Dynamic Human Anatomy

SECOND EDITION

William C. Whiting, PhD
California State University, Northridge

HUMAN KINETICS

Library of Congress Cataloging-in-Publication Data
Names: Whiting, William Charles, author.
Title: Dynamic human anatomy / William C. Whiting.
Other titles: Dynatomy
Description: Revised edition. | Champaign, IL : Human Kinetics, [2019] |
Preceded by Dynatomy / William C. Whiting, Stuart Rugg. c2006. | Includes
bibliographical references and index.
Identifiers: LCCN 2017047888 (print) | LCCN 2017049718 (ebook) | ISBN
9781492549864 (e-book) | ISBN 9781492549871 (print)
Subjects: | MESH: Anatomy | Movement | Biomechanical Phenomena
Classification: LCC QP301 (print) | LCC QP301 (ebook) | NLM QS 4 | DDC
612.7/6--dc23
LC record available at https://lccn.loc.gov/2017047888

ISBN: 978-1-4925-4987-1 (print)

This book is a revised edition of *Dynatomy: Dynamic Human Anatomy,* published in 2006 by Human Kinetics, Inc.

The web addresses cited in this text were current as of February 2018, unless otherwise noted.

Senior Acquisitions Editor: Joshua J. Stone; **Developmental and Managing Editor:** Amanda S. Ewing; **Copyeditor:** Joy Hoppenot; **Proofreader:** Karla Walsh; **Indexer:** Ferreira Indexing; **Permissions Manager:** Dalene Reeder; **Graphic Designer:** Sean Roosevelt; **Cover Designer:** Keri Evans; **Cover Design Associate:** Susan Rothermel Allen; **Photograph (cover):** Getty Images/Henrik Sorensen; **Photographs (interior):** © Human Kinetics, unless otherwise noted; **Photo Asset Manager:** Laura Fitch; **Photo Production Manager:** Jason Allen; **Senior Art Manager:** Kelly Hendren; **Illustrations:** © Human Kinetics, unless otherwise noted; **Printer:** Walsworth

Printed in the United States of America 10 9 8 7 6 5 4 3 2 1

The paper in this book was manufactured using responsible forestry methods.

Human Kinetics
P.O. Box 5076
Champaign, IL 61825-5076
Website: www.HumanKinetics.com

In the United States, email info@hkusa.com or call 800-747-4457.
In Canada, email info@hkcanada.com.
In the United Kingdom/Europe, email hk@hkeurope.com.

For information about Human Kinetics' coverage in other areas of the world, please visit our website: **www.HumanKinetics.com**

E7051

With everlasting love to my wife, Marji; sons, Trevor and Tad;
and daughter, Emmi, for giving special meaning to my life.

In loving memory of two colleagues and good friends,
Jennifer Romack and Shane Frehlich, who left us too soon.

Contents

Part IV Movement Applications 215

Preface

No earthly creation is as intricate, versatile, or mysterious as the human body. Our bodies perform countless functions that ensure our existence. Of all the body's functions, movement is arguably the most essential, for without movement, the human body could not survive. With movement, the body flourishes. At one end of the movement spectrum, the capacity for purposeful movement allows us to maintain fundamental physiological processes; at the other end, it allows us to pursue the limits of athletic and artistic expression.

For those involved in the many areas of human health and performance, knowledge of the body's structure and function is essential. Students in these areas undoubtedly have taken introductory courses in human anatomy. All too often, however, these courses (which attempt to cover all of the body's systems in a single academic term) provide students with the basics of structural anatomy but offer limited exposure to the elegance and complexity of the body's functional movement anatomy.

From my experience teaching human anatomy, kinesiology, and biomechanics over the past three decades, I can attest to the unfortunate fact that some students emerge from their introductory anatomy experience with limited competency in applying their anatomical knowledge to human movement problems. These students have learned what I refer to (not too euphemistically, and only somewhat kiddingly) as "dead anatomy."

The purpose of this book is to bring anatomy to life by exploring the marvelous potential of the human body to express itself through movement and to provide students with the information and skills they need to appreciate and assess dynamic human anatomy.

Organization

This book, the successor edition to *Dynatomy: Dynamic Human Anatomy* (Whiting & Rugg, 2006), has been developed primarily for students who already have taken, or currently are taking, an introductory course in human anatomy and who need a more detailed exposure to concepts of human *movement* anatomy. These students include those in the fields of exercise science and human movement studies, kinesiology, biomechanics, physical education, coaching, athletic training, ergonomics, and the health sciences (e.g., medicine, physical therapy).

Given the amount of material presented and memorized in introductory anatomy courses, it is neither unusual nor unexpected that students forget many of the details. Thus, part I (Anatomical Foundations) provides a concise review of relevant anatomical information and neuromechanical concepts. Part I begins in chapter 1 (Introduction to Human Anatomy and Movement) with an introduction to the dynamics of human movement and essentials of anatomical structure. The chapter presents a brief overview of motor behavior, including life-span motor development, motor control, and motor learning, and it establishes a contextual framework for later discussion of specific movements and populations.

Chapter 2 (Osteology and the Skeletal System) describes the microstructure and macroscopic morphology of bone and the organization of the skeletal system. It also identifies specific bones and bony landmarks.

Part I continues in chapter 3 (Joint Anatomy and Function) with a discussion of joint structure, function, and motion, along with a description of major joints of the extremities and spine. Chapter 4 (Skeletal Muscle) begins with a discussion of the structure and function of skeletal muscle and the basic physiology and mechanics of muscle action. The chapter continues with exploration of muscle actions; voluntary, reflex, and stereotypical movements; nervous system control of muscle action; and factors (e.g., muscle length and velocity, fiber type, and muscle architecture) that affect muscle force production. It concludes with a presentation of the primary muscles responsible for the production and control of human movement.

With the requisite anatomical and neuromechanical foundations in place, part II (Biomechanics and Movement Control) provides the essentials of a dynamic approach to movement. This part begins in chapter 5 (Biomechanics) with a review of mechanical concepts essential to an understanding of human movement. Chapter 6 (Muscular Control of Movement and Movement Assessment) explains the muscle control formula, a set of simple steps used to identify the muscles and types of muscle actions responsible for producing or controlling human movements. Numerous examples, using simple movement patterns, demonstrate the utility of the formula. Mastery of the muscle control formula provides a valuable tool for independent assessment of muscle action across a broad spectrum of human movements. Chapter 6 concludes with consideration of several important topics relevant to movement assessment. These include single-joint versus multijoint movements, coordination, movement efficiency, and assessment of movement.

Part III (Fundamentals of Movements) begins in chapter 7 (Posture and Balance) with a discussion of the types and functions of posture and balance, mechanisms of postural control, postural alterations and perturbations, developmental considerations, life-span aspects of balance, and postural dysfunction. Chapter 8 (Gait) explores various aspects of walking and running gait. Part III concludes with chapter 9 (Basic Movement Patterns), which examines the fundamental movement patterns of jumping, kicking, lifting, throwing, and striking.

With the anatomical, biomechanical, and basic movement pattern foundations in place, part IV (Movement Applications) discusses movement-related aspects of strength and conditioning applications (chapter 10), sport and dance applications (chapter 11), clinical applications (chapter 12), and ergonomics applications (chapter 13).

One of the primary goals of this book has been to include concepts not found in many traditional anatomy texts, concepts that emphasize function and application. An understanding of the material covered will enhance your preparation for work in the many areas of human movement. To assist you is a comprehensive glossary that defines hundreds of important terms and concepts. Terms boldfaced in the text can be found in the glossary. Each chapter also includes objectives and critical thinking questions to guide and challenge students, as well as suggested readings for further study.

Updates to the Second Edition

The second edition has been updated to include recent research and citations. Additional sidebars have been included to present applications of concepts and movement-related research in biomechanics. Three new chapters have been added to expand on applications in the areas of strength and conditioning (chapter 10), clinical issues (chapter 12), and ergonomics (chapter 13). In addition, chapter 11 (Sport and Dance Applications) has been expanded to cover movement mechanics in baseball and softball, basketball, American football, golf, soccer, tennis, and volleyball.

Student Resources

Students have access to a web study guide that provides several learning aids:

- Summary of articulations for the spine and upper and lower extremities
- Tables that provide the origin, insertion, action, and innervation for all major muscle groups
- Interactive practice problems that allow students to apply the muscle control formula discussed in chapter 6
- Critical thinking questions

The web study guide is available at www.HumanKinetics.com/DynamicHumanAnatomy.

Instructor Resources

New to this edition is a full array of instructor resources:

- **Presentation package:** More than 270 slides present the key concepts from the chapters, including select figures and tables. These slides can be updated to fit unique lecture and classroom needs.
- **Image bank:** The image bank includes most of the figures and tables from the text. These images can be used to supplement PowerPoint slides, lectures, student handouts, and so on.
- **Instructor guide:** The instructor guide provides an introduction on how to use the ancillaries, suggestions for student activities, a sample syllabus, and chapter-specific files that include summaries and lecture outlines. It also includes the answers to the critical thinking questions in the web study guide.
- **Test package:** The test package includes 330 questions that instructors can use to build quizzes to test student understanding.

The instructor ancillaries are available at www.HumanKinetics.com/DynamicHumanAnatomy.

Final Thoughts

Please keep in mind that the material in this text is only a beginning. Consideration of any one of the many topics presented could be expanded to fill a chapter, or even a book, of its own. I make no pretense of having exhaustively covered any of the topics, but rather have sought to present a concise introduction to many of the fascinating areas of human movement.

This book will have served its purpose well if you emerge with an appreciation for human movement that is more wondrous and less mysterious than when your journey began. In the words of British author Laurence Sterne, "So much of motion, is so much of life, and so much of joy" (1980, p. 354).

Acknowledgments

Although only one name appears on its cover, this book could not have been completed without the contributions of many people. I extend my thanks to the many friends and colleagues who have contributed to its completion. I am especially grateful to Stuart Rugg, my coauthor on the first edition of this book. His valuable contributions helped lay the foundation for this second edition.

Special thanks to Bob Gregor and Ron Zernicke for sharing their knowledge and enthusiasm for biomechanics and helping me develop my love of teaching. Thanks also to Judi Smith for her pioneering work in establishing the human anatomy program in the then–Department of Kinesiology at UCLA that helped provide me with my foundation in anatomy and my belief in the remarkable power of human movement.

My thanks to the staff at Human Kinetics, especially Josh Stone and Amanda Ewing, for their outstanding assistance, patience, support, and belief in this project.

And finally, and most importantly, I thank my family for their unwavering support and love, during the course of this project and always!

PART I

Anatomical Foundations

The four chapters of part I present a concise review of human anatomy and a conceptual framework for our discussion of human movement. Chapter 1 (Introduction to Human Anatomy and Movement) introduces anatomical and movement concepts that form the foundation for understanding details presented in subsequent chapters. Chapter 2 (Osteology and the Skeletal System) reviews the organization of the skeletal system and details of bone structure and function. Using the information presented in chapter 2, our discussion continues in chapter 3 (Joint Anatomy and Function) with examination of the articular system and joint motion, including review of the major joints involved in human movement. Chapter 4 (Skeletal Muscle) explores the structure and function of skeletal muscle, the tissue that provides the force needed to produce and control our movements. In this chapter, we also examine factors that affect muscle force production. Chapter 4 concludes part I with a presentation of the muscles of movement and a summary of the specific muscles responsible for movement control.

The information presented in part I is essential to our understanding of dynamic human anatomy. These chapters set the stage for our application of dynamic human anatomy to a wide variety of human movements.

1

Introduction to Human Anatomy and Movement

Objectives

After studying this chapter, you will be able to do the following:

- Define the terms *anatomy* and *physiology* and describe the relationship between these two areas of study
- Explain the importance of movement in our daily lives
- Understand how movement changes across the life span
- Describe the importance of understanding motor behavior and its subareas of motor control, motor learning, and motor development
- Explain movement considerations in younger and older populations
- Describe the importance of differences in movement ability among individuals
- Explain the anatomical concepts of complexity, variability, individuality, adaptability, connectivity, and asymmetry
- Describe the levels of structural organization and primary tissue types
- Explain the structure and function of muscle and connective tissues
- Define and explain anatomical terms describing body regions, body positions, movement planes, axes of rotation, joint positions, and movements
- Appreciate the multidisciplinary perspective necessary for understanding human movement

In this introductory chapter, we establish the conceptual framework for understanding the details of human anatomy and movement in the chapters that follow. In building this framework, we take a life-span approach and focus on general anatomical terminology and concepts and an overview of human structure. We emphasize the elements of anatomy (e.g., bones and muscles) directly related to human movement.

Introduction to Human Anatomy

Anatomy, simply defined, is the study of the structure of organisms. The term derives from the Greek *anatome*, which means "to dissect" or "to cut." The organism considered here is the human organism. Historically, dissection has been a primary means of discovery about the structure of the human organism, and it remains a useful educational tool in training professionals in the health sciences (e.g., medicine, physical therapy). However, in recent decades, technological advances of many kinds (e.g., MRI, CAT scans) have allowed us to better understand the body's structure without cutting it open.

Anatomy is an expansive science that encompasses many branches, or subdivisions. Some of these subdivisions are described in table 1.1. In establishing our anatomical foundation for study of human movement, we focus on gross (macroscopic) anatomy, systemic anatomy (especially the skeletal, muscular, and nervous systems), regional anatomy as it relates to specific movement patterns, and, most importantly, functional anatomy.

Although anatomy is our focus, the related area of **physiology**, which deals with the *functions* of the body parts and systems, deserves mention. Anatomy and physiology are closely related, and study of one without consideration of the other makes little sense. We begin by presenting, in elementary terms, the origin, development, and structure of the body's components and then show how the parts are both structurally and functionally interconnected.

We begin with an overview of human movement, followed by an introduction to concepts of functional anatomy. These anatomical concepts form a necessary foundation for your understanding of dynamic human anatomy.

TABLE 1.1 Subdivisions of Anatomy

Anatomy subdivision	Description
Gross (macroscopic) anatomy	Study of structures without the aid of a microscope
Microscopic anatomy	Study of structures using microscopic techniques
Systemic anatomy	Study of specific body systems (e.g., skeletal, muscular, nervous, respiratory, cardiovascular)
Regional anatomy	Study of specific body regions (e.g., head, extremities)
Functional anatomy	Study of how body systems (e.g., skeletal, muscular, nervous) work cooperatively to perform various functions
Surface anatomy	Study of landmarks on the surface of the body
Developmental anatomy	Study of structural development across the life span, from fertilization to death
Embryology	Study of development from fertilization through the 8th week in utero
Histology	Microscopic study of tissue structure
Cytology	Microscopic study of cell structure

Applying the Concept

Vesalius

Some of the most admired anatomical treasures of the Renaissance were drawn by Andreas Vesalius (1514-1564). His master work, *De Humani Corporis Fabrica* (1543), stands as one of the great volumes in the history of modern science. In this work, Vesalius blended text with picture into a truly integrated presentation of his understanding of human anatomy.

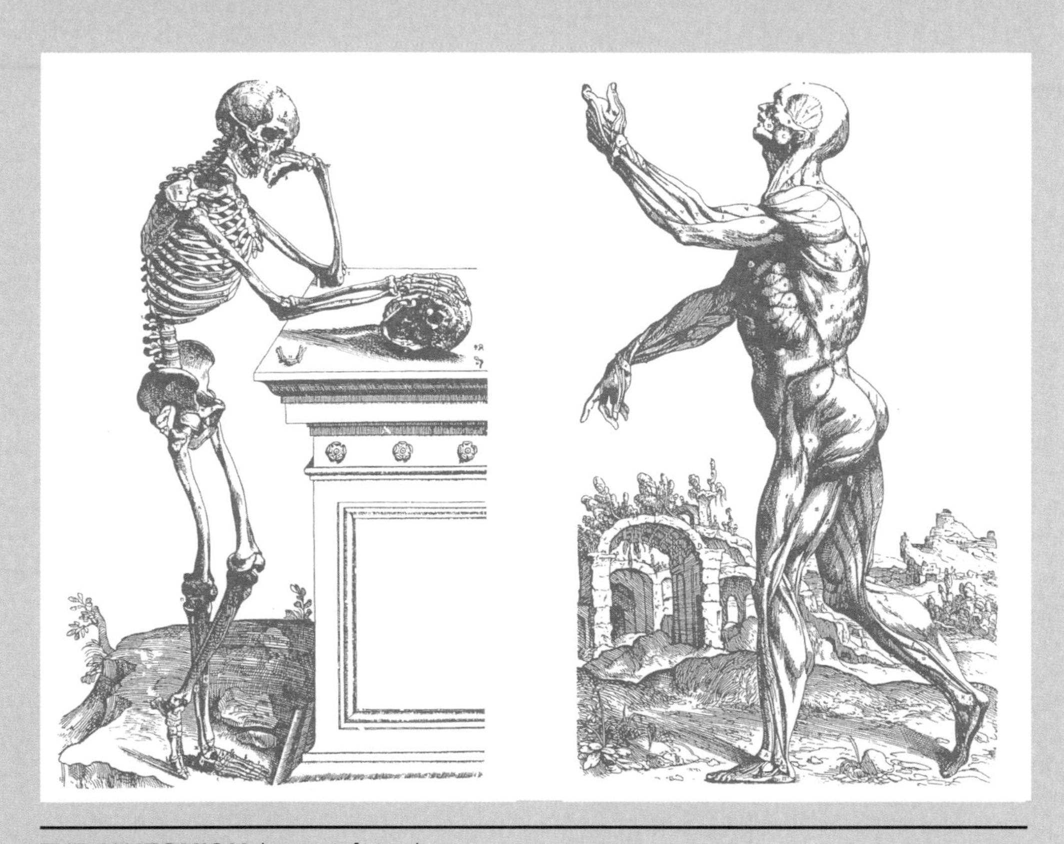

THE ANATOMICAL legacy of Vesalius.

Introduction to Human Movement

Movement is a fundamental behavior essential for life itself. Life processes such as blood circulation, respiration, and muscle contraction require motion, as do activities such as walking, bending, and lifting. The human organism seeks, consciously or not, to move. Children, in particular, provide clear evidence of the inherent nature of humans to move. They never seem to stop. Even as we age and slow down, movement remains an essential part of our lives. As noted by French scientist and philosopher Blaise Pascal (1623-1662), "Our nature consists in motion; complete rest is death" (Pascal & Krailsheimer, 1995, p. 211).

At one end of the spectrum, the capacity for purposeful movement supports basic physiological processes; at the other end, movement allows us to explore the limits of athletic and artistic expression. Limited movement, such as when a person is bedridden or elects a sedentary lifestyle, can contribute to adverse health conditions such as cardiovascular disease, diabetes, and cancer. Thus our ability to move, or the choice to limit movement, may contribute, either directly or indirectly, to our susceptibility to disease and injury.

The study of human movement across all its dimensions is known as **kinesiology** (Gk. *kinesis*, fr. *kinein* "to move"). Kinesiology is a broad discipline that encompasses both the science and art of human movement. It draws from many related disciplines, including anatomy, physiology, biomechanics, motor behavior, and psychology, together with clinical and applied disciplines such as medicine, physical therapy, engineering, and physical education. A primary goal of studying kinesiology is to identify the underlying mechanisms and consequences of human movement.

The range of human movements is broad and complex. Some movements, such as throwing, consist of a single episode; others, such as walking, involve repetitive movement cycles. And none of us moves in the same way, even while doing the same thing. Each person, for example, adopts a unique walking pattern based on individual structure, purpose, and style. This can be seen in the common experience of viewing a person's silhouette in the distance and knowing who it is by the way she walks, long before we recognize her by facial or body features. Something in the way she moves tells us who she is.

Movement Across the Life Span

As we navigate life's journey, each of us becomes aware of our own movement capabilities and limitations. As children, we progress from immobility to crawling, walking, running, and jumping. With age, we grow larger and stronger, and typically become more proficient in our movement patterns throughout most of our lives. Our later years may be characterized by movement pattern changes due to muscle strength decreases, fatigue, nervous system declines, postural changes, injury, disease, and environmental factors. One thing is clear—our ability to move changes across the life span. These changes can enhance movement potential, as when a child grows and develops, or limit movement, as in the case where injury, disease, or age-related functional declines cause a deterioration in movement capacity. Given the myriad changes to our movements from cradle to grave, we emphasize a life-span perspective throughout this book.

Anatomical changes that occur from day to day are barely detectable, if at all, but as days turn to weeks, weeks to months, and months to years, changes become evident. These changes may be due to growth (e.g., an adult is larger and stronger than a child); physical conditioning (e.g., an athlete performs better after a strength or endurance training program); injury or disease (e.g., the growth of a child's bone may be affected by a serious fracture); or a variety of environmental, sociological, psychological, or cultural influences.

Each of us changes in countless unique ways throughout our lives. These changes directly affect our potential and capacity to move. The chapters that follow take a life-span approach to functional anatomy and human movement and show many ways in which movement is affected by our time and place along life's journey. The acquisition and refinement of movement skills across the life span are examined through the study of motor behavior and its subareas: motor control, motor learning, and motor development.

Motor Behavior

In the context of human movement, the term **motor** describes things related to a muscle, nerve, or brain center involved in producing or controlling movements. **Motor behavior** is an umbrella term that encompasses several specialized areas related to neuromuscular control of movement: motor control, motor learning, and motor development. Each of these areas contributes to our understanding of human movement.

Each area of motor behavior typically encompasses a distinctive time frame. Motor control covers events with very short time intervals; motor learning involves times of hours, days, and weeks; and motor development addresses events over months, years, and even decades. For example, to control the muscles involved in throwing a ball, the nervous system sends signals to the shoulder and arm. These signals may last for a fraction of a second. Learning to throw may take days or weeks. And to see developmental changes in throwing from childhood to adulthood, we need to observe the task across many years. Though these areas are clearly related, we consider each in turn.

Motor Control

Motor control is the study of the neural, physical, and behavioral aspects of movement (Schmidt & Lee, 2011). More specifically, motor control refers to how the body's systems organize and control muscles involved in movement. The system primarily involved in control is the nervous system, which consists of the brain, spinal cord, and peripheral nerves emerging from the cord.

Research in Mechanics

The Legacy of Borelli

In the history of science, the name Giovanni Alfonso Borelli (1608-1679) usually does not appear on lists with such giants as Galileo and Sir Isaac Newton. In the history of movement science, however, Borelli deserves a prominent place. His definitive treatise, *De Motu Animalium,* serves as the seminal work on the science of animal movement.

This richly illustrated volume includes discussion of musculoskeletal anatomy and muscle physiology, along with application of mechanical principles to movement across a broad spectrum of animal motion. Borelli modeled the leverage of muscles, tendons, and bone and applied such mechanical constructs to the analysis of human gait, as well as to the locomotor patterns of horses, insects, birds, and fish.

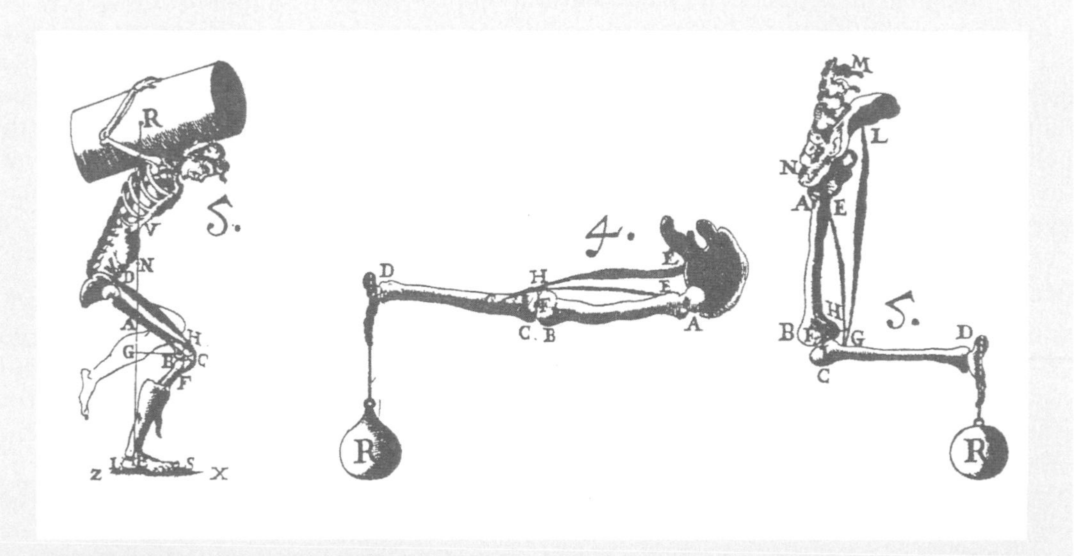

THE LEGACY of Giovanni Borelli.

In many ways, Borelli was ahead of his time. He was restricted—by limited technology and yet-to-be-formalized theories of mechanics—from fully researching the concepts he clearly embraced. Regrettably, Borelli never saw his *De Motu Animalium* in published form. Most of Borelli's work was published only after his death (Cappozzo & Marchetti, 1992).

See the references for the full citation:

Cappozzo & Marchetti, 1992.

The degree of motor control clearly depends on the developmental, or maturational, stage of the individual. A 2-year-old child, for example, cannot be expected to have the same degree of movement control as a 10-year-old or an adult. As the systems involved in movement (nervous, muscular, skeletal, cardiovascular, respiratory) develop, the body is better able to control movements and develop a wider variety of potential movement skills.

The study of motor control originated with neurophysiologists who explored how the nervous system worked with muscles to produce and control movement. Over the decades, various models and theories have attempted to explain the mechanisms of motor control, including the equilibrium-point theory, first proposed by Feldman more than 50 years ago, the uncontrolled manifold hypothesis, and the principle of abundance. More recent efforts (e.g., Martin et al., 2009; Latash et al., 2010) have attempted to combine these three models into a unified theory of motor control "to create a formal description, operating with exactly defined variables, of the physical and physiological processes that make coordinated voluntary movements possible" (Latash et al., 2010, p. 391).

Motor Learning

Experience and the environment can teach us a lot, and so can others who share their knowledge and experience. **Motor learning** deals with how we learn to move and develop motor skills. A **motor skill** is a voluntary movement used to complete a desired task action or achieve a specific goal. When we first begin to learn a new motor skill, our movements usually are awkward, uncoordinated, and inefficient. With practice, we learn to refine the movement until it becomes relatively permanent and proficient.

The process of motor learning involves several stages, or phases. These phases, as first presented by Fitts (1964), are the cognitive phase, associative phase, and autonomous phase. In the initial cognitive phase, a person must devote considerable conscious thought to the movement task and try various strategies. With repetition, the mover retains the strategies that best accomplish the task and discards those that do not. In the associative phase, movements are less variable, and the mover determines the best movement strategy. The movement requires less cognitive involvement, and the mover can focus on refining and perfecting the movement. Finally, in the autonomous phase, the movement becomes automatic, or instinctive, and the mover can attend to other factors such as environmental conditions that may affect task performance and movement.

Motor Development

Motor development explores the changes in movement behavior that occur as one progresses through the life span from infancy until death. The changes typically are continuous, sequential, and age related. An infant, for example, increases her mobility by progressing from crawling and creeping to walking, and then from walking to running. These developments occur continuously, in a predefined sequence, and obviously are age related.

Movement changes are directed by numerous individual, environmental, and task constraints (Newell, 1986). Individual constraints may be either structural (e.g., body size and strength) or functional (e.g., motivation). Environmental constraints include elements outside the body and can be physical (e.g., weather or surface conditions), sociological, or cultural (e.g., societal norms or traditions that favor one group's participation over another's). Task constraints refer to external factors inherent to the task at hand. A participant in a wheelchair basketball game has obvious constraints imposed by his own wheelchair and those of others in the game, along with the challenges of avoiding his opponent's defensive efforts.

One prominent approach to exploring motor development employs a dynamic systems theory, which views development as a complex, multicausal process (Smith & Thelen, 2003) with a goal of decreasing intraindividual variability (Deutsch & Newell, 2005). In the words of Smith and Thelen (2003), "Development is about creating something more from something less, for example, a walking and talking toddler from a helpless infant" (p. 343).

Movement Considerations in the Young

Comparison of movement between younger and older people readily shows that children are *not* just miniature adults. Infants and children are anatomically, physiologically, psychologically, and emotionally different from adults, and their structure and movement behaviors reflect these differences. For example, infants' body proportions differ from adults', as can be seen in the fact that infants have proportionally shorter limbs and larger heads than adults. This difference in weight distribution plays an important role in movement dynamics of infants compared with older children and adults.

Near-universal similarities exist among people in terms of human developmental progression. However, the progressive changes do not occur at the same age, or at the same rate, for every person. Nor are the developmental changes the same for everyone. Averages exist across a given population, but within the population, considerable variability is apparent. As expected, average values are reported for developmental measures, but many people fall above or below the average. The individual nature of development should always be kept in mind.

Similarly, because children reach developmental milestones at different ages, movements should be viewed and selected using developmentally appropriate standards rather than age. The average age for a baby's first walking steps, for example, is 11.7 months. The age of walking onset ranges, however, from as early as 9 months to as late as 17 months. Clearly, not all infants and children reach developmental movement milestones at the same age.

Developmental factors include genetic, environmental, maturational, and cultural influences, none of which alone determines the course of development. Many factors combine to determine the course of motor development and the eventual movement capacity of each individual.

Movement Considerations in Older People

Movement capability typically declines with age. Older people tend to move more slowly, with more limited range of motion, and with greater hesitancy than do younger, healthy adults. Older people are more likely to be afflicted with certain diseases (e.g., heart disease,

In Others' Words

Charles Darwin

Under a transport of joy or of vivid pleasure, there is a strong tendency to various purposeless movements, and to the utterance of various sounds. We see this in young children, in their loud laughter, clapping of hands, and jumping for joy; in the bounding and barking of a dog when going out to walk with his master; and in the frisking of a horse when turned out into an open field. Joy quickens the circulation, and this stimulates the brain, which reacts again on the whole body. . . .

It is chiefly the anticipation of a pleasure and not its actual enjoyment, which leads to purposeless and extravagant movements of the body, and to the utterance of various sounds. We see this in children when they expect any great pleasure or treat.

Moreover, the mere exertion of the muscles after long rest or confinement is in itself a pleasure, as we ourselves feel, as we see in the play of young animals. Therefore on this latter principle alone we might expect, that vivid pleasure would be apt to show itself conversely in muscular movements. (Darwin, 1998, pp. 80-81)

See the references for the full citation:

Darwin, 1998.

diabetes) and suffer from injuries (e.g., hip fractures). Other factors, such as slower reaction times, reduced muscle strength and balance, and fear of falling may also contribute to slowed movements. Subsequent chapters examine many of these factors in detail.

Undeniably, some slowing happens as we age. But do we slow down because we age, or do we age because we slow down? To some extent, the latter is true, largely because of the adoption of sedentary lifestyles. The movement capacity of older people who get little, if any, regular physical activity typically declines, often markedly. These individuals may deteriorate to the point where simple activities of daily living (ADLs) become a challenge. ADLs include tasks such as bowel and bladder control, personal grooming, toilet use, feeding, bathing, mobility, transfer, and climbing stairs. Taken together, ADLs may be used as an index of functional independence. A low cumulative index score may indicate a loss of independence and a lowered quality of life. That's the bad news.

The good news is that research clearly shows that the body can respond to training throughout the life span and that it is never too late to begin restorative training programs aimed at improving cardiorespiratory function, muscle strength, and balance. In a now-classic study, Fiatarone and colleagues (1990) noted strength gains of 174% after just 8 weeks of high-intensity **resistance training** in frail, institutionalized people with an average age of 90 years. The researchers demonstrated that high-resistance weight training promoted gains in muscle strength, size, and functional mobility in people up to 96 years of age!

Although some decline in function may be inevitable, with proper lifestyle choices—and a little luck—most of us have the potential to maintain productive movement well into our later years.

Differences in Movement Ability

Each of us demonstrates movement abilities falling somewhere along what can be termed an *ability continuum*. An elite athlete may exhibit high ability in several movement forms, such as the ability of a professional basketball player to jump high, run fast, and control his body while in the air. A professional dancer may show exceptional ability to control her arm and leg movements and maintain balance throughout a performance. At the other end of the ability continuum, a person with physical disabilities may have difficulty walking or controlling limb movements.

Many different movement forms exist, including gross motor skills and fine motor skills. **Gross motor skills** involve moving and controlling the limbs, as when someone walks, runs, or jumps. **Fine motor skills**, in contrast, involve small movements (e.g., manipulating objects with the fingers), such as sewing, writing, or typing. Good manual dexterity (i.e., ability to move the fingers) is essential for many occupations and activities.

Some activities require a combination of gross and fine motor skills for successful performance. A wide receiver in American football, for example, needs good gross motor skills to run down the field and elude defenders but must also have sufficient fine motor skills and eye–hand coordination to catch the ball thrown by the quarterback.

Few people have high abilities across all movement forms. Someone with good gross motor skills, such as a soccer player, may have difficulty with fine motor tasks. In contrast, a concert pianist with exceptional fine motor skills may be much less capable of performing gross movements.

Many factors determine motor abilities, including natural genetic potential, practice, level of physical conditioning, motivation, disease, and injury. A gifted athlete, for example, may enjoy the good fortune of having athletic parents who passed favorable movement potential on to their child. That movement potential may be fully realized through proper training and conditioning, or it may never be fulfilled because of lack of motivation or debilitating injury. A person not as genetically blessed may nonetheless excel through hard work, perseverance, and strong motivation.

Dysfunctional, or disabled, movement refers to limited or compromised ability to move. Many factors can cause movement disabilities, including genetics (e.g., an infant born with a

Applying the Concept

Exceptional Gross and Fine Motor Skills

Most of us have a functional balance of gross and fine motor skills adequate for accomplishing tasks of daily living. Some people have notable skills in certain movement domains. Very few people have truly exceptional motor skills. Olympic gold medalist Paul Anderson showed incredible gross motor skills and brute strength when he back lifted an astonishing 6,270 lb (2,844 kg) on June 12, 1957. In contrast, exceptional fine motor skill is exhibited by virtuoso pianists, who can strike more than 10 piano keys per second.

congenital deformity), disease (e.g., shaking limbs seen in people with Parkinson's disease), or injury (e.g., a fractured bone). Other factors may also lead to movement dysfunction. In describing pathological walking, Perry & Burnfield (2010) identifies four functional factors that can weaken gait: deformity, muscle weakness, impaired control, and pain. A functional deformity such as excessive tightness in a muscle, tendon, or ligament can limit joint range of motion. Muscle weakness due to paralysis or muscular disease can compromise the muscle's ability to move the arms or legs. Impaired control refers to a diminished ability of the nervous system to provide sensory input and tell the muscles what to do. Finally, pain can discourage a person from performing movement throughout a full range of motion—or from moving at all.

Anatomical Concepts

The study of dynamic human anatomy involves many details and considerable memorization. Although knowing the details is important, it is equally important not to lose sight of the forest for the trees. Keeping the larger context in mind is essential, and consideration of several anatomical concepts provides a useful framework for the study of the details. These concepts include anatomical complexity, variability, individuality, adaptability, connectivity, and asymmetry.

Complexity

We inarguably are complex organisms. The human body is composed of more than 200 bones and some 600 muscles, hundreds of tendons and ligaments, miles of circulatory vessels, and countless nerve cells. These tissues are connected to one another and interact in many and wonderful ways. We have learned much about the human body's structure and function, but much more remains a mystery. Thousands of researchers spend their professional lives trying to unravel these mysteries and enhance our knowledge of the human organism. Due to the complexity of the human organism, most problems and challenges related to the human body are multifactorial (i.e., multiple factors must be considered).

Variability and Individuality

Humans come in all shapes and sizes. The basic components, for the most part, are the same in each of us. With few exceptions, each of us has the same number of bones, muscles, and internal organs. However, the size, shape, composition, and function of these components can vary greatly. At a gross level, some individuals exceed 7 feet (2 m) in height, while others are less than 4 feet tall (just over 1 m). Body weights vary greatly. Skin color spans the spectrum from light to dark. Some individuals have large, strong, and dense bones. Others have fragile bones that may easily fracture. The chapters to follow present the basics of anatomy and function, but it is important to keep in mind that because we are not all built the same, we will not function alike. Certain structural characteristics may favor one individual over another when it comes to movement proficiency and performance capability.

The importance of anatomical variation cannot be overstated. As noted in the preface of the *Compendium of Human Anatomic Variation*,

> Most modern textbooks of anatomy are more or less devoid of information on variations that so commonly appear in the dissecting room and, more importantly, in practice. Variations must be considered to be normal and thus must be anticipated and understood. It has been repeatedly stated in the literature that textbook descriptions are accurate . . . in only about 50-70% of individuals. From the standpoint of utilization of anatomic information in a clinical setting, textbooks are not only inadequate but may be dangerously misleading as well. (Bergman, Thompson, Afifi, & Saadeh, 1988, p. v)

The importance of anatomic variability was confirmed by the establishment of the online journal *International Journal of Anatomical Variations* in 2008 and the publication of *Bergman's Comprehensive Encyclopedia of Human Anatomic Variation* (2016).

Another important anatomical concept closely related to structural variability is individuality. Each of us has individual characteristics and circumstances that affect our anatomy and physiology. These characteristics include gender, age, environment, genetics, ethnicity, cultural background, and family and individual history.

Adaptability

People have a remarkable ability to adapt. Human body adaptations can be physiological or structural; they can be advantageous or detrimental. Muscles, for example, can increase in size (**hypertrophy**) and strength in response to resistance training or decrease in size (**atrophy**) when disused as a result of disease, injury, or a sedentary lifestyle (figure 1.1). The cardiovascular system can improve its ability to transport oxygen following endurance training; in contrast, its functional capacity can be compromised by physical inactivity. Tissues such as bone can structurally adapt to the forces imposed on them over time. The nervous system (brain, spinal cord, nerves) has the capacity to modify its structure and function as it learns or is injured. This ability to adapt to environmental stimuli and events is termed **plasticity**.

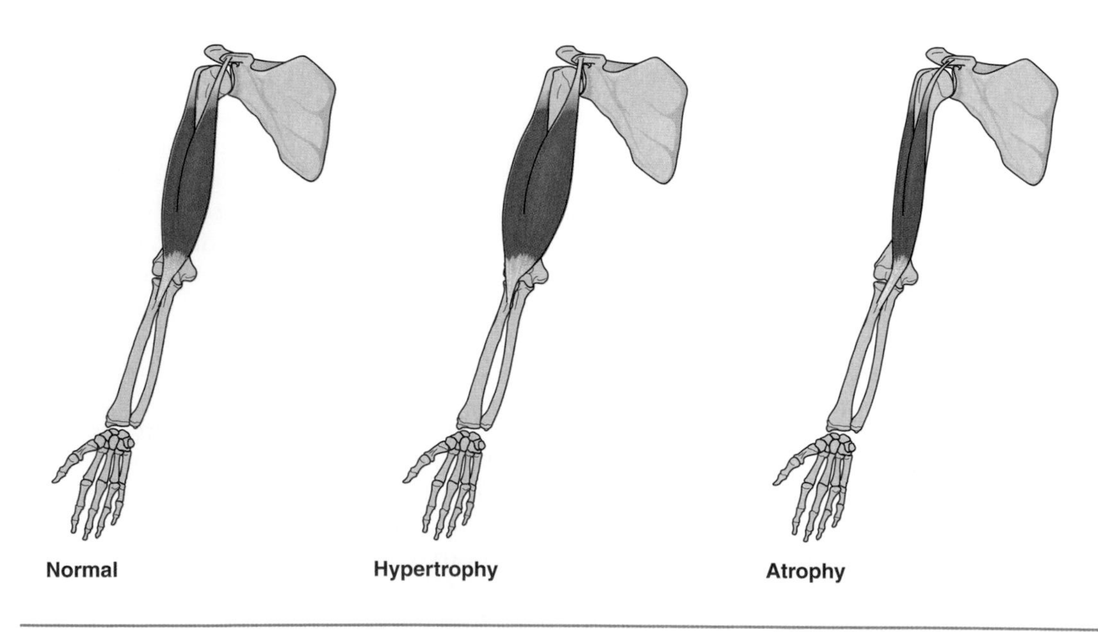

FIGURE 1.1 Examples of normal muscle, muscle hypertrophy, and muscle atrophy.

All the body's tissues and systems can adapt to imposed demands. This characteristic of adaptability is described by the **SAID** (specific adaptation to imposed demands) **principle**, which states that tissues and systems in the human body can adapt, either structurally or physiologically, in a way that is specific to the mechanical or metabolic demands placed on them. In general, there is a level of demand, or stimulus, that allows for optimal adaptation. A stimulus that is too small (e.g., from physical inactivity) reduces the body's performance ability and structural integrity, while one that is too great can result in tissue injury, such as when bones are overloaded and fracture.

Connectivity

Traditional approaches to studying anatomy logically begin with examination of thc individual structural components (e.g., bones, muscles, tendons). Although this is an appropriate starting place, these tissues are connected to one another in the body, and the coordinated function of the whole organism depends on the connective characteristics. For example, muscle and tendon are joined at the **musculotendinous** (also **myotendinous**) **junction**. Here the muscle fibers intertwine with collagen fibers of the tendon in a structural arrangement that enhances surface area for attachment and strengthens the site.

In general, forces applied to the body are transmitted from one tissue to another through their connections, and failures to the structural units (e.g., bone–ligament–bone) frequently occur at attachment sites. Thus, we need to keep in mind the connectivity of the body's structure and the important role it plays in effective performance, tissue adaptation, and injury causation and prevention. In general, structural units will fail at their weakest link, the place where the tissue is most vulnerable given the current environment and loading conditions.

Asymmetry

At first glance, the human body appears to be designed symmetrically: two each of arms, legs, eyes, ears, and so on. From the outside, our left and right sides, generally speaking, appear to be mirror images of each other. Further exploration, however, reveals that in many ways we are not designed symmetrically. The heart is not located along the body's midline. The liver is on the right side and the spleen is on the left. The right lung has three sections (lobes), whereas the left lung has two.

Joints in the body also are not symmetrical. The knee joint, for example, has a number of structural asymmetries. The supporting ligaments on the sides of the joint (collateral ligaments) are arranged differently. The ligament on the inner (medial) side of the knee structurally blends with the joint capsule that surrounds the joint. The ligament on the outer

Applying the Concept

Situs Inversus

An interesting example of anatomical asymmetry is seen in the congenital condition known as *situs inversus* (also *situs transversus* or *situs oppositus*), in which the internal organs of the thorax are reversed, or placed in mirror image, compared to normal. In situs inversus, the heart, for example, is located to the right of the body's midline instead of its normal placement to the left of the midline. The liver is found on the left side of the body (instead of the right side). And so goes the remainder of the internal organs. The prevalence of situs inversus varies but has been estimated at less than 1 in 10,000 (<0.01%). Most people with situs inversus can lead normal healthy lives. Complications can arise in the medical diagnosis of unrelated problems (e.g., appendicitis) or in cases of organ transplants.

(lateral) side of the joint is structurally distinct and is located outside the capsule. The size and shape of the bony contact surfaces on the medial and lateral sides are not the same, nor are the cartilage discs (menisci) that help the knee function properly.

Anatomical asymmetry has allowed the body to develop into the complex organism it is, with all of the necessary organs neatly but asymmetrically packaged inside. Finding the genes responsible for determining what goes where is but one of the many anatomical challenges facing developmental geneticists. Structural asymmetries often have functional advantages, and it is important to keep this anatomical concept in mind as we learn more about the details of human anatomy and movement.

Levels of Structural Organization

Early study of the human body and its structure was limited to what could be observed by the unaided eye. Only when microscopic techniques were developed could the body's smaller structures be investigated. We now know that the larger (macroscopic) structures are composed of smaller structures, which in turn are made of even smaller structures. To view the spectrum of structural complexity, we describe the body's levels of structural organization (figure 1.2).

The most basic organizational level is the chemical, or molecular, level. The human body is composed of more than a dozen elements, but four of these elements (hydrogen, oxygen, nitrogen, carbon) account for nearly all (>99%) of the atoms in the body. Atoms can be combined to form small molecules, such as water (H_2O) and oxygen (O_2), or larger molecules (macromolecules). The molecular composition of the body consists of water (67%) and macromolecules categorized as proteins (20%), lipids (10%), and carbohydrates (3%).

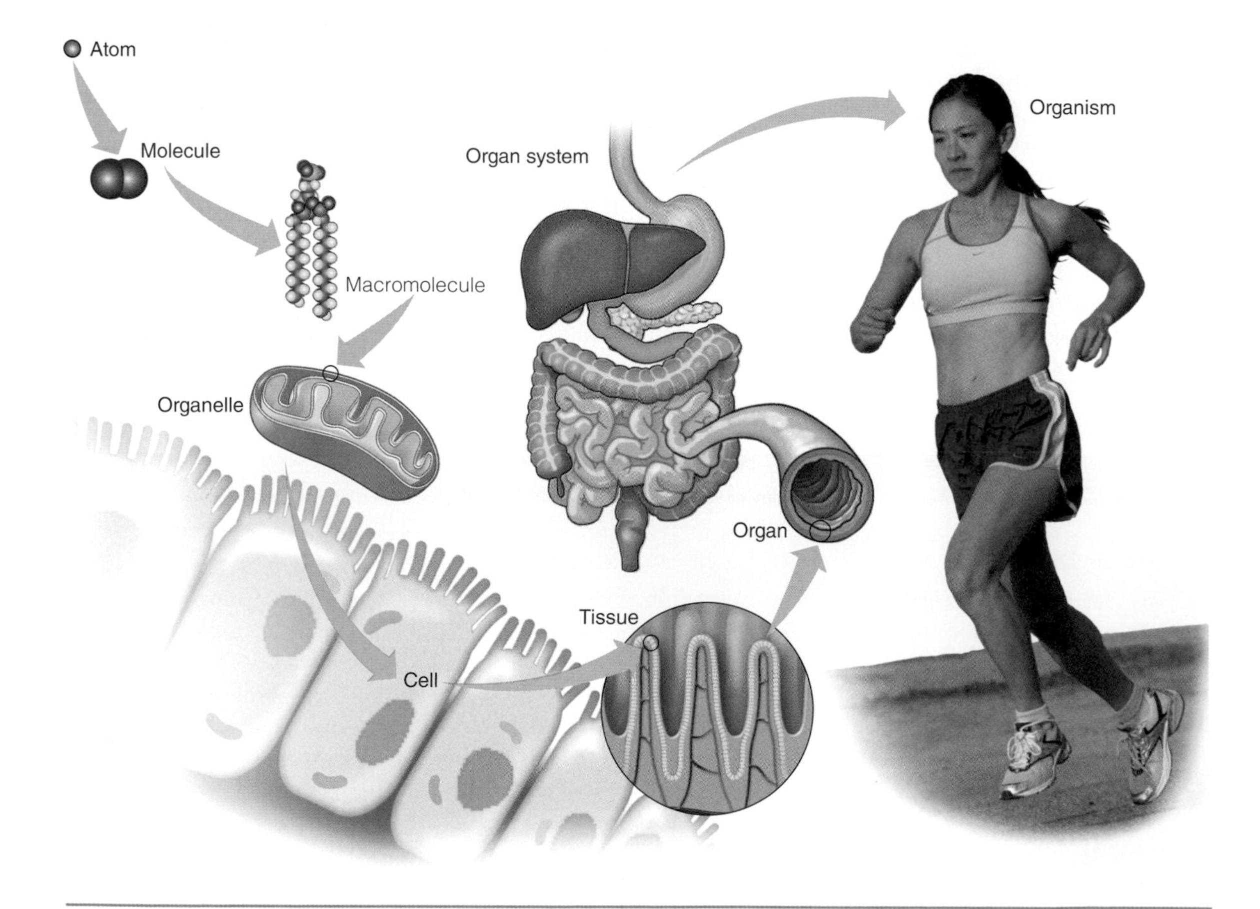

FIGURE 1.2 Levels of structural organization.

The next level of organization consists of cells, the smallest living units in the body. The human body contains trillions of cells that come in many different shapes and sizes. Specific cell types develop from undifferentiated cells known as **mesenchymal cells**, or **stem cells**.

At a higher level, cells combine to form tissues. **Tissues** are composed of similar cells (with specialized functions) and surrounding noncellular material that gives each tissue its structure and determines, in large part, its mechanical properties. The study of tissue structure is known as **histology**. Four primary tissue types exist: epithelial, muscle, nervous, and connective.

- *Epithelial tissue* is found throughout the body. It covers and lines body surfaces both inside and out, provides protection for adjacent tissues and organs, and helps regulate secretion and absorption.
- *Muscle tissue*, unlike any other tissue in the body, has the unique ability to produce force. The body uses these forces, for example, to move the arms and legs, pump blood, and facilitate transport of materials in the digestive and cardiovascular systems.
- *Nervous* (also *nerve*) *tissue*, found in the brain, spinal cord, and peripheral nerves, acts as the body's communication system. It receives sensory signals (e.g., pain, heat), integrates this sensory information at various levels of the brain and spinal cord, and then sends instructions to muscles and glands throughout the body.
- *Connective tissue* serves multiple functions. In general, connective tissues connect, bind, support, and protect. The noncellular matrix of connective tissues is especially important because it determines each connective tissue's mechanical properties.

Organs are structures composed of two or more tissues that have a definite form and function. The heart, for example, is an organ containing muscle tissue (to pump blood), nervous tissue (to control electrical activity), connective tissue (to hold the tissues together), and epithelial tissue (to protect the heart). An **organ system** consists of a group of organs that work together to perform specific functions. The body has 11 organ systems (see table 1.2).

The highest organizational level is the organismic level, which combines all of the organ systems to form an organism, or a single living individual. Organisms are distinct from nonliving things in their ability to perform certain physiological processes. These include metabolism (i.e., the chemical processes that occur in our bodies), responsiveness (e.g., to stimuli such as pain), growth and differentiation, reproduction, and movement.

Muscle Tissue

Just as an automobile contains an engine to make it go, the human body has its own engines that provide the power needed to allow us to move. These engines are our muscles. They have a unique ability among all of the body's tissues to generate force and to contract. Their specialized cells allow muscle to produce tensile forces (sometimes simply referred to as *tension*) and to change their shape by shortening, or contracting.

The neural, biochemical, and mechanical details of how our muscles generate force are quite elegant and complex and are considered in detail in chapter 4. As a prelude, here we consider some of the basic characteristics of muscle tissue.

Muscle tissue is classified as one of three types: cardiac, smooth, or skeletal. Some of the structural distinctions between the three muscle types are shown in figure 1.3.

Cardiac muscle tissue (*myocardium*) is located only in the heart and is responsible for generating the forces required for the heart to pump blood out to the lungs and the rest of the body. The cardiac cells are tightly connected to one another and function, in normal situations, under the coordinated control of pacemaker cells. When the pacemaker cells are dysfunctional, the cardiac cells act randomly in a pattern known as fibrillation. If an external defibrillator is not used to shock the cells back into a coordinated pattern of contraction, death may result.

Smooth muscle is found in and around structures in the circulatory, respiratory, digestive, urinary, and reproductive systems. Smooth muscle facilitates movement of substances through tracts in those systems.

TABLE 1.2 Organ Systems

Organ system	Components	Functions
Cardiovascular	Heart, blood, blood vessels	Transports nutrients and oxygen to cells and tissues and carries away waste products
Digestive	Mouth, esophagus, stomach, intestines, teeth, tongue, salivary glands, liver, pancreas, rectum, anus	Breaks down and processes food through mechanical and chemical actions
Endocrine	Pituitary gland, pineal gland, hypothalamus, ovaries, testes, thyroid gland	Communication within the body using hormones
Integumentary	Skin, hair, nails, fat, sweat glands	Prevents dehydration, stores fat, produces vitamins and hormones, protects internal structures
Lymphatic/Immune	Lymph nodes and vessels, lymph, thymus, spleen, leukocytes	Protects the body from disease-causing toxins, transfers lymph (fluid) between tissues and blood stream
Muscular	Muscles	Enables movement, maintains posture, produces heat, protects underlying structures
Nervous	Brain, spinal cord, nerves	Communication system that collects, transfers, and processes sensory information, responds to external environment, monitors and coordinates organ functions
Reproductive	Female: ovaries, uterus, fallopian tubes, vagina, mammary glands. Male: testes, scrotum, penis, vas deferens, prostate	Production of offspring
Respiratory	Nose, pharynx, larynx, trachea, bronchi, lungs, diaphragm	Provides oxygen to the body by way of gas exchange from external environment and blood gases
Skeletal	Bones, joints, ligaments, cartilage, tendons	Supports, protects, provides framework for movement
Urinary	Kidneys, bladder, urethra, ureters	Excretion of urine, fluid and electrolyte balance

Note: Some sources combine the cardiovascular and lymphatic systems under a heading of *circulatory system* and therefore list 10 organ systems in the human body. Some sources identify an *excretory system* that includes combined functions of the integumentary, digestive, respiratory, and urinary systems.

From a movement perspective, the most important muscle type is **skeletal** (also **striated**) **muscle**. The human body contains more than 600 skeletal muscles, but many of these muscles are relatively small and play little or no role in human movement. Fewer than 100 pairs of skeletal muscles account for the vast majority of movement production and control. The most important of these muscles are discussed in chapter 4.

Connective Tissue

Connective tissue, the most abundant tissue type in the body, encompasses an array of individual tissues with vast differences in structure, function, and mechanical characteristics. Collectively, connective tissues provide support and protection, serve as a structural framework, and help bind tissues together. They also fill the space between cells, tissues, and organs;

produce blood cells; store fat; protect against infections; and help repair damaged tissues. No single connective tissue, however, performs all these functions.

As with all tissues, connective tissues contain cellular and noncellular components. The cells are primarily responsible for each tissue's physiological functions. Connective tissues contain specialized cells unique to the tissue's specific functional requirements (e.g., osteocytes in bone, chondrocytes in cartilage).

The noncellular portion (also termed **extracellular matrix** or **intercellular substance**) largely determines the mechanical characteristics of each connective tissue. The noncellular mineral component of bone, for example, gives bone its hardness and resistance to fracture. Connective tissues are distinguished by a relative abundance of extracellular matrix as compared with other tissues. The matrix is composed of fibers and ground substance.

The extracellular **ground substance** is the primary determinant of whether a connective tissue is solid (e.g., bone), fluid (e.g., blood), or somewhere in between (e.g., tendons, ligaments). In bone, the ground substance is a combination of primarily calcium phosphate and, to a lesser extent, calcium carbonate. The rigidity of these calcium salts is complemented by the flexibility of collagen fibers in bone's matrix to produce a structure that combines the best of both worlds: a tissue that is fracture resistant and remarkably well designed to meet the demands of everyday living. At the other end of the mechanical spectrum is blood, whose ground substance is its plasma. Blood contains cells (red blood cells, white blood cells, platelets) surrounded by a liquid (plasma) that facilitates the transport of materials through the cardiovascular system.

Intercalated disc
Nucleus
Cardiocytes
Striations
Cardiac muscle

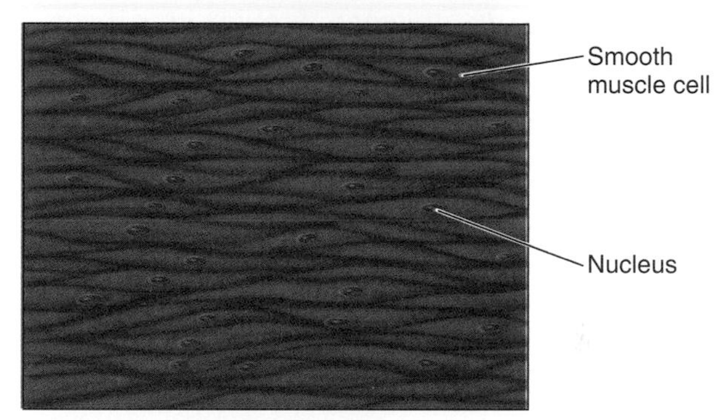

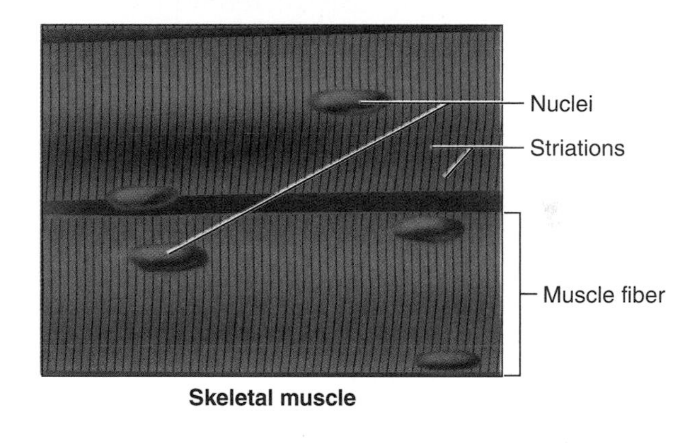

FIGURE 1.3 Different types of muscle tissue.

We now provide some details of several connective tissues (tendons, ligaments, cartilage) that play an important role in the anatomy of human movement. A detailed examination of bone is presented in the next chapter.

Tendons

A **tendon** is a cordlike connective tissue that connects skeletal muscle to bone (figure 1.4). The cellular component of tendons consists of many fibrocytes embedded in an extracellular matrix composed primarily of collagen fibers. The function of tendon is to transmit forces generated by the muscle to the bone to produce and control movement. To do this, the tendon's collagen fibers are aligned parallel to the tendon's line of action. The body also uses broad tendonlike sheets, known as **aponeuroses**, to cover a muscle's surface or to connect the muscle to another muscle or other structures.

The connective structure of a tendon creates three structural zones: the body of the tendon, the connective region between tendon and bone (**osteotendinous junction**), and the connective region between tendon and muscle (**myotendinous junction**, or **musculotendinous junction**). The osteotendinous junction contains microscopic transition zones of progressively stiffer tissues between the tendon and the bone. This structural feature distributes the load being transmitted from the muscle to the bone, thus reducing the chance of injury.

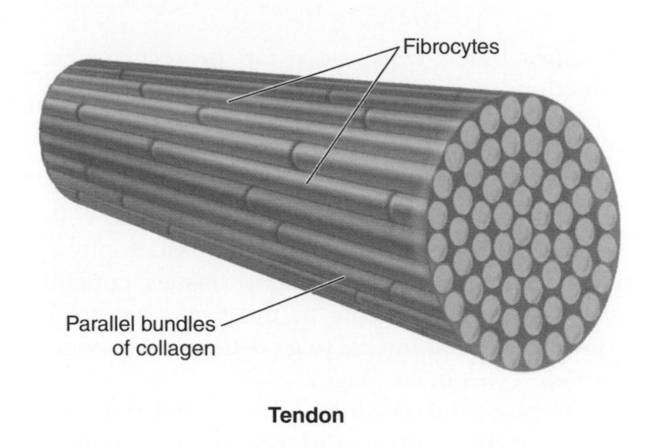

FIGURE 1.4 Tendon structure. Note the parallel arrangement of collagen fibers.

The myotendinous junction similarly attempts, through its structure, to distribute forces passing through it. The collagen fibers of the tendon are intertwined with folded cell membranes of the muscle. In normal situations, the tendon performs its force-transmission function well. If the forces being transmitted exceed the strength of the fibers (or the whole tendon), however, the tendon may be injured. Chronic overuse of tendon may result in an inflammatory response, or **tendinitis** (also *tendonitis*). This inflammation may affect the tendon itself or related structures such as the tendon's outer covering (sheath) or associated **bursa** (fluid-filled sacs that help cushion impact or reduce friction) adjacent to the tendon.

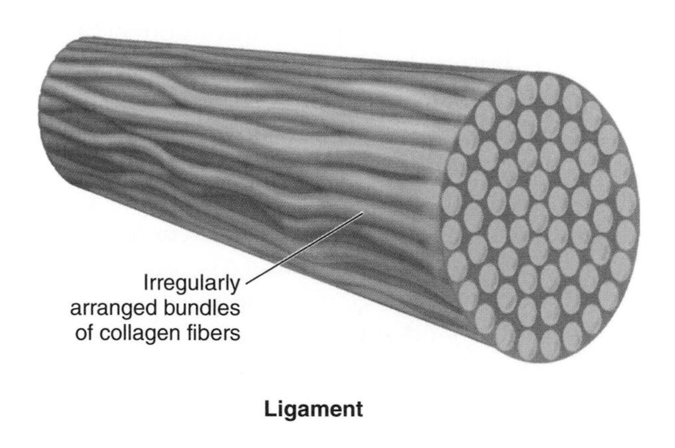

FIGURE 1.5 Ligament structure. Note the irregular arrangement of collagen fibers.

Tendons also contain sensory structures, such as Golgi tendon organs (GTOs), that detect changes in muscle force and tension. GTOs are located near the myotendinous junction.

Ligaments

A **ligament** is a connective tissue that joins bone to bone. Its primary function, like that of tendons, is to resist tensile forces (figure 1.5). Ligaments protect bone-to-bone connections by resisting excessive movements or dislocation of the bones. In that role, ligaments are sometimes referred to as *passive joint stabilizers*. A ligament's fibers (primarily collagen with some elastic and reticular fibers) are oriented according to their function. The fibers may be arranged parallel, obliquely, or even in a spiral formation. In addition, some ligaments are easily identifiable, isolated structures. Others appear as indistinct thickenings of the fibrous joint capsule that surrounds certain joints.

Joint ligaments contain a variety of sensory receptors capable of providing the nervous system with information about movement, position, and pain. The exact role of these sensory structures remains somewhat controversial and is the subject of ongoing study.

Ligament injury (**sprain**) occurs when applied forces exceed the tensile strength of the ligament's extracellular matrix. The severity of injury can range from mild to severe. Mild (also type I) sprains are characterized by local tenderness, no visible injury, and no loss of joint stability. Moderate (also type II) sprains involve a partial tearing of the ligament, visible swelling, marked tenderness, and some loss of stability. Severe (also type III) sprains result in complete tearing of the ligament with gross swelling, considerable pain, and joint instability.

Cartilage

Cartilage is a stiff connective tissue whose ground substance is nearly solid. Cartilage cells (chondrocytes) have the ability to produce both the fibers and ground substance of the extracellular matrix. The types and amount of matrix constituents distinguish three kinds of cartilage: hyaline cartilage, fibrocartilage, and elastic cartilage. None of these cartilage types has intrinsic blood, lymph, or nerve supply. The absence of circulatory structures makes it necessary for cartilage to receive its nutrients and remove metabolic waste through diffusion.

Hyaline cartilage gets its name from its glassy appearance (*hyalos* = glass) and is the most common cartilage type. It is found on the articular surfaces of most joints, at the anterior portions of the ribs, and in components of the respiratory system (e.g., trachea, nose, bronchi). Hyaline cartilage also serves as the precursor tissue to bone in the developing fetus (as discussed in chapter 2).

Both strong and flexible, **fibrocartilage** is functionally well suited to the role it plays at stress points in the body where friction could be a problem. Fibrocartilage also serves as a filler material between hyaline cartilage and other connective tissues and is found near joints, ligaments, and tendons and in the intervertebral discs.

Elastic cartilage, as its name suggests, has a great deal of flexibility. Its matrix contains elastic and collagen fibers. The matrix of elastic ligaments appears more yellowish in color than hyaline or fibrocartilage does because of its higher proportion of elastic fibers. Elastic cartilage is found in the external ear, nose, and portions of the respiratory system.

Anatomical References and Terminology

Although knowing and understanding the structural details of the human body are essential, so too is communicating this knowledge with others in the field. To facilitate communication, scientific disciplines share a system or set of terms (**nomenclature**). The anatomical nomenclature uses terms to identify and describe body regions, body positions, movement planes, axes of rotation, joint positions, and movements. These terms allow us to specify the location of the body or any of its segments at a point in time and to describe joint movements. This common nomenclature allows professionals in many areas of kinesiology and related fields to communicate effectively. Note, however, that there is *not* universal agreement on all anatomical terms and their use. You are encouraged to be flexible to occasional differences in nomenclature.

Body Positions and Body Planes

The standard body reference position is called **anatomical position** (figure 1.6). The body is erect with the head facing forward and the arms hanging straight down, palms facing forward. While anatomical position may not be one people regularly assume in everyday life, it nonetheless provides us with a suitable reference point. In this position, all joints are said to be in a **neutral position**. Joint positions or ranges of joint motion are measured starting from these reference positions.

In anatomical position, the human body can be sectioned into three mutually perpendicular flat surfaces, or planes, as shown in figure 1.7. Sectioning the body in this way is useful for describing movements of the body and its segments. The three principal, or primary, planes are the sagittal (also median), frontal (also coronal), and transverse planes. Each plane is at right angles to the other two. Movement occurring parallel to a given plane has that as its plane of action.

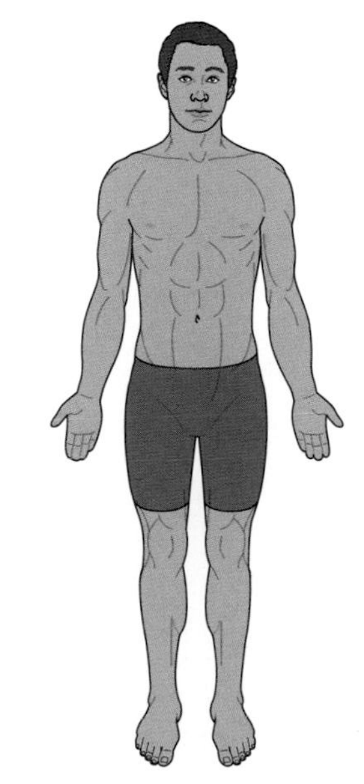

FIGURE 1.6 Anatomical position.

Sagittal planes, in general, divide the body into left and right sections. If a particular sagittal plane is placed along the midline, dividing the body in half, it is termed a **midsagittal** or **median plane**. Any sagittal plane offset from the midline to one side or the other is called a **parasagittal plane**. Body movements are said to occur in the sagittal plane if they are made parallel to the midsagittal plane. For example, forward bending of the trunk (figure 1.8) and flexion of the elbow (from anatomical position) occur in the sagittal plane. The **frontal plane** sections the body into anterior and posterior parts. Moving the arms away from the midline, leaning to one side or the other, and jumping jacks all occur in the frontal plane. The third principal plane, the **transverse plane**, divides the body into superior and inferior parts. Rotational movements such as twisting the head from side to side occur in the transverse plane.

Although it is helpful to define the three principal planes and describe movements in these planes, many human movements do not occur in a single plane and instead cross two or more planes. Description of these more complex movements often can

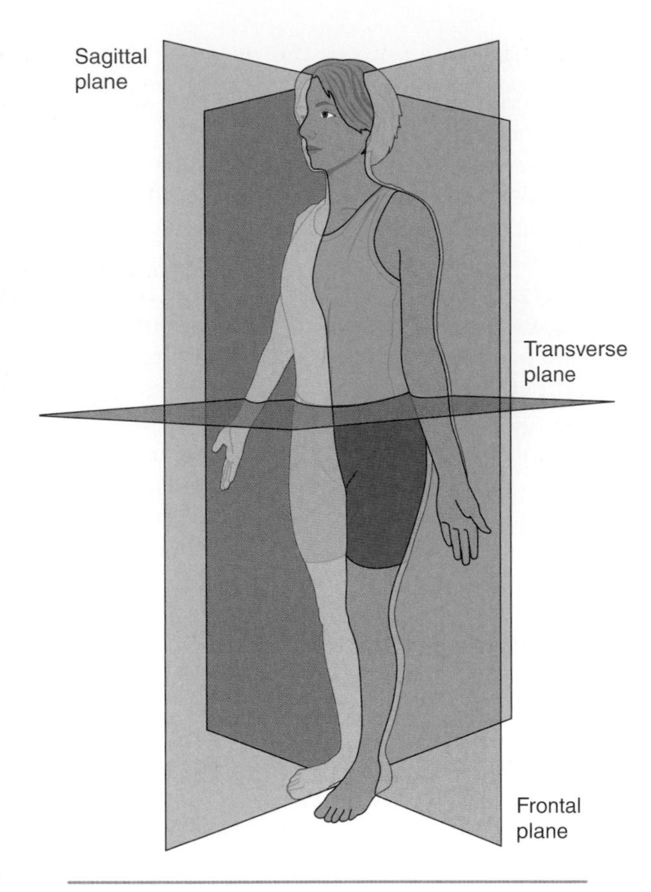

FIGURE 1.7 Principal body planes.

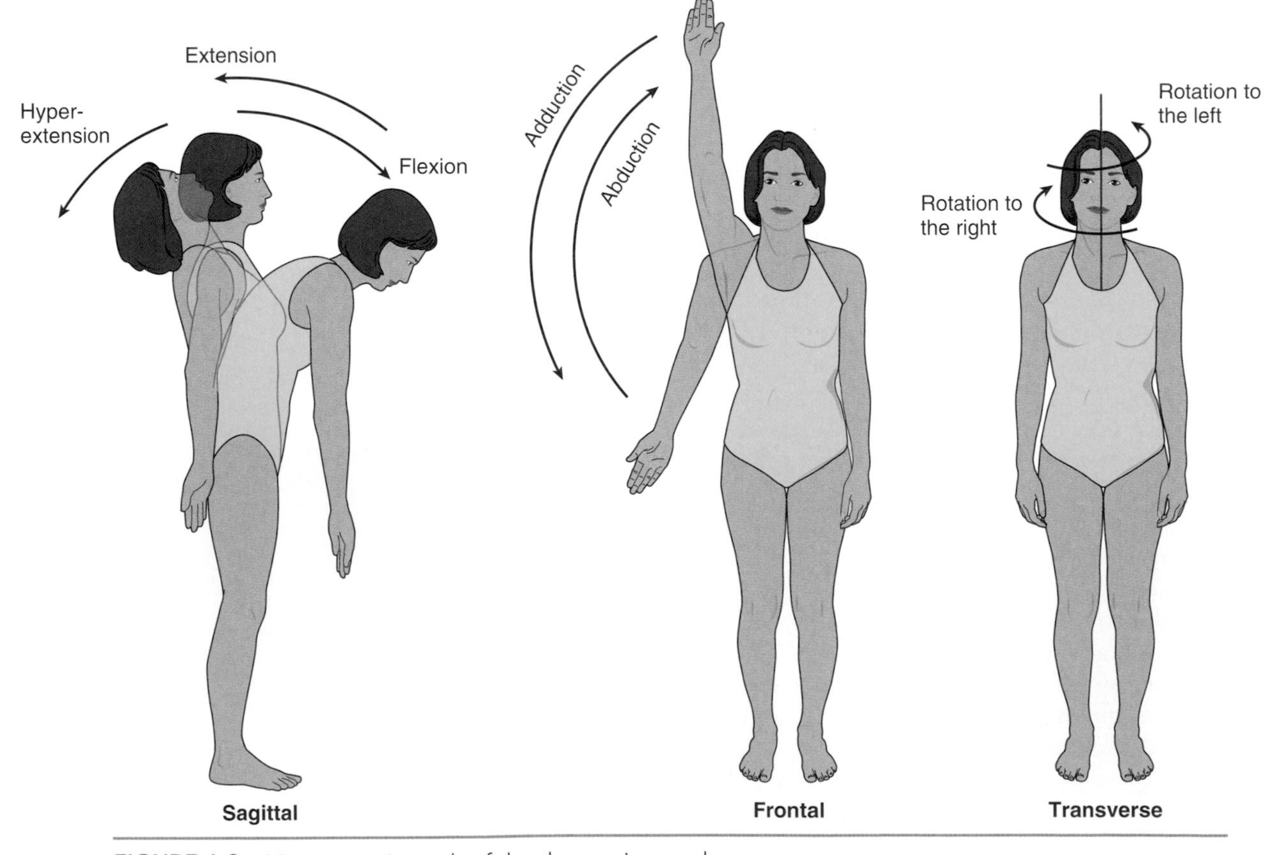

FIGURE 1.8 Movement in each of the three primary planes.

be challenging. If our movements were constrained to a single plane, they would appear robotic and unnatural. We present the planar description and analysis of movement as a starting point, with no presumption that such a scheme will be able to describe all movements.

Directional Terms and Axes of Rotation

Anatomical terms of direction describe the relative position, orientation, or direction of body surfaces, planes, or segments. Common terms are summarized in table 1.3. These terms typically occur in pairs (e.g., anterior and posterior), with each member of the pair having a meaning opposite of the other.

Rotational, or angular, movements of body segments happen around specific axes of rotation. An **axis of rotation** is an imaginary line about which rotation occurs. The axis is always perpendicular (i.e., at a right angle) to the plane of action. For example, lifting (rotating) the arm away from the midline occurs in the frontal plane. The axis of rotation is a line through the center of the shoulder joint, with an anterior–posterior orientation perpendicular to the frontal plane. The axes of rotation for joint-specific movements are detailed in chapter 3.

TABLE 1.3 Anatomical Terms of Direction

Term	Description
Anterior (ventral)	In the front or toward the front
Anteroinferior	In front and below
Anterolateral	In front and to one side (usually outside)
Anteromedial	In front and toward the inner side or midline
Anterosuperior	In front and above
Bilateral	On two, or both, sides
Contralateral	On the opposite side
Deep	Away from the surface of the body
Distal	Farther away from the axial skeleton (trunk)
Inferior (caudal)	Toward the feet or ground, or below relative to another structure
Ipsilateral	On the same side
Lateral	Away from the midline of the body or structure, or farther from the midsagittal plane
Medial	Toward the midline of the body or structure, or closer to the midsagittal plane
Posteroinferior	Behind and below, in back and below
Posterior (dorsal)	Behind, in back, or toward the rear
Posterolateral	Behind and to one side (usually outside)
Posteromedial	Behind and toward the inner side or midline
Posterosuperior	Behind and above
Prone	Body lying on the belly with face downward
Proximal	Closer to the axial skeleton (trunk)
Superficial	Toward or nearer to the surface of the body
Superior (cephalic)	Toward the head, or above relative to another structure
Supine	Body lying on the back with face upward
Unilateral	On one side

Study of Human Movement

The study of human movement requires a multidisciplinary perspective that considers anatomical, mechanical, physiological, psychological, environmental, sociological, and cultural factors. Each of these areas offers unique insights into the complexities of human movement. Only by integrating information from all these areas, however, can we reach a comprehensive understanding and appreciation of human movement.

Anatomy, as described earlier, is the study of the body's structure and how the different structural components functionally relate to one another. Biomechanics studies the mechanical aspects of movement and explores the role of force, time, and distance. Physiology deals with the functions of the body parts and systems and, in our current context, focuses on movement initiation and control. Although our discussions in this book focus mostly on these three areas of study, movement is also greatly affected by psychological, environmental, sociological, and cultural variables.

Psychology deals with the mind and behavior. Psychological factors such as perception and motivation play an important role in how we move. An athlete in the throes of intense competition, for example, can move in ways that are not possible while in a less-motivated state.

Human movement does not occur in isolation, but rather is influenced by environmental factors. These external influences include weather, surface characteristics, apparel, and equipment. The walking pattern of a person struggling across a hot desert terrain, for example, would be much different from that of someone cautiously negotiating an icy sidewalk in the middle of winter.

Sociology is the study of social institutions and social relationships. Sociological factors play an essential role in movement, such as in team settings when one person's performance can be enhanced or restricted by interactions with teammates and opponents. Movement can also be influenced by a variety of cultural norms and traditions.

We emphasize that each of these ways of studying human movement offers valuable information and insights, but only by taking a multidisciplinary approach can we truly understand movement.

Concluding Comments

The goal of this first chapter is to establish a conceptual framework and perspective of anatomy and human movement that will enhance our understanding of the human body and how it works. In his treatise on the biophysical dynamics of movement and sport, John Jerome wrote, "Analyze us finely enough and in the end we are nothing more than electrochemical soup" (1980, p. 43). In one sense, Jerome is correct because we *are* a complex composite of chemical compounds. But in so many ways we are much more than just soup. Our bodies are composed of countless molecules arranged hierarchically to form cells, tissues, and organs. The organs are gathered in organ systems that collectively form the human organism. The integrated functioning of all the body's systems allows us to perform the myriad tasks essential to life, in particular the ability to move.

In the following chapters, you will have the opportunity to go beyond mere memorization of details and grow in your understanding and appreciation of this most marvelous of creations, the human body, unparalleled in its complexity, variability, individuality, and adaptability. Keep in mind, too, that the human body nonetheless is imperfect, both in its structure and its function, and subject to influences that can limit movement proficiency.

Go to the web study guide to access critical thinking questions for this chapter.

Suggested Readings

Behnke, R.S. (2012). *Kinetic anatomy* (3rd ed.). Champaign, IL: Human Kinetics.

Cappozzo, R., Marchetti, M., & Tosi, V. (1992). *Biolocomotion: A century of research using moving pictures.* Rome: Promograph.

Cech, D.J., & Martin, S. (2011). *Functional movement development across the life span* (3rd ed.). St. Louis: Elsevier Saunders.

Floyd, R.T. (2018). *Manual of structural kinesiology* (20th ed.). New York: McGraw-Hill.

Hale, R.B., & Coyle, T. (1988). *Albinus on anatomy.* Mineola, NY: Dover.

Hale, R.B., & Coyle, T. (2000). *Anatomy lessons from the great masters.* New York: Watson-Guptill.

Jenkins, D.B. (2008). *Hollinshead's functional anatomy of the limbs and back* (9th ed.). Philadelphia: Saunders.

Keele, K.D. (1983). *Leonardo da Vinci's elements of the science of man.* New York: Academic Press.

MacKinnon, P., & Morris, J. (1994). *Oxford textbook of functional anatomy* (Vol. 1, rev. ed.). Oxford: Oxford University Press.

Martini, F.H., Timmons, R.J., & McKinley, M.P. (2014). *Human anatomy* (8th ed.). New York: Pearson.

O'Malley, C.D., & Saunders, J.B. de C.M. (1997). *Leonardo da Vinci on the human body.* Avenel, NJ: Wings Books.

Saunders, J.B. de C.M., & O'Malley, C.D. (1973). *The illustrations from the works of Andreas Vesalius of Brussels.* Mineola, NY: Dover.

Thelen, E., & Smith, L.B. (2000). Dynamic systems theories. In W. Damon (Ed.), *Handbook of child psychology* (5th ed., pp. 563-634). New York: Wiley.

Trew, M., & Everett, T. (2001). *Human movement: An introductory text* (4th ed.). Edinburgh: Churchill Livingstone.

2

Osteology and the Skeletal System

Objectives

After studying this chapter, you will be able to do the following:

- Describe the functions of the skeletal system
- Name and describe the function of bone cells
- Explain the macroscopic and microscopic structure of bone
- Describe skeletal system organization, including bones of the lower body, upper body, and trunk
- Explain the events involved in bone modeling, growth, and development
- Explain the processes of bone remodeling
- Explain the primary factors involved in bone health, including exercise, diet, and aging

In light of the skeletal system's importance in supporting us, protecting us, and allowing us to move, it is both ironic and unfortunate that the image of a skeleton often is associated with fear and death. From the skeletons that abound during the frightful Halloween season to the stark picture of skeletal remains strewn across a desert landscape in a Hollywood movie, bones often are used to portray a feeling of lifelessness and dread. In reality, bones are dynamic organs that elegantly blend form with function and serve many useful purposes.

Functions of the Skeletal System

The skeletal system plays an integral role in mechanical and physiological functions essential for everyday life. Mechanically, the skeleton provides structural support for the body, acts as a lever system to permit movement of body segments and joints, and protects organs of the body's other systems. The bones of the skeletal system also act as a storehouse for minerals and contain tissues responsible for blood cell production. These five functions, varied as they are, all play critical roles in a comprehensive understanding of how the skeletal system facilitates human movement.

- *Structural support.* The skeletal system consists of 206 bones that collectively form a supporting framework and give shape to the body (figure 2.1). This framework bears applied loads with limited deformation, supports organs and soft tissues, and provides attachment sites for many muscles. Without bones, we might be structurally likened to an intelligent jellyfish. The bones also give shape, or form, to the body. Bones of the skull, for example, determine the shape of the head. The size and shape of bones often are related to the functions they perform. Women, for example, have a relatively wider pelvis than do men to allow for passage of a newborn during delivery. Similarly, the long bone (tibia) of the lower leg is shaped like a column to support the body's weight.
- *Movement.* Bones function as mechanical levers. When muscles (attached through tendons) pull on a bone, they change the magnitude and direction of forces and create rotational effects at joints (e.g., elbow, knee). The joints then rotate about an imaginary line known as the axis of rotation. Joint movements can range from the subtle motion of a violinist's fingers to complex, multijoint movements such as a vertical jump in which simultaneous extension of the hip, knee, and ankle powerfully propel the body upward.
- *Protection.* The hardness of bone makes it an ideal material to protect other more vulnerable organs in the body. For example, the skull protects the brain and eyes; the vertebrae shelter the spinal cord; the ribs protect the heart, lungs, and other visceral organs; and the pelvis shields organs of the urinary and reproductive systems. Without the protection provided by the skeletal system, the human body would be dramatically more susceptible to injury.
- *Mineral storehouse.* The calcium salts found in bone provide an important mineral reserve used by the body to maintain normal concentrations of calcium and phosphorus ions in body fluids. The body contains approximately 1 to 2 kg of calcium, 98% of which is contained in the bones in the form of calcium phosphate crystals. The small amount of remaining calcium is distributed throughout the body and is essential to physiological processes such as muscular contraction and nerve conduction.
- *Blood cell production.* The cavities and spaces in bone house a loose connective tissue known as **marrow**. The two types of marrow are named according to their color. *Yellow marrow* contains large numbers of fat cells (**adipocytes**) that produce its characteristic color. It commonly resides in the marrow cavity in the shaft of long bones and provides a reserve source of energy. *Red marrow* consists of a mixture of red and white blood cells and platelets, along with the precursors, or stem cells, that produce them. The stem cells form blood cells in a process called **hematopoiesis** (also *hemopoiesis*). Important sites of blood cell formation include the proximal ends of the femur and humerus, sternum, ribs, and vertebrae.

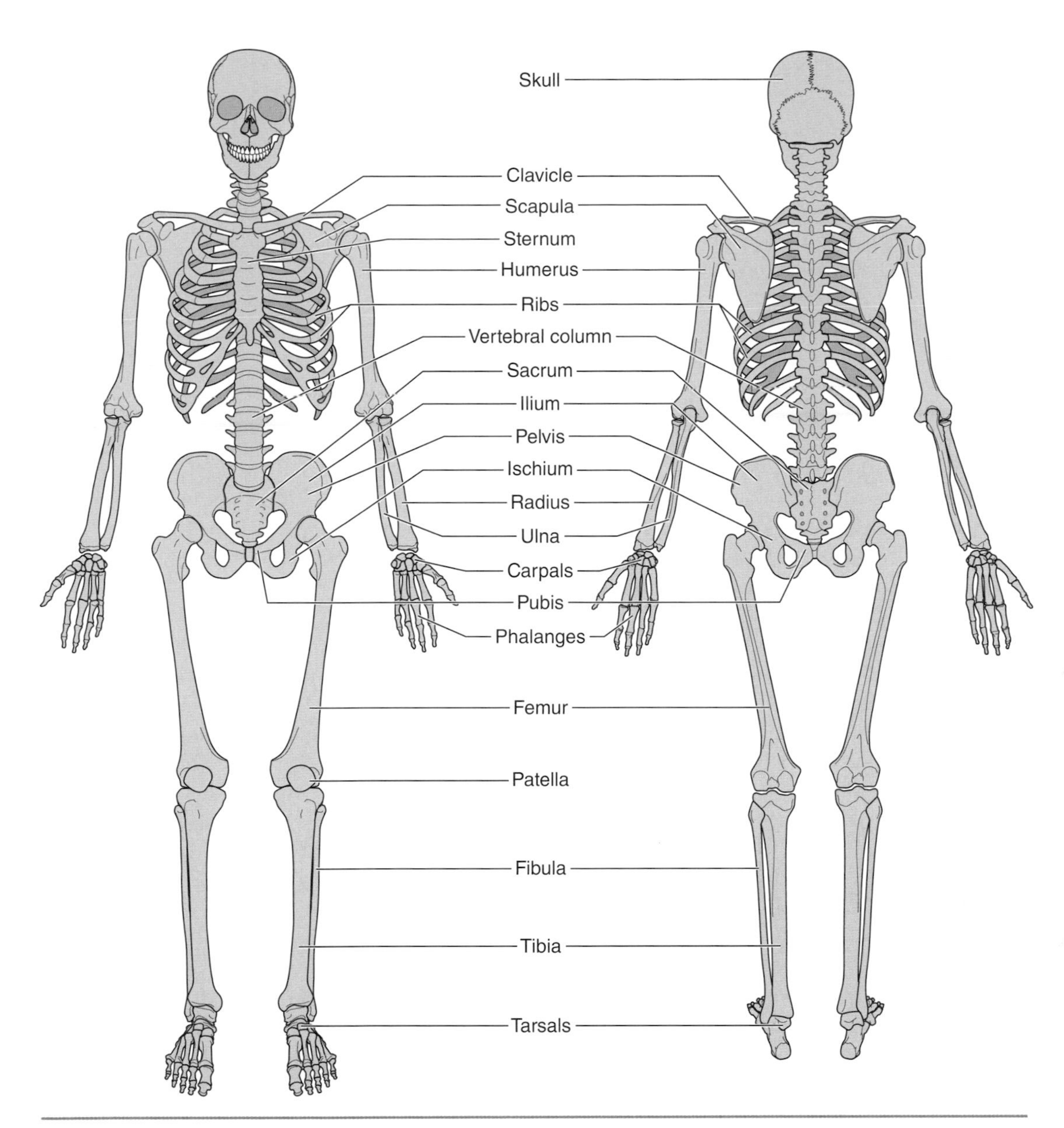

FIGURE 2.1 Anterior and posterior views of the human skeletal system.

Applying the Concept

Stronger Than Steel?

Is bone stronger than steel? In simple terms, the answer is yes. Weight for weight (or pound for pound), bone is up to five times stronger than steel. But that does not tell the whole story. Bone is generally stronger than steel, in relative terms, when resisting compressive (i.e., pushing together) forces. If the forces are tensile (i.e., pulling apart) or create torsion (i.e., twisting), the answer is more complex. Also, the rate (i.e., how fast) the force is applied plays an important role. Nonetheless, human bones are marvels of nature, strong enough to resist most forces, yet light enough to allow us to move freely.

Bone Histology and Composition

As do all tissues, bone has a cellular component and a noncellular (extracellular) component. At the cellular level, specialized bone cells produce new bone and monitor and maintain their surrounding matrix. The extracellular component, or matrix, composed of mineral salt crystals and collagen fibers, gives bone its rigidity and strength. Because bones form the levers that allow us to move, healthy bones are essential for effective movement. Bone health relies on the action of bone cells and their surrounding matrix.

Bone Cells

The cellular component of bone is composed of four cell types, namely osteoprogenitor (mesenchymal) cells, osteoblasts, osteocytes, and osteoclasts. The **osteoprogenitor cells** are relatively few in number and are undifferentiated mesenchymal cells with the ability to produce daughter cells that can differentiate to become osteoblasts. The differentiation of mesenchymal cells into osteoblasts involves a 2- to 3-day process that seems to be triggered by mechanical stresses applied to the tissue.

The resulting **osteoblasts** are mononuclear (single nucleus) cells described as "producers" or "formers" because they produce the organic portion (**osteoid**) of the extracellular matrix. In performing this function, the osteoblasts are responsible for the materials required for new bone formation.

As an osteoblast matures, it becomes smaller and less metabolically active, finally becoming encapsulated by the calcified osteoid in a small pocket known as a **lacuna** (pl. *lacunae*). The osteoblast's role changes from forming new bone to monitoring and maintaining the bony matrix. These smaller mature cells are now called **osteocytes**.

Osteoclasts are large, multinucleated cells formed by the fusion of monocytes that originate in the red bone marrow. The osteoclasts are described as "resorbers" because their primary role is the **resorption**, or breakdown, of bone. These cells produce and release acids that demineralize the bone and enzymes that dissolve collagen fibers in the matrix.

Bone Matrix and Bone Structure

The extracellular matrix of bone consists of water, calcium phosphate (hydroxyapatite) crystals [$Ca_5(PO_4)_3(OH)$ or $Ca_{10}(PO_4)_6(OH)_2$], collagen, and small amounts of ground substance. These elements are divided into organic and inorganic portions. The organic components (collagen and ground substance) are primarily produced by the osteoblasts and to a much lesser extent by the osteocytes. The collagen fibers account for about 30% of the bone's dry weight and provide resistance to tension, bending, and twisting.

The structure of bone can be considered at several levels. Human bones come in a wide variety of shapes and sizes and are classified at a gross level according to their shape (figure 2.2):

- Long, such as the femur
- Short, such as the carpal
- Flat, such as the sternum and manubrium
- Irregular, such as a vertebra
- Sesamoid, such as the patella

Another bone classification is sutural, found in the skull. Bone shape plays an essential role in determining how much movement is allowed at a given joint, as is explained in the next chapter.

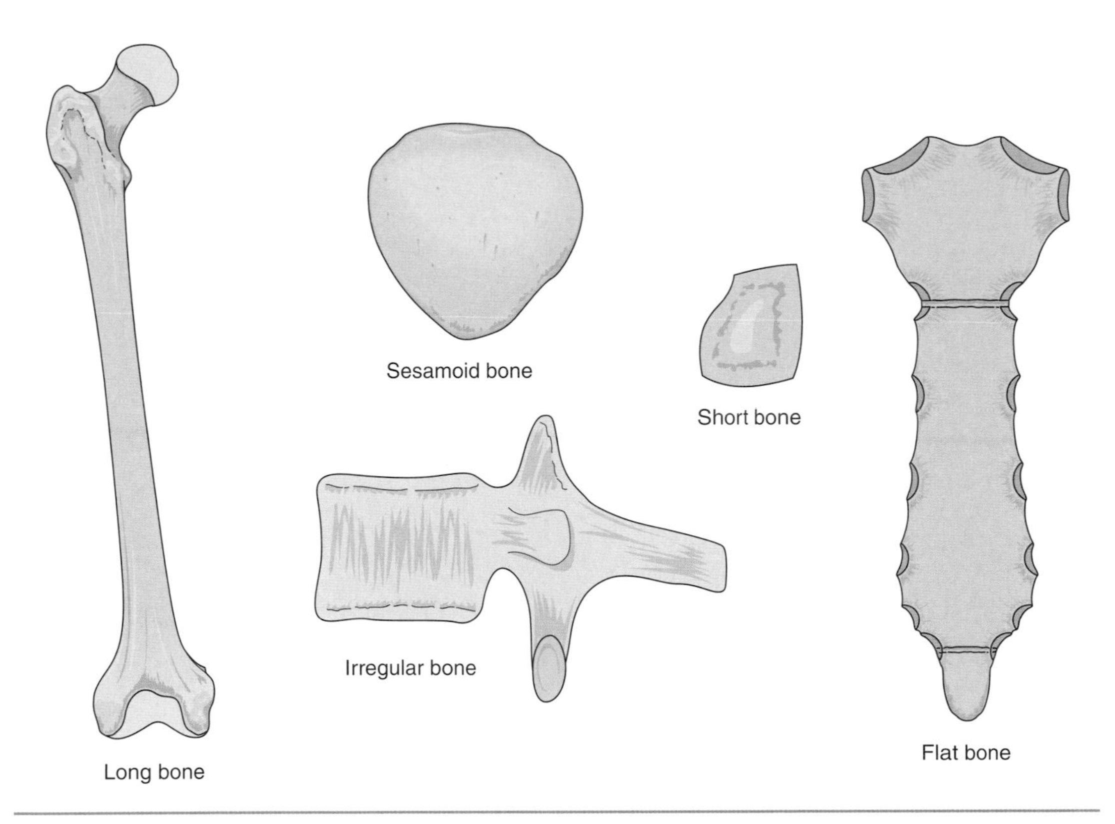

FIGURE 2.2 Bone shapes.

Examination of bone's gross structure reveals two types of bone, based on the density and structural arrangement of its components. At the microscopic level, each of these two bone types exhibits distinct structural characteristics.

Macroscopic Bone Structure

A cross section of long bone (figure 2.3) reveals two types of bone, distinguished by the tissue's **porosity**. Porosity is measured by the amount of soft tissue in the bone. Bone with little soft tissue has low porosity, while bone with a greater amount of soft tissue has high porosity. From the perspective of bone **density** (i.e., the inverse of porosity), measured by the amount of hydroxyapatite crystals per unit volume, bones with high density have low porosity. Conversely, bones with low density have high porosity. In theory, bone porosity could range anywhere along a continuum from 0% to 100%. In reality, however, bones typically have either high or low porosity.

Bone with low porosity (5%-10%) is called **compact** (also **cortical) bone**. Compact bone is found in the shaft **(diaphysis)** of long bones and forms the hard outer covering **(cortex)** of all bones. This protective outer bony surface layer sometimes is described as a **cortical shell**. The cortical shell is in turn covered by a fibrous connective tissue, known as the **periosteum**, that covers the entire bone except at the articular (joint) surfaces, which are covered by a thin (~1-3 mm) protective layer of hyaline cartilage, referred to, based on its location, as *articular cartilage*.

Bone with high porosity (75%-95%) is termed **spongy** (also **trabecular** or **cancellous) bone**. Spongy bone is found under the cortical shell in vertebrae, flat bones, and the **epiphyses** of long bones. The bony matrix of spongy bone consists of plates of bone tissue known as trabeculae. Red bone marrow fills the space between the trabeculae.

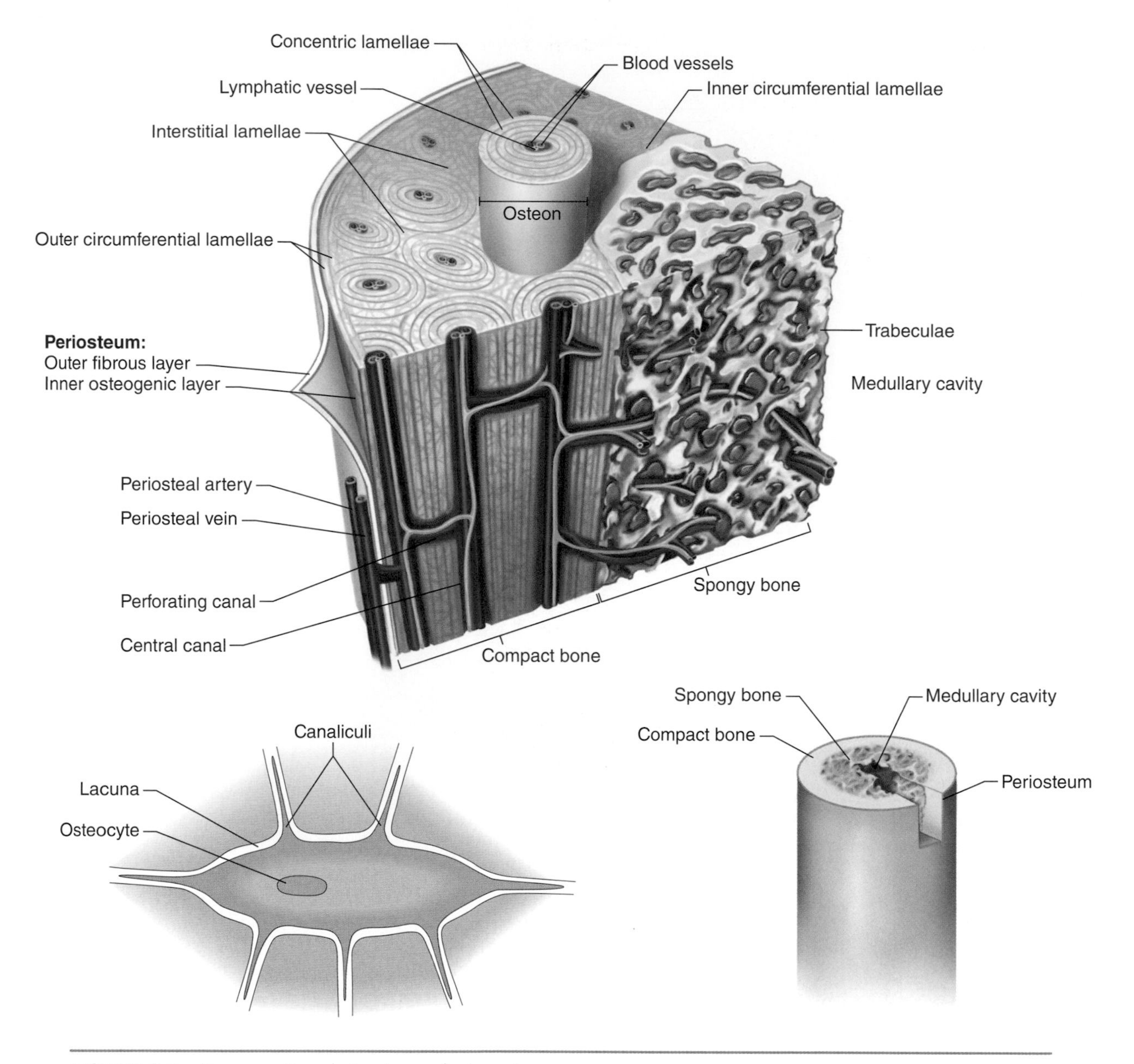

FIGURE 2.3 Cross-sectional structure of a long bone.

Note that spongy and compact bone contain the same cellular and extracellular components; the difference is in the structure and density, not in the constituent materials.

Microscopic Bone Structure

Compact bone, when viewed by the unaided eye, appears as a solid, dense mass of bone. Microscopic evaluation, however, reveals a complex structure with a variety of connecting canals (see figure 2.3). The fundamental structural unit of compact bone is the **osteon** (also *Haversian system*). Each osteon consists of concentric layers (rings) of bone arranged around a central canal that houses blood vessels. The osteon's rings of bone, known as concentric lamellae, are configured like the rings around the bull's-eye of a target or the age rings in the cross section of a tree. Larger rings of bone, called circumferential lamellae, form the inner and outer boundaries of the bone's surface (figure 2.4).

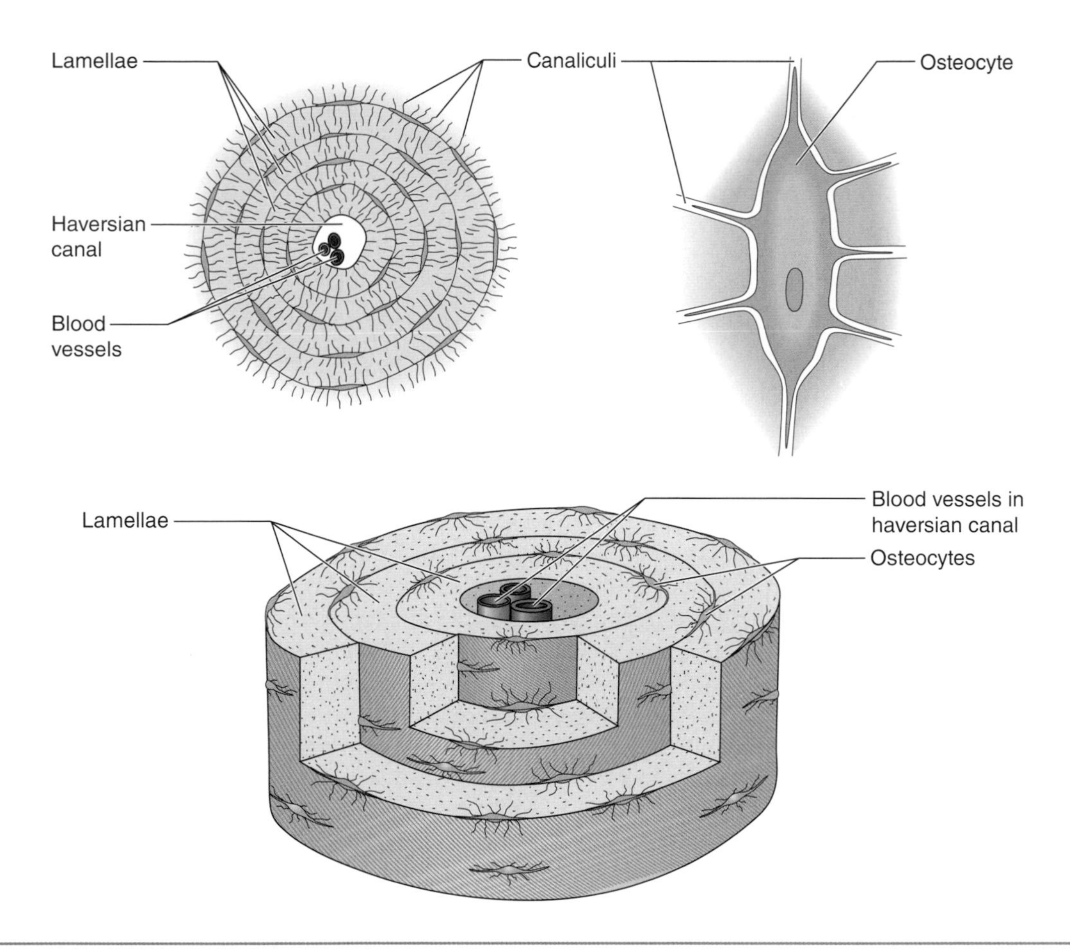

FIGURE 2.4 Structure of an osteon.

The capillaries of the central canal receive their blood supply from perforating canals (Volkmann's canals) that run transversely between adjacent osteons. These canals also connect osteons and facilitate inter-osteon nutrient supply and communication, as well as supply blood to the periosteum on the bone's outer surface.

In contrast to compact bone, spongy bone is much less organized. At first glance, it appears to be arranged randomly. Spongy bone is virtually devoid of osteons. However, the arrangement of spongy bone's bony plates (trabeculae) is far from random. The trabeculae are oriented to accept the forces applied to the bone.

Bone Modeling, Growth, and Development

Modeling refers to the formation (addition) of new bone. Bone modeling starts prior to birth. Skeletal development begins when mesenchymal cells from the mesodermal layer of the fetus condense and differentiate into osteoblasts. In some bones (e.g., cranium and facial bones and, in part, the ribs, clavicle, and mandible), the cellular condensations form fibrous matrices that subsequently ossify directly in a process known as *intramembranous ossification*.

In most limb and axial bones, mesenchymal condensations form a cartilaginous model (*anlage*) of the bones, rather than proceeding directly to calcification and ossification. The anlage, made of hyaline cartilage, then proceeds through a process known as *endochondral ossification*. Beginning in the middle of the anlage, in an area known as the *primary ossification center*, the hyaline cartilage undergoes a series of changes and eventually is replaced by

Applying the Concept

The Short and Tall of It All

Consistent with the anatomical concept of variability discussed in chapter 1, humans come in all shapes and sizes. Here are the verified tallest and shortest people of all time.

Tallest woman: Zeng Jinlian (People's Republic of China); 8'1" (2.48 m)
Tallest man: Robert Wadlow (United States); 8'11"(2.72 m)
Shortest woman: Pauline Musters (Netherlands); 1'11" (0.58 m)
Shortest man: Chandra Bahadur Dangi (Nepal); 1'9.5" (0.55 m)

bony material. This replacement proceeds from the bone's center toward the ends (*epiphyses*) of the bone. Subsequently, similar changes happen in the epiphyses, with replacement of the hyaline cartilage by bony material, at what are termed *secondary ossification centers*. The cartilage remaining between the bony regions of the diaphysis and epiphysis is called the *growth plate*. The growth plate eventually disappears as the bony regions fuse together in an event known as *closure*. At this point, the bone is done growing longitudinally (i.e., the bone will not get any longer). Once all long bones reach closure, a person will not grow any taller.

Bone is, however, not done with its growth and development. The modeling process continues and is characterized by continuous bone deposition on any bony surface that produces a net gain in bone. During modeling, osteoblasts and osteoclasts are not active along the same surface; resorption occurs along one surface while deposition occurs along another. For example, in the development of a long bone, even after closure, osteoblasts continuously add bone to the periosteal (outer) surface, while osteoclasts enlarge the central canal by resorbing bone on the endosteal (inner) surface.

Skeletal System Organization

The generally accepted number of bones in the skeletal system is 206. This number may, however, vary from person to person. Some individuals lack certain bones, while others have extra ones (e.g., small **sesamoid bones** that develop in tendons to reduce friction where tendons pass over bony prominences). The 206 bones are divided into two subsystems, or divisions: the **axial skeleton** and the **appendicular skeleton** (table 2.1). The primary purpose of the axial skeleton (skull, spinal column, and thoracic cage) is to protect and support internal organs and provide sites for muscle attachment. The appendicular skeleton, in contrast, consists of the bones and supporting structures of the upper and lower limbs and is involved in movements used in everyday life (e.g., walking, lifting).

Each of our bones has its own unique shape, size, and physical characteristics that match its function. We highlight here bones of the vertebral column, thorax, pectoral girdle, upper limbs, pelvic girdle, and lower limbs, because these bones are the ones that support us and allow us to move from place to place and interact with our immediate environment.

Trunk: Vertebral Column, Spinal Curvatures, and Thorax

The vertebral column (or spine) consists of 26 bones (vertebrae) that span from the base of the skull down to the pelvis (figure 2.5). The spine is divided into five regions: cervical region (7 vertebrae), thoracic region (12 vertebrae), lumbar region (5 vertebrae), sacrum, and coccyx. The sacrum is a single bone formed during development by the fusion of 5 sacral vertebrae. The coccyx is formed by the fusion of (usually) 4 vertebrae.

TABLE 2.1 Bones of the Adult Skeletal System

Division of the skeleton	Structure	Number of bones
Axial skeleton	Skull	
	Cranium	8
	Face	14
	Hyoid	1
	Auditory ossicles	6
	Vertebral column	26
	Thorax	
	Sternum	1
	Ribs	24
		Subtotal = 80
Appendicular skeleton	Pectoral (shoulder) girdles	
	Clavicle	2
	Scapula	2
	Upper limbs (extremities)	
	Humerus	2
	Ulna	2
	Radius	2
	Carpals	16
	Metacarpals	10
	Phalanges	28
	Pelvic (hip) girdle	
	Hip, pelvic, or coxal bone	2
	Lower limbs (extremities)	
	Femur	2
	Fibula	2
	Tibia	2
	Patella	2
	Tarsals	14
	Metatarsals	10
	Phalanges	28
		Subtotal = 126
		Total = 206

Each vertebra is identified by shorthand notation that specifies the spinal region and the number of the vertebra within that region. The most superior vertebra in each region is number 1. For example, the most superior vertebra in the cervical region is labeled C1. Underneath C1 is C2 and so on through C7. Similarly, the thoracic vertebrae are numbered T1 to T12, and the lumbar vertebrae are designated as L1 to L5. The superior surface of the fused sacral bone is identified as S1.

Most vertebrae have a common structural plan, consisting of an anteriorly located vertebral body (or centrum), a vertebral arch (also neural arch) that extends posteriorly, and various processes that project from the vertebra and serve as muscle and ligament attachment sites or as articulation sites with adjacent vertebrae or ribs (figure 2.6). The vertebral body is typically oval shaped (as viewed from above) and relatively thick. One of its primary functions is to accept and transfer compressive forces acting through the spine. Moving caudally (inferiorly) from the head, each successive vertebral body must support increasing weight and therefore is thicker and stronger.

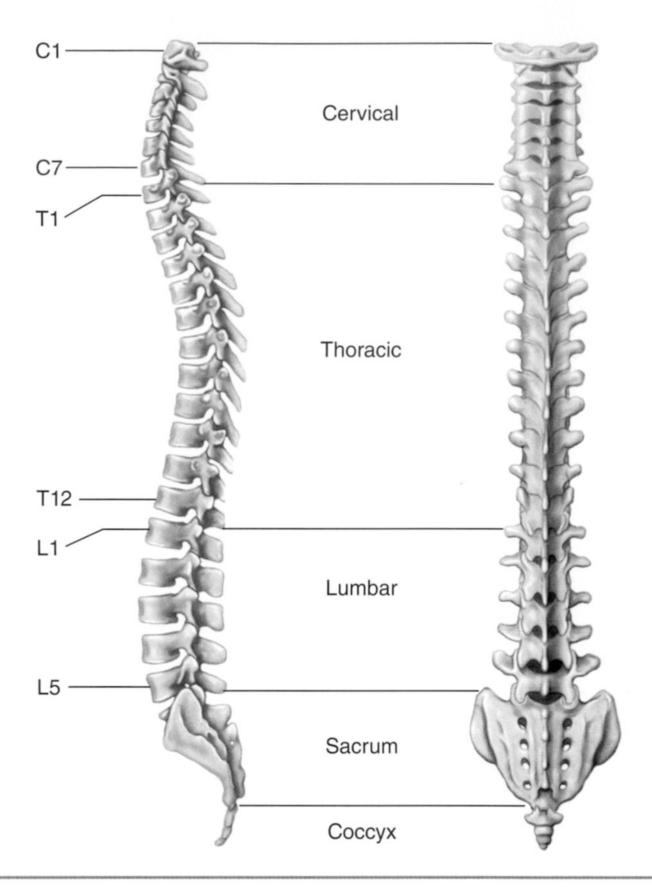

FIGURE 2.5 Lateral and posterior views of the vertebral column.

The vertebrae also are integrally involved in movements of the trunk and spine. Each vertebra moves relative to its adjacent ones; the summation of these movements determines the overall movement of the spine.

The orientation of the vertebral facets varies from region to region. In the cervical spine, the facets are inclined at about 45° above the horizontal plane and parallel with the frontal plane. Facets in the thoracic region are inclined at about 60° above the horizontal plane and deviate 20° behind the frontal plane. In the lumbar region, facets are inclined about 90° above the horizontal plane and deviate 45° behind the frontal plane. These regional changes in facet orientation play an essential role in determining movement potential between adjacent vertebrae in each region.

When viewed from the front or back, the normal spine is straight. From the side, however, the spine has characteristic curvatures (see figure 2.5). In the fetus, there is a single anteriorly concave curvature. About 3 months after birth, when the infant begins to hold her head erect, a posteriorly concave curvature develops in the cervical region. Similarly, when the infant begins to walk, a curvature appears in the lumbar region. The thoracic and sacral regions retain their original curvatures, hence the term **primary curvatures**. Because they develop later, the oppositely directed curves of the cervical and lumbar regions are called **secondary curvatures**. The spinal curvatures serve important mechanical functions, including balance, shock absorption, and movement facilitation.

The sternum and ribs form skeletal components of the thorax (chest; see figure 2.7). The sternum (breastbone) is a centrally located flat bone made up of three sections (manubrium, body, xiphoid process). The manubrium and body join each other to form the sternal angle.

The manubrium also articulates with the proximal end of the clavicle and the first and second ribs. Ribs 2 to 10 attach directly or indirectly to the sternal body, with rib 2 also attaching to the manubrium.

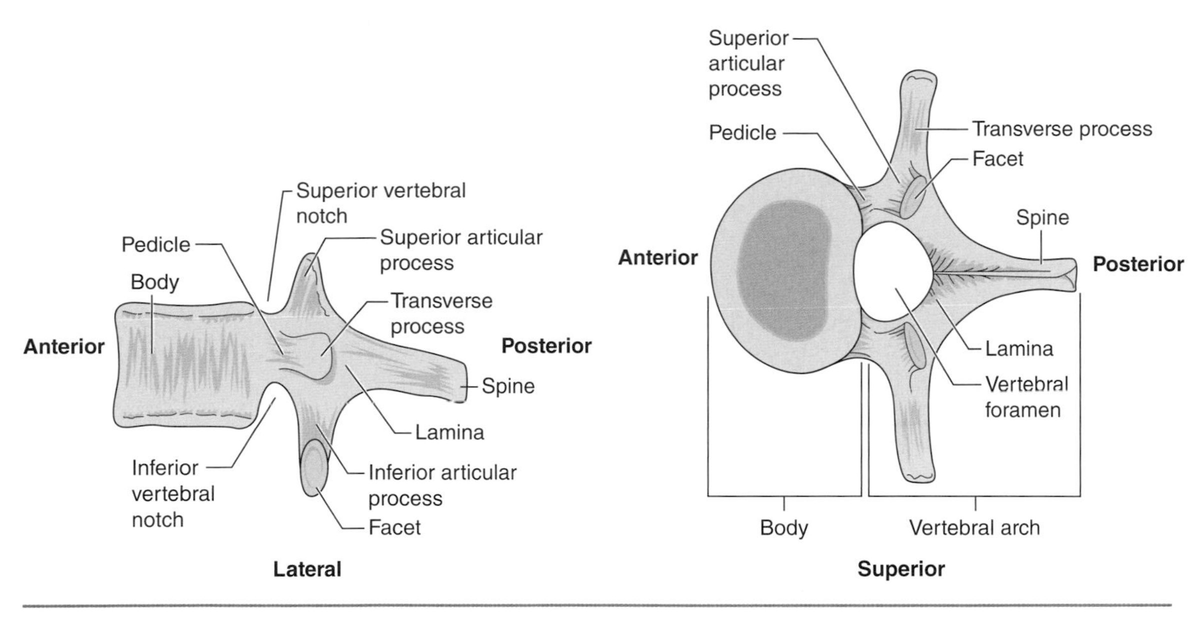

FIGURE 2.6 Lateral and superior views of a typical vertebra.

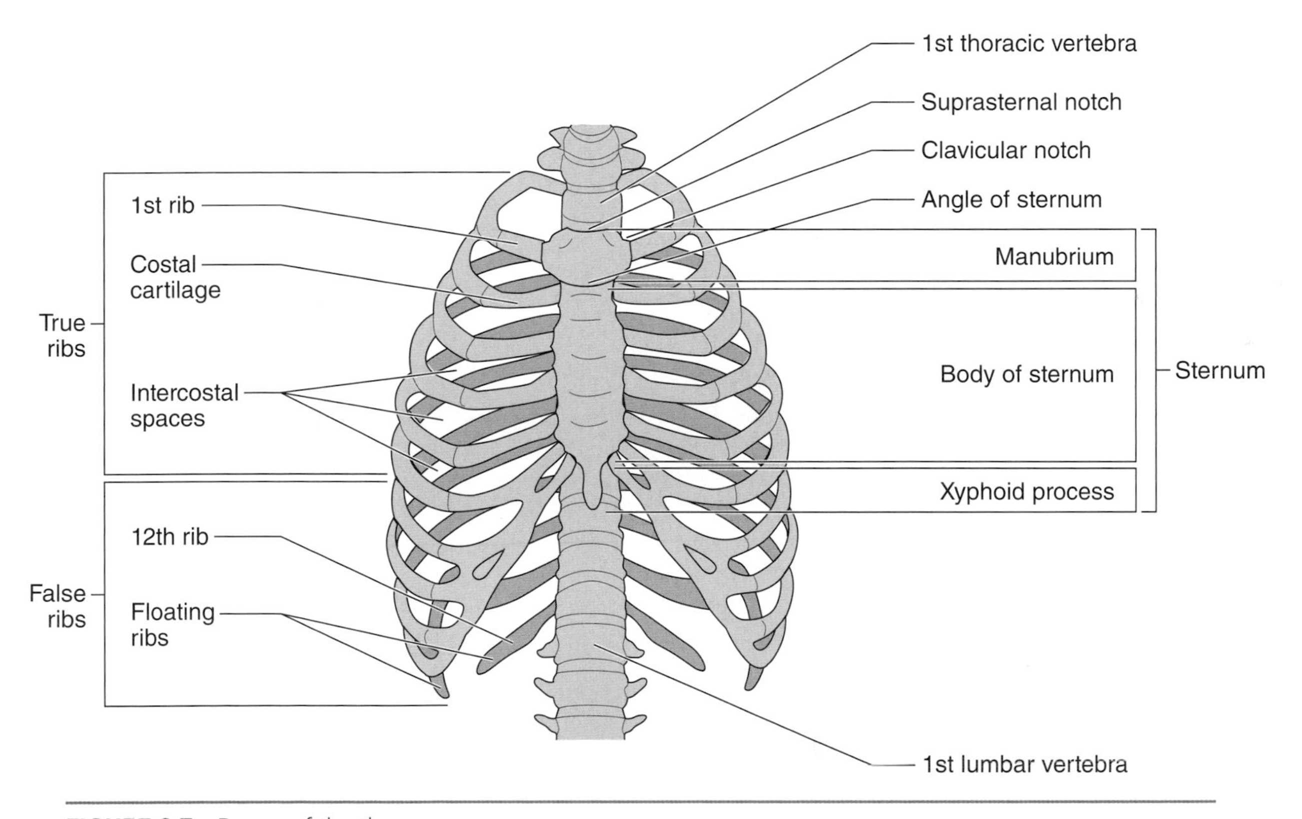

FIGURE 2.7 Bones of the thorax.

The thorax contains 12 rib pairs that collectively encase and protect important internal organs such as the heart and lungs. The first 7 rib pairs articulate directly to the manubrium or sternal body and are thus called *true ribs*. The next 3 pairs attach indirectly through costal cartilage to the sternum. Because these ribs lack direct connection to the sternum, they are termed *false ribs*. The cartilage of ribs 8 to 10 (vertebrochondral ribs) merges with the cartilage of rib 7, which then connects with the sternum. The remaining two rib pairs (11-12) have no connection at all with the sternum and are therefore called *floating ribs*.

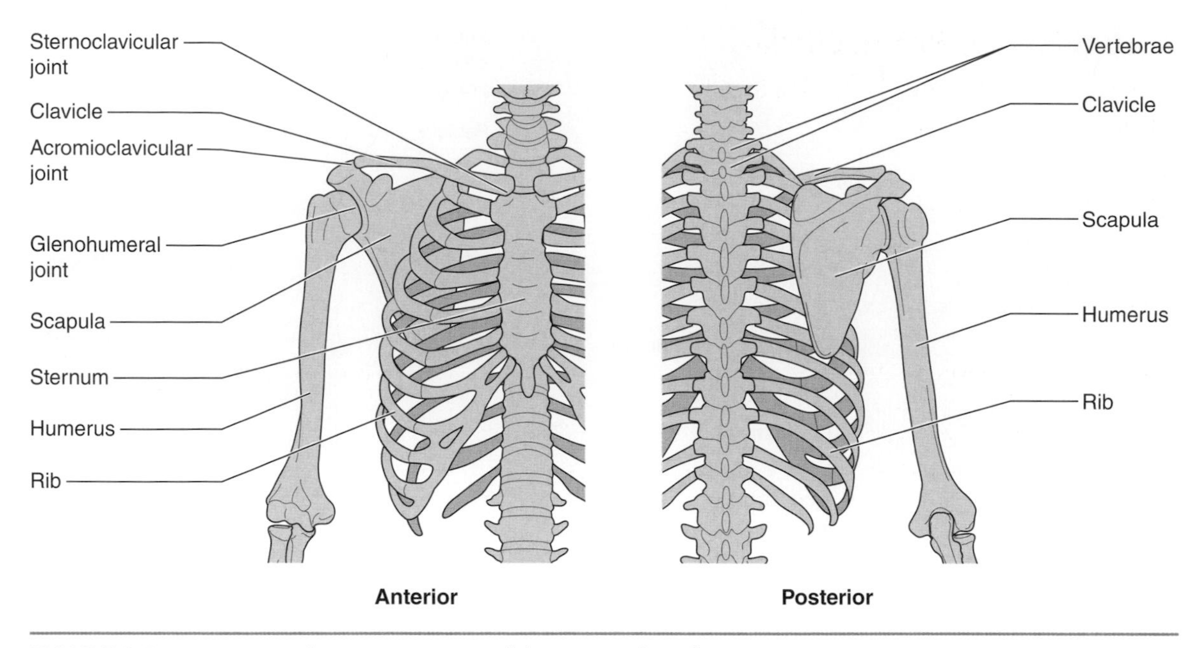

FIGURE 2.8 Anterior and posterior views of the pectoral girdle.

Upper Body: Pectoral Girdle and Upper Limbs

The **pectoral girdle**, or shoulder girdle, consists of two bones, the clavicle and scapula (figure 2.8). These two bones work in a coordinated fashion to facilitate the complex and necessary movements of the pectoral girdle and the arm. The most prominent joint of the pectoral girdle is the glenohumeral (shoulder) joint where the head of the upper arm bone (humerus) articulates with the scapula. The three-dimensional nature of the glenohumeral joint allows exceptional range of motion for activities such as throwing and lifting.

Each upper extremity, or limb (figure 2.9), is made up of one bone (humerus) in the brachium, or upper arm (sometimes called just the arm), and two bones (radius, ulna) in the antebrachium, or forearm. The wrist and hand contain 27 bones, including 8 carpal bones, 5 metacarpal bones, and 14 phalanges (figure 2.10). The bones and associated joints of the upper limbs provide for both large-scale (gross) movements such as pushing, pulling, and throwing, and small-scale (fine) movements such as writing, gripping, and grasping.

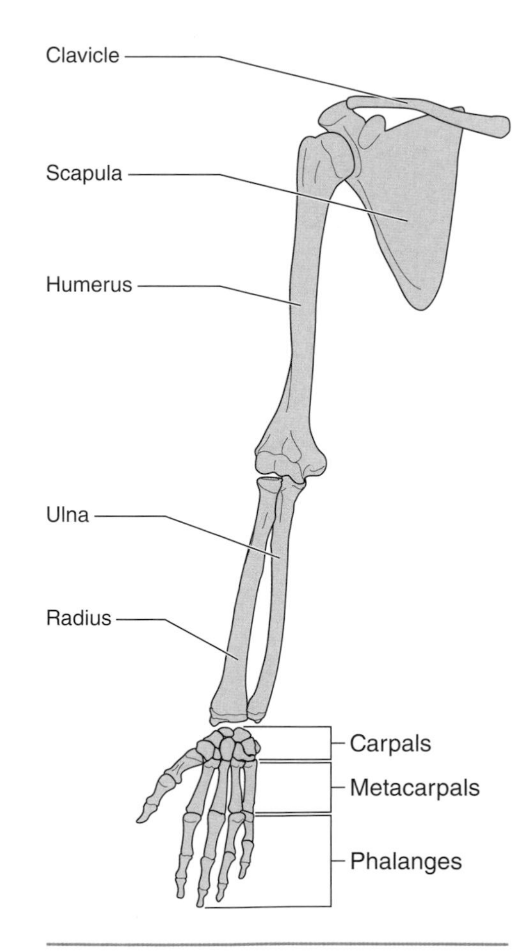

FIGURE 2.9 Bones of the upper extremity (anterior view).

Lower Body: Pelvic Girdle, Lower Limbs, and Arches of the Foot

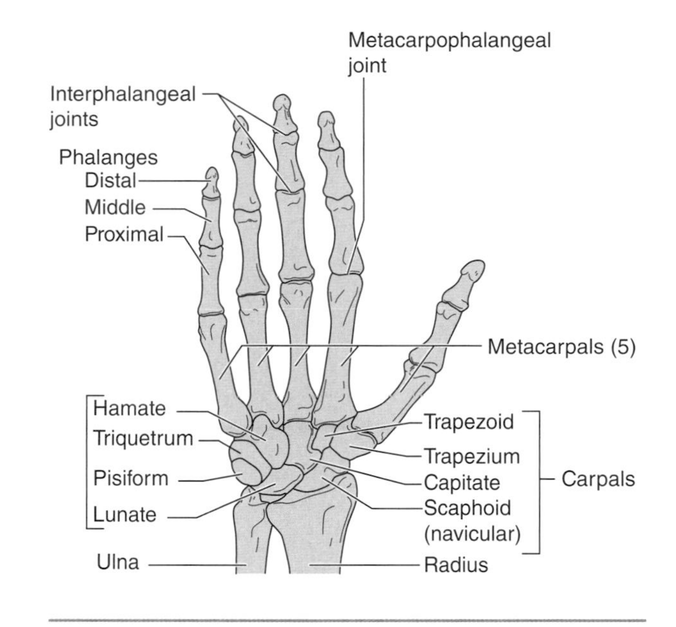

FIGURE 2.10 Bones of the wrist, hand, and fingers.

The **pelvic girdle** is similar to the pectoral girdle in that it connects limbs of the appendicular skeleton to the axial skeleton. However, distinct differences between the two girdles exist. The pelvic girdle, because of its role in weight bearing and locomotion, is much larger and heavier than the pectoral girdle, similar to the way that bones of the lower extremities are larger and stronger than their counterparts in the upper extremities. Collectively, the pelvic girdle, sacrum, and coccyx form the pelvis (figure 2.11).

Each lower extremity, or limb (figure 2.12), is made up of one bone (femur) in the thigh and two bones (tibia, fibula) in the lower leg (also shank). The foot contains 26 bones, including 7 tarsal bones, 5 metatarsal bones, and 14 phalanges (figure 2.13).

The bones of the foot are arranged to form two primary arches: the longitudinal arch, running from the calcaneus to the distal ends of the metatarsals, and the transverse arch, which extends from side to side across the foot (figure 2.14). The longitudinal arch is divided into a medial portion that includes the calcaneus, the talus, the navicular, three cuneiforms, and the three most medial metatarsals. The lateral portion is much flatter and is in contact with the ground during standing. The transverse arch is formed by the cuboid, cuneiforms, and bases of the metatarsals.

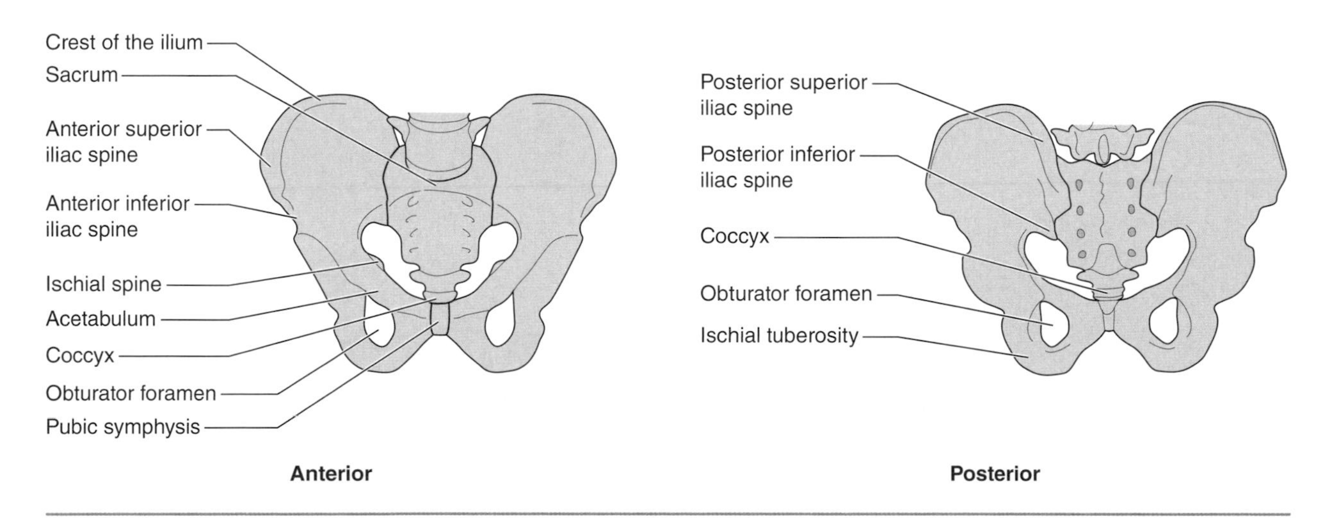

FIGURE 2.11 Pelvic girdle, anterior and posterior views.

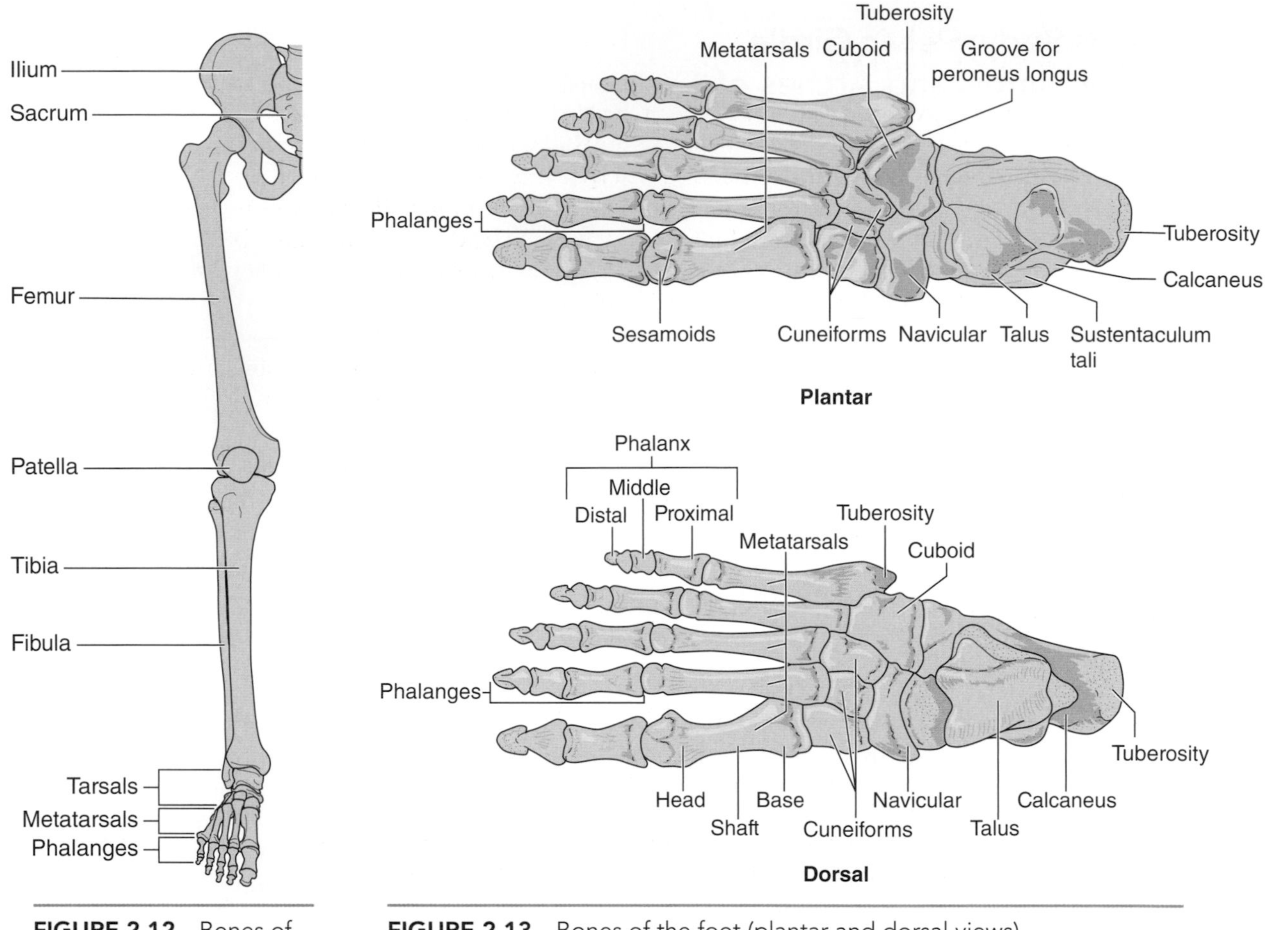

FIGURE 2.12 Bones of the lower extremity (anterior view).

FIGURE 2.13 Bones of the foot (plantar and dorsal views).

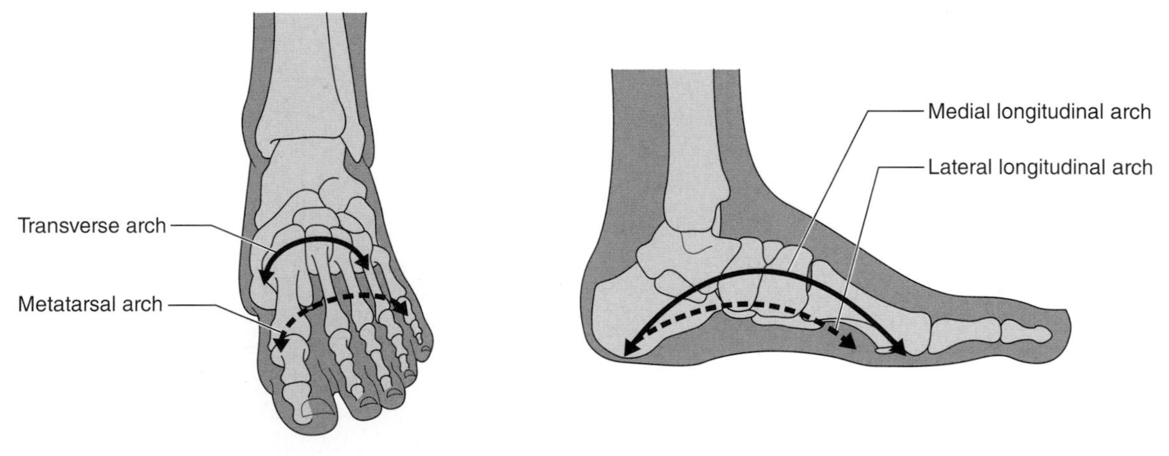

FIGURE 2.14 Arches of the foot.

During weight-bearing activities (e.g., walking, running, jumping), the arches compress to absorb and distribute the load. Several ligaments assist in this force distribution, including the plantar calcaneonavicular ligament ("spring ligament"), the short plantar ligament, and the long plantar ligament. The integrity of the arches and their ability to absorb loads are maintained by the tight-fitting articulations between foot bones, the action of intrinsic foot musculature, the strength of the plantar ligaments, and the plantar aponeurosis (plantar fascia).

Bone Adaptation

Bone has an unfortunate reputation of being an inert, lifeless structure. Nothing could be further from the truth. Bone is a remarkably dynamic tissue that adapts to multiple internal factors (e.g., hormone levels, calcium concentrations) and external factors (e.g., mechanical loads). Adaptation is a modification of an organism or its parts that makes it more fit for existence under the conditions of its environment. This definition concisely characterizes the adaptability of bone and how it meets the changing needs imposed by activity levels and environmental demands.

Remodeling refers to the resorption and replacement of existing bone. Solid evidence exists that the remodeling of bone is, in fact, triggered by microdamage to the bone. In remodeling, the action of the osteoblasts and osteoclasts is tightly coupled, and the overall effect on bone is determined by the *net* activity of these cells. If osteoblast activity surpasses osteoclast activity, a net increase in bone mass results. Conversely, if osteoclast activity predominates, bone mass will decrease. This remodeling process is essential for maintaining the bone strength required to support the body during weight-bearing movements such as walking, running, and jumping.

Bone Health

Bone health is affected by numerous genetic, environmental, nutritional, hormonal, and mechanical factors. The complex interaction of these factors ultimately determines the physiological and biomechanical integrity of bone tissue. Bone research before the mid-20th century emphasized the physiological function of bone. In more recent decades, exploration of bone function and adaptation has widened to include other areas of scientific inquiry and clinical application. Although much has been learned about the requisites for bone health, much still remains a mystery. One thing that is *not* a mystery is the fact that bones need to experience forces to remain healthy. Human movements that involve weight-bearing forces are essential for the maintenance of bone health. Activities such as running and jumping, for example, subject bones of the lower extremities to high compressive forces. These forces stimulate the remodeling process and strengthen bones.

Applying the Concept

Wolff's Law

The capacity of bone to adapt its structure to imposed loads has become known as **Wolff's law**. Although Julius Wolff, a 19th-century surgeon, is given credit for this idea, he was not the first to observe bone's adaptive abilities, did not define many of the concepts now associated with his law, and was mistaken in some of his observations. Nevertheless, Wolff did recognize the adaptive capacity of bone, and his name has become inseparably associated with bone adaptation. Wolff was quoted by Keith (1918) as follows:

> Every change in the form and function of . . . bone[s] or of their function alone is followed by certain definite changes in their internal architecture, and equally definite secondary alterations in their external conformation, in accordance with mathematical laws. (Martin et al., 215, p. 275)

Among the concepts that arose in the 19th century that are now incorporated into Wolff's law are the optimization of bone strength with respect to bone weight, trabecular alignment with the lines of principal stress, and the self-regulation of bone structure by cells responding to mechanical stimuli (Martin et al., 2015).

Exercise

There is little doubt that exercise and physical activity stimulate bone remodeling. The details of the phenomenon are extremely complex and involve the interaction of many variables, including exercise type and intensity, skeletal maturity, bone type, anatomical location, and hormone levels.

The primary relationships between physical activity and bone mass can be summarized as follows (Whiting & Zernicke, 2008):

1. Growing bone responds to low- to moderate-intensity exercise through significant deposition of new cortical and trabecular bone.
2. Bone responds negatively, by suppressing normal modeling activity, when certain activity thresholds are exceeded.
3. Moderate- to high-intensity physical activity can elicit modest increases (1%-3%) in bone mineral content in men and premenopausal women and may generate modest and site-specific increases in bone mass in postmenopausal women.
4. Long-term benefits of exercise on bone health are maintained only by continuing exercise.
5. Bone mass gains appear to depend on initial bone mass (i.e., individuals with very low bone mass may show greater gains from exercise than those with only moderately reduced bone mass).

Exercise-related bone changes are usually specific to the region being affected (i.e., mechanically loaded) and are not systemic. For example, a person who runs on a regular basis would expect increased bone density in the legs but not in the arms. A softball pitcher would likely experience bone adaptation in her throwing-arm side but not in the other arm.

The dynamics of bone remodeling are important to keep in mind because the timing of the resorption–deposition cycle may leave the bone vulnerable to injury. If excessive loads are applied during the time between resorption and deposition, injury may occur. Athletes therefore are well advised to increase the intensity of their workouts gradually and progressively. If they do too much, too soon, they may end up on the sidelines tending to a fractured bone.

Too little exercise can also damage bone. People who choose a sedentary lifestyle or who are bedridden because of disease or injury are likely to lose bone mass and strength. Bone is deposited where it is needed and removed from where it is not. Thus, skeletal tissue not subjected to periodic mechanical loading from physical activity (e.g., walking or running) will atrophy. This net bone loss weakens the tissue and increases the chance of fracture.

One group at particular risk of bone loss is astronauts. Detectable bone loss has been measured after only a few days in space, and extended stays in space (i.e., weeks or months) can significantly compromise bone quality (see sidebar). Appropriate exercises while in space may reduce the negative effects of hypogravity on astronauts' skeletal systems.

Diet

Nutrition profoundly affects bone health. Bone growth and remodeling are integrally dependent on having the proper kinds and quantities of nutrients required for building healthy bone. The body's normal bone mineral balance is typically well regulated by the synergistic actions of vitamin D, parathyroid hormone, and calcitonin. These substances control the absorption of dietary calcium, bone mineral deposition and resorption, and the renal (kidney) secretion and resorption of calcium and phosphorus. Adequate dietary calcium intake is essential because the body excretes calcium on a continual basis throughout the day. Calcium absorption is affected by vitamin D, dietary protein, phosphorus, fiber, and fat.

Insufficient levels of vitamin D may compromise calcium absorption and result in bones with reduced mineral content. These bones are softer and more bendable than healthy bones. Children with vitamin D deficiency may develop a condition called rickets, in which their more pliable bones, subjected to repeated forces, become bent. This may result in a bowlegged deformity because the bones lack the rigidity to withstand the forces of walking.

Research in Mechanics

Bone Health in Astronauts

Numerous research studies have explored the physiological effects of short-duration and prolonged space flight on astronauts. Grimm and colleagues (2016) published a review of the research specifically related to the effects of microgravity on bone in humans. They reviewed more than 180 research articles that examined changes in calcium, sodium, and bone metabolism of astronauts; mechanisms of bone loss; models to simulate the effects of microgravity on bone on Earth; and countermeasures used in space to counteract the effects of microgravity.

Among the many effects reviewed and reported by Grimm and others is that microgravity results in demineralization of bone, increased resorption, and decreased bone formation, all of which may lead to eventual osteoporosis. A variety of countermeasures have been tested over the past few decades to minimize the negative effects of microgravity. While some of these have been ineffective, others have shown promise. Recently, for example, an advanced resistance exercise device (ARED) used on the International Space Station has proved beneficial for bone health in astronauts (Smith et al., 2012).

See the references for the full citations:

Grimm et al., 2016.

Smith et al., 2012.

Adequate intake of vitamin C also is essential for bone health. Insufficient intake of this vitamin results in a disease called scurvy. In these cases, the lack of vitamin C compromises the synthesis of collagen. Low collagen levels result in brittle bones that may fracture easily.

Dietary protein levels affect the handling of calcium in the urine. Protein deficiency has been identified as a factor in the onset of bone mass reduction. Too little protein can lead to reduced calcium absorption in the intestine and increased urinary calcium levels. At the other extreme, too much dietary protein can lead to greater renal calcium loss and a negative calcium balance.

High levels of dietary fat and sugar have a negative effect on the absorption of calcium in the intestines. Therefore, low to moderate intake of fat and sugar is recommended for optimal bone health.

Aging

Many physiological, anatomical, and psychological changes accompany aging. Among these many changes are bone-specific alterations that affect the quality of bone tissue and consequently our ability to move effectively. During our growth period (from conception to approximately 25 years), the body normally increases its bone mass as osteoblast activity far exceeds the work of osteoclasts. This results in a net gain in bone as our skeletal system grows and develops. At some point, which varies across individuals and specific bones, we reach a level at which we have the greatest amount of bone we will ever have. This is referred to as **peak bone mass**. The level of peak bone mass is affected by many factors, including heredity, diet, and exercise.

Our middle years (25-50) are characterized by a relatively constant level or slight decrease in bone mass; osteoblasts and osteoclasts live in relative harmony in maintaining our skeletal integrity. As we move into our later years (50 plus), osteoblast activity decreases. This results in net bone loss and reduced skeletal strength. Postmenopausal women experience a more rapid rate of bone loss than do men, primarily because of reduced production of estrogen, a hormone that inhibits osteoclast activity.

One of the keys to bone health is accumulation of considerable bone mass in our early years of bone growth and development. If the inevitable decline in bone mass begins from a higher peak bone mass, the likelihood of eventually reaching a bone mass near the fracture threshold is greatly reduced.

What can we do to optimize bone mineral acquisition during the growing years? We can do many things, including making a lifelong commitment to weight-bearing exercise, engaging in a variety of vigorous short-duration activities, increasing muscle strength through resistance training, and avoiding immobility and prolonged sedentary periods. In addition, bone health can be enhanced by proper nutrition, adequate rest, limited alcohol intake, and not smoking. The benefits of these lifestyle choices are certainly not restricted to bone. All of the body's systems work best with a healthy combination of diet and exercise.

Osteopenia and Osteoporosis

Some degree of bone loss is a natural part of aging. Excessive bone loss, however, can have catastrophic consequences. **Osteopenia** refers to general bone loss. More serious bone loss with increased risk of fracture is termed **osteoporosis**. Osteoporotic bone exhibits excessive porosity and accompanying structural changes that greatly increase fracture risk. Spongy bone seems most prone to osteoporosis. All bones theoretically are at risk, but those with high levels of spongy bone (e.g., vertebrae and the proximal femur) are at particular risk.

Osteoporosis is a disease marked by reduced bone mineral mass and changes in bone geometry, leading to an increased probability of fractures, primarily of the hip, spine, and wrist. Progressive loss of bone mass can be a function of the normal aging process or can be caused by other disease processes. Many individuals are unaware of the existence of their osteoporosis, especially in its early stages.

Both men and women experience some loss of bone mass as part of normal aging, but osteoporosis progresses much more rapidly in postmenopausal women. In women, bone loss increases significantly for about five years after menopause and then slows to a more gradual loss.

Postmenopausal women have the highest risk of osteoporotic fracture. However, other populations also are at considerable risk for osteoporosis. Young female athletes, for example, with low body fat who train at high intensities commonly experience menstrual dysfunction. They may have no or very few menstrual cycles (**amenorrhea**) or irregular menstrual cycles (**oligomenorrhea**); as a result, they have reduced estrogen levels. This creates an imbalance in the bone deposition–resorption ratio that can eventually lead to osteoporosis. A comparison between amenorrheic elite athletes and a group with normal menstrual function found that the athletes with amenorrhea had up to 25% lower vertebral mineral mass than the control group (Marcus et al., 1985). Their problems are compounded by the fact that these athletes (e.g., runners, gymnasts) subject their bodies to excessive loading that further stresses the bones and increases the chance of fracture.

Older individuals are at particular risk for falls and bone fractures. The statistics associated with the incidence of falls are sobering. For example, in the United States alone, more than 300,000 hip fractures happen annually. These statistics are particularly alarming in light of the fact that these injuries all too often are the first event in a chain that leads to eventual incapacity, dependence, and even death. Many of the characteristics associated with aging magnify fall risk. These include decreases in bone and muscle strength, impaired cardiorespiratory function, compromised visual acuity, slower reaction times, and reduced balance. The important issues related to the effects of aging are considered in greater detail in chapter 7 (Posture and Balance).

Given that people now live longer, current demographic trends point to an exploding population of older people. This fact, coupled with more sedentary lifestyles, indicates that the problems associated with osteoporosis will grow as a major health issue.

Applying the Concept

Osteoporosis-Related Fractures: A Worldwide Problem

Thousands of research studies have been conducted on various aspects of osteoporosis. The International Osteoporosis Foundation (IOF) summarized more than 200 of these studies. Among the many facts reported by the IOF are the following:

- Osteoporosis causes nearly 9 million fractures annually worldwide.
- An estimated 200 million women worldwide are affected by osteoporosis.
- Worldwide, 1 in 3 women and 1 in 5 men over the age of 50 will suffer osteoporotic fractures.
- By 2050, the worldwide incidence of hip fracture in men and women is projected to increase by 310% and 240%, respectively, compared to fracture rates in 1990.
- Women are about three times more likely to experience hip fractures than men.
- The most common sites of osteoporotic fractures are the hip, wrist or forearm, and vertebrae.

Concluding Comments

The skeletal system plays an important role in many processes needed for our survival and effective functioning, including structural support, protection, movement, blood cell production, and mineral storage. Healthy bones allow the skeletal system to perform all these functions well. Unhealthy bones, however, can dramatically affect our ability to perform even ordinary tasks of daily living. Many factors such as diet, hormones, activity level, and genetics interact to make bone health a multifactorial problem to which we have only some of the answers.

Go to the web study guide to access critical thinking questions for this chapter.

Suggested Readings

Alexander, R.M. (2000). *Bones: The unity of form and function.* New York: Basic Books.

Benyus, J.M. (2002). *Biomimicry: Innovation inspired by nature.* New York: Harper Perennial.

Martin, R.B., Burr, D.B., Sharkey, N.A., & Fyhrie, D.P. (2015). *Skeletal tissue mechanics* (2nd ed.). New York: Springer.

Nordin, M., & Frankel, V.H. (2012). Biomechanics of bone. In M. Nordin & V.H. Frankel (Eds.), *Basic biomechanics of the musculoskeletal system* (4th ed., pp. 26-55). Philadelphia: Lippincott Williams & Wilkins.

3

Joint Anatomy and Function

Objectives

After studying this chapter, you will be able to do the following:

- Describe joint structure and classification
- Classify synovial joints according to their structure and function
- Explain the concepts of joint stability and mobility
- Describe movement planes and joint motion
- Describe types of joint movement
- Identify movements of the hip and pelvis, knee, ankle, and foot
- Identify movements of the shoulder, elbow, forearm, wrist, and hand
- Identify movements of the head, neck, and spine
- Describe spinal deformities (scoliosis, kyphosis, lordosis)

Our ability to move depends on the body's joints. In this chapter, we present information on general joint structure and movement terminology, followed by details of the major joints involved in movement. Our approach emphasizes the functional aspects of joint structure, thereby providing you with the foundation needed to fully appreciate and understand the complexities of human movement in subsequent chapters.

Joint Structure and Classification

Our infinite capacity for movement depends on a musculoskeletal design that includes well-functioning articulations, or joints. Each joint's structure is well matched with its function. Some articulations allow considerable range of movement, or joint **mobility**. Others are built for joint **stability** and vigorously resist movement. The amount of joint mobility, or **range of motion**, is determined by both the structural congruity of the bones (i.e., how well they fit together) and the amount of support provided by tissues surrounding the joint (periarticular tissues). The shoulder (glenohumeral) joint, for example, has remarkable range of motion allowed by the relatively poor fit between the head of the humerus and the shallow glenoid fossa of the scapula. At the other extreme, suture joints of the skull are formed by bones that interlock in much the same way as pieces of a jigsaw puzzle fit tightly together and are thus immovable.

Joints can be classified in a number of ways. Most commonly, joints are categorized structurally according to the type of tissue that binds the joint together (binding tissue). Joints may also be classified functionally by the type or extent of movement they allow. Functional designations include joints that allow no movement (**synarthroses**), limited movement (**amphiarthroses**), or free movement (**diarthroses**). Logically, synarthroses and amphiarthroses tend to predominate in the axial skeleton, while diarthroses are more commonly found in the appendicular skeleton.

Structurally, joints fall into one of three categories based on the tissues that support the joint and bind the bones together and whether or not the joint contains a synovial cavity between the bones. **Fibrous joints** do not have a joint cavity and are bound by connective tissues composed primarily of collagen fibers. **Cartilaginous joints** also are void of a joint cavity but have cartilage as their binding tissue. The most common and complex articulations are **synovial joints**, which are distinguished by a fibrous joint capsule that surrounds and encapsulates the joint.

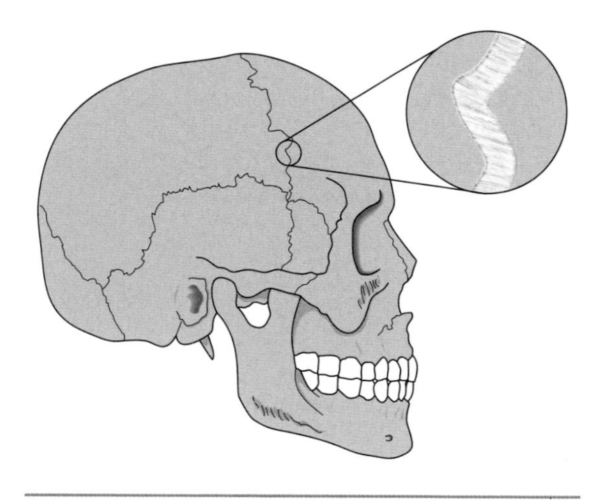

FIGURE 3.1 Suture joint.

Fibrous Joints

The common features of fibrous joints include the absence of a synovial joint cavity and the presence of fibrous (collagenous) tissue to reinforce the bone junction. The two most common types of fibrous joints are sutures and syndesmoses. The amount of movement allowed at fibrous joints varies, but usually little or no movement is evident.

Suture joints connect bones of the skull. These interlocking bones are bound by a dense fibrous connective tissue that renders the joints immovable (figure 3.1). Suture joints therefore are functionally classified as synarthroses.

Syndesmoses are joints bound by ligaments, collagenous structures that connect bone to bone. The amount of joint motion may be limited, as in the articulation between the radius and ulna, which are bound together by a collagenous **interosseous membrane** (figure 3.2). The tibiofibular joint is functionally an amphiarthrosis. In contrast, considerable movement is allowed between the radius and ulna of the forearm. Here, the longer fibers of the interosseous membrane permit more extensive bone excursion and result in the radioulnar joint's functional classification as a diarthrosis (i.e., freely movable).

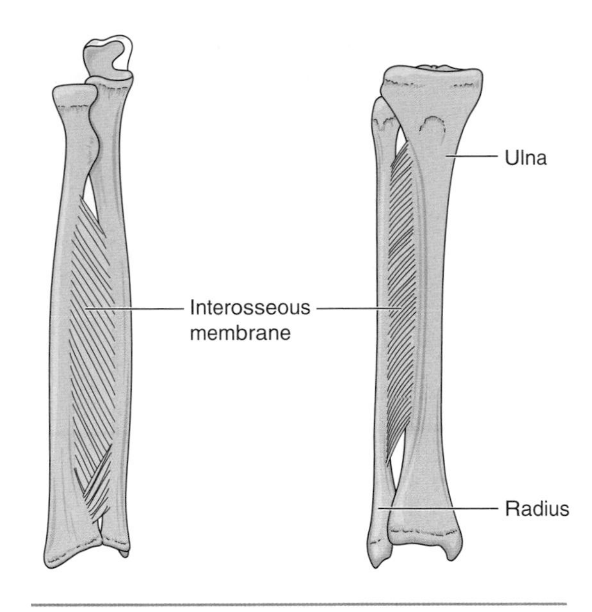

FIGURE 3.2 The interosseous membrane between the radius and the ulna is a fibrous synarthrotic joint.

Cartilaginous Joints

Of the three types of cartilage, only hyaline cartilage and fibrocartilage are found as the binding tissue at cartilaginous joints. Elastic cartilage is never used to reinforce joints. Hyaline cartilage joints, or **synchondroses**, are found in the mature skeleton, for example, between the first rib and the manubrium (sternocostal joint) and temporarily in the developing skeleton between the diaphysis and epiphysis in the form of the **epiphyseal growth plate**, or **growth plate**.

Symphyses are joints between bones separated by a fibrocartilage pad interposed between two joint surfaces that are covered by hyaline cartilage. The pubic symphysis, for example, joins the two pubic bones (figure 3.3). Normally, this joint is relatively immovable, except during late pregnancy and birth when hormonal changes allow greater tissue extensibility and joint movement. Other examples of symphyses include the sternomanubrial joint (joining the manubrium with the sternal body) and intervertebral joints (between adjacent vertebrae).

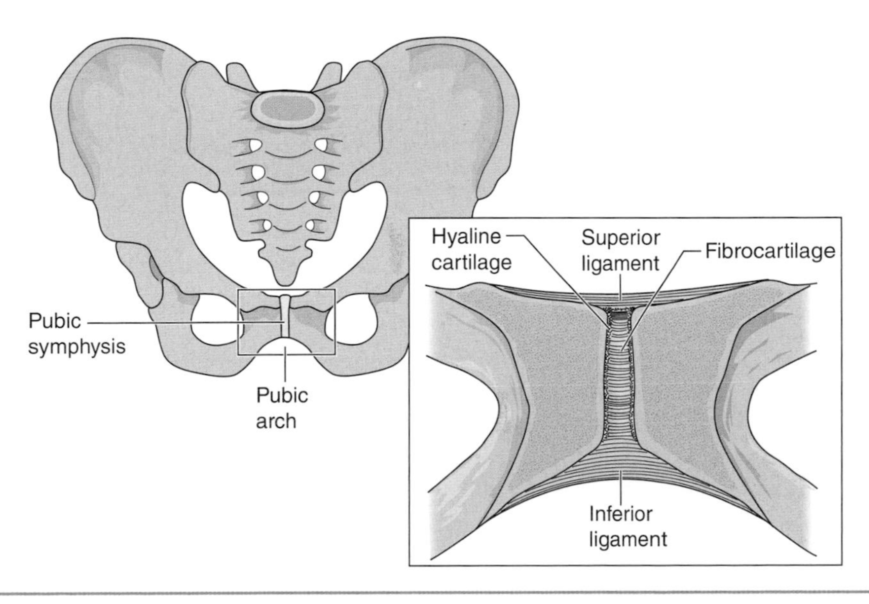

FIGURE 3.3 Cartilaginous joint at the pubic symphysis.

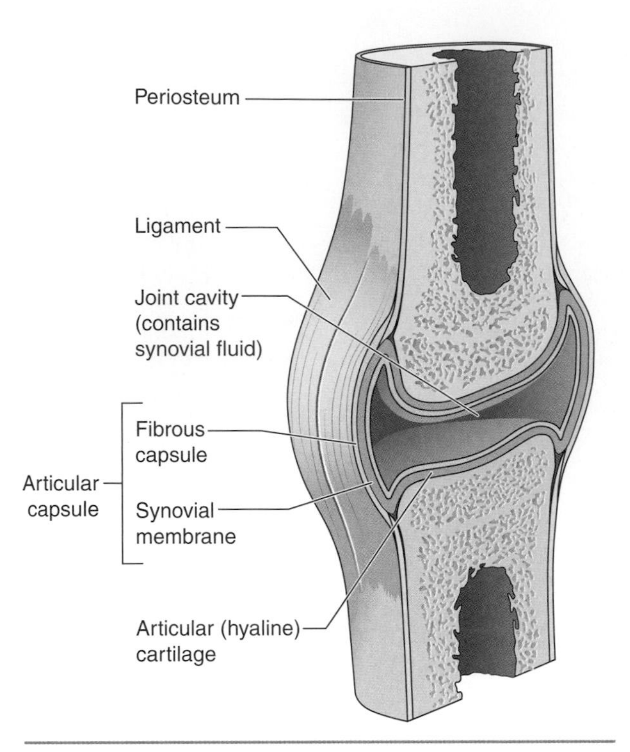

FIGURE 3.4 Structure of a synovial joint.

Synovial Joints

Most movements of our extremities occur at major joints such as the hip, knee, ankle, shoulder, elbow, and wrist. All of these joints, and many others, are synovial joints.

Synovial joints share common structural elements, including a **synovial joint cavity** encapsulated by a **fibrous joint capsule**. The joint cavity is filled with **synovial fluid** produced by the **synovial membrane**, a thin membrane that lines the inner surface of the joint capsule. A smooth, shiny layer of **articular cartilage** (composed of the hyaline type of cartilage) covers the joint surfaces of the articulating bones. This structure allows the joints to move freely. Synovial joints therefore are functionally classified as diarthroses.

The fibrous joint capsule consists of dense connective tissue that blends with the periosteum of the articulating bones (figure 3.4). The capsule forms the outer boundary of the synovial joint and provides structural support. A thin synovial membrane covers the inner surface of the capsule. Although not providing any structural support, the synovial membrane is nonetheless important for overall joint function because it secretes a thick synovial fluid that lubricates the joint. As a lubricant, the fluid reduces friction between the articular surfaces and assists with the absorption of compressive loads applied to the joint. The synovial fluid also provides nutrients for the articular cartilage. The hyaline cartilage on the joint surfaces is avascular (i.e., lacks an inherent blood supply) and therefore must rely on the synovial fluid to provide nutrients through diffusion. Synovial joints contain a relatively small amount of this important fluid. Large joints such as the knee, for example, contain a scant 3 ml or less of synovial fluid.

When irritated, as in joint trauma, the synovial membrane may produce excess synovial fluid. This excess fluid causes the joint to swell. A severely twisted knee, for example, may swell in response to injury and be diagnosed as so-called water on the knee. In fact, the swelling results from overproduction of synovial fluid by an irritated synovial membrane.

Synovial joints may also contain other structural components that enhance their function. These accessory structures include ligaments, tendons, bursae, tendon sheaths, fibrocartilage pads (menisci), and fat pads.

The tensile strength of ligaments makes them ideal structures for reinforcing synovial joints and enhances their resistance to dislocation. Three ligament types are found in and around synovial joints. The most common type, the capsular ligament, blends with the joint capsule and appears as a capsular thickening. Extracapsular ligaments also are common and lie completely outside the joint capsule. Much less common are intracapsular ligaments that lie entirely within the capsule and attach directly to the bone.

Tendons, although technically not a part of the joint itself, can lend structural support through their action across or around joints. Muscle forces transmitted through the tendon to the bone can reinforce the joint and increase its stability. The contribution of muscle action to joint stability is sometimes termed **active support**, in contrast to the **passive support** provided by noncontractile (i.e., non-force-producing) tissues such as ligaments and the joint capsule.

Bursae are sacs, or pockets, filled with synovial fluid that serve as spacers between a tendon or ligament and the underlying bone. They reduce friction and pressure and provide shock

Research in Mechanics

Joint Cracking

Some joints, particularly those whose joint surfaces are congruent (i.e., fit well together), can produce a cracking or popping sound when pulled apart or quickly forced toward their end range of motion. Two questions often arise. First, what causes the popping sound? Second, does the action causing cracking injure the joint?

Over the past 70 years, various mechanisms have been suggested to explain the cause of cracking or popping joints. In 1947, Roston and Wheeler Haines proposed that joint cracking was caused by the rapid distraction (i.e., pulling apart) of two bones and that this quick separation was responsible for the cracking sound. In contrast, Unsworth and colleagues (1971), based on an investigation of cavitation in the metacarpophalangeal joint, concluded that the sound was caused by a change in hydrostatic pressure in the fluid inside synovial joints. When the bones are first pulled, a negative pressure develops in the synovial fluid and a gas bubble (e.g., nitrogen, carbon dioxide) forms. As the bones separate, the pressure within the joint causes the bubble to collapse and make the popping sound. This explanation was generally accepted for more than 40 years. However, Kawchuk and others (2015) challenged the popping bubble hypothesis. Based on their study using magnetic resonance imaging (MRI), the authors concluded, "Our results offer direct experimental evidence that joint cracking is associated with cavity inception rather than collapse of a pre-existing bubble" (p. 1). Further, Fryer and colleagues wrote, "At this time, it remains a mystery whether sound generation occurs before, after, or simultaneously to the time of cavity formation" (2017, p. 38).

As for the second question, there is no evidence that the hydrostatic pressure change and popping sound will cause injury. Contrary to myth, joint cracking will not lead to osteoarthritis. That said, rapid and repeated stretching of articular structures theoretically might lead to damage if the tissues are repeatedly deformed to the extremes of joint motion.

See the references for the full citations:

Fryer, Quon, & Vann, 2017.

Kawchuk, Fryer, Jaremko, Zeng, Rowe, & Thompson, 2015.

Roston & Wheeler Haines, 1947.

Unsworth, Dowson, & Wright, 1971.

absorption. Tendon sheaths are elongated, tube-shaped bursae that surround tendons and facilitate tendon movement by providing a lubricated surface. Joints are further supported and cushioned by fat pads that act as fillers around the joint.

Articular discs, or **menisci**, are fibrocartilage pads interposed between bones. These pads provide joint cushioning and act as wedges to improve the bony fit. They typically are ring- or C-shaped and attach to one of the articular surfaces. The menisci of the knee (figure 3.5), for example, attach to the joint surfaces of the proximal tibia and are necessary for effective knee joint function.

Classification of Synovial Joints

Bones come in a wide variety of shapes and sizes. The bones of each joint are uniquely configured to permit certain movements and restrict others. Synovial joints are structurally classified according to their respective bony fit and movement potential. Each of the classification types has mechanical correlates, as illustrated in figure 3.6.

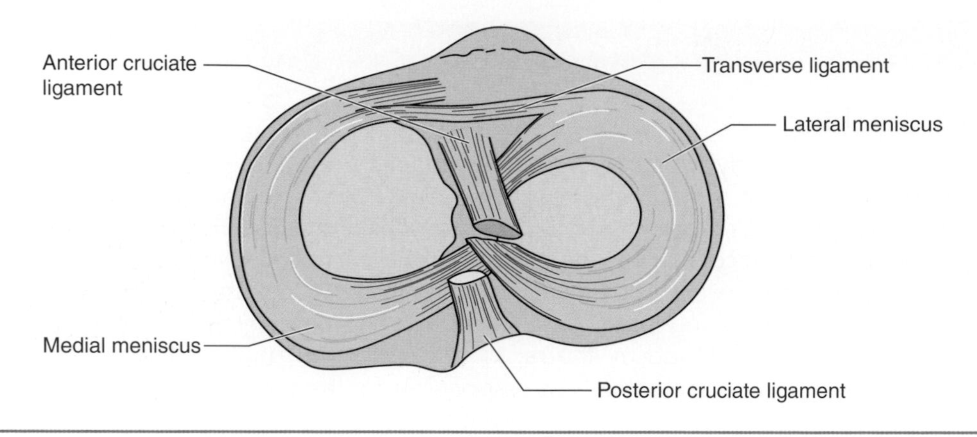

FIGURE 3.5 Lateral and medial menisci of the knee.

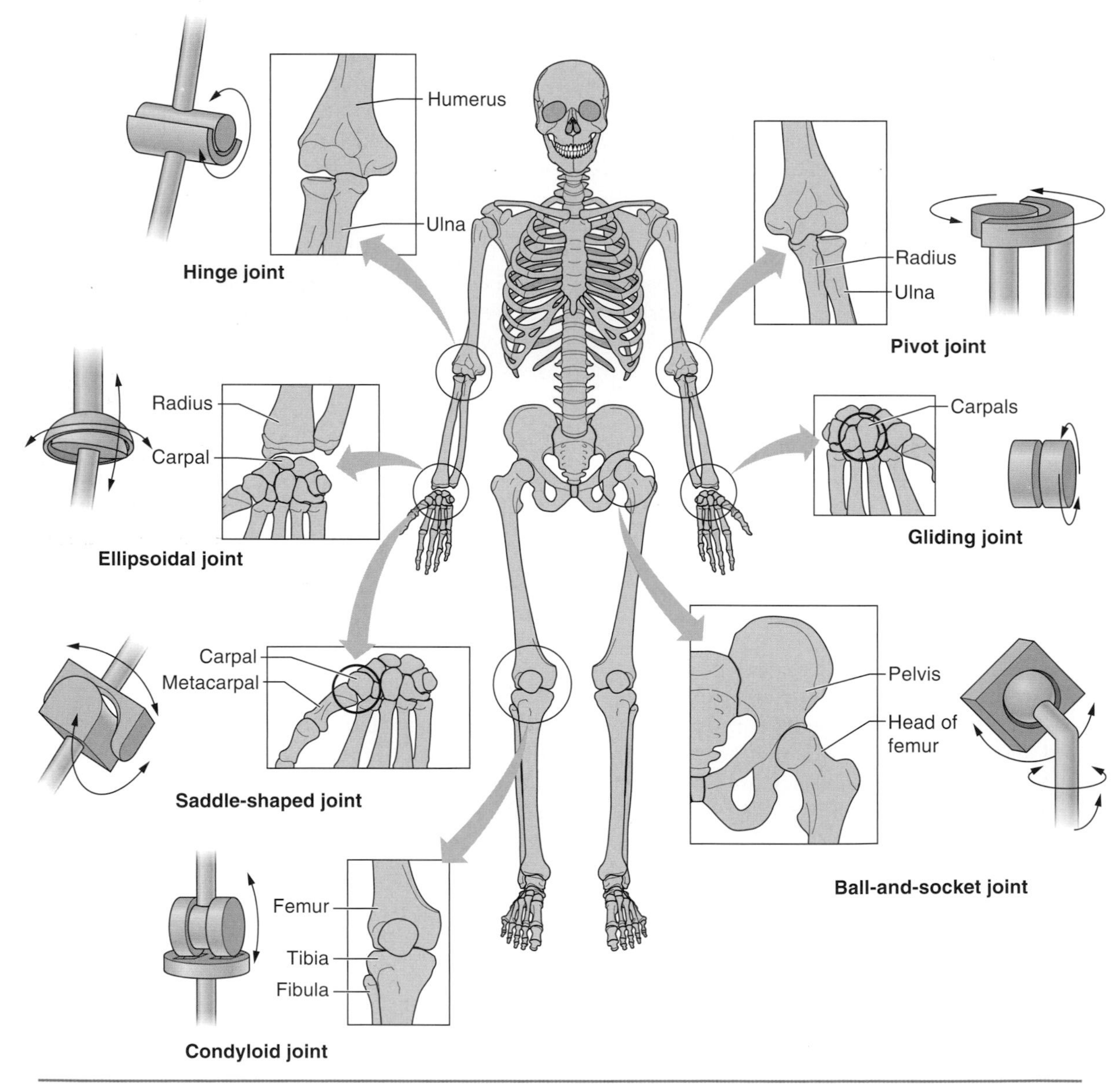

FIGURE 3.6 Mechanical correlates of synovial joint types.

Applying the Concept

Arthritis

Arthritis refers to inflammation of a joint. The term *arthritis* encompasses many conditions that have either primary or secondary inflammatory involvement. Among the major types are those resulting from chronic and excessive mechanical loading (e.g., **osteoarthritis**), systemic disease (e.g., **rheumatoid arthritis**), or biochemical imbalances (e.g., **gouty arthritis**).

Osteoarthritis (OA), also known as degenerative joint disease, is the most common form of arthritis. It is characterized by deterioration of the hyaline articular cartilage and by bone formation on joint surfaces. Osteoarthritis results from mechanical trauma and accompanying chemical process alterations. Given its mechanical etiology, OA most often affects the load-bearing joints of the lower extremities (e.g., hip and knee).

Rheumatoid arthritis (RA) is an autoimmune condition in which the body's immune system malfunctions and attacks its own joint tissues. RA often affects the joints of the hand and fingers and is characterized by considerable joint swelling, pain, and limited range of motion. In severe cases, the bones can become fused together. This eliminates joint motion altogether.

Gouty arthritis (or gout) is caused by excess production of uric acid in the form of uric acid crystals. These crystals are carried by the circulatory system and become embedded in synovial joint structures. The crystals irritate the joint and initiate an inflammatory response.

- *Gliding (planar) joints* have opposing flat or slightly curved surfaces that permit limited sliding between bones. These joints have no axis of rotation, and their movements are described as planar. Gliding joints include those formed by the articular processes of adjacent vertebrae, intertarsal joints, and intercarpal joints.

- *Hinge joints* exhibit angular motion about a single fixed axis of rotation, much like that of a door hinge. Movements are restricted to a single plane (uniplanar) and a single axis (uniaxial) by the joint's bony configuration. These joints (e.g., ankle, elbow, interphalangeal) are relatively stable, largely because of their tight bony fit.

- *Pivot joints* are uniaxial (like hinge joints) but are distinguished by an axis that runs longitudinally along a bone. The pivot joint between the atlas and axis (in the neck) permits rotation of the head from side to side. Similarly, the proximal radioulnar joint pivots to allow the radius to rotate (or "roll") over the ulna in pronation of the forearm.

- *Condyloid (ellipsoid) joints* are relatively unstable joints formed by articulation of a shallow convex surface of one bone with the concave surface of another. The metacarpophalangeal and wrist joints provide good examples of condyloid joints and their ability to move in two planes (biplanar). The articulation between the distal femur and the proximal tibia at the knee forms a double condyloid (or bicondyloid) joint. Given the instability of this configuration, the knee is reinforced by numerous ligaments and other supporting structures. (Note that the knee sometimes is classified as a hinge joint; this is technically incorrect. Although the primary movement at the knee involves flexion or extension, the knee also can rotate when in a flexed position and hence is biplanar. This violates the uniplanar requirement of a hinge joint.)

- The name *saddle joint* is given to articulations where one bone sits on another as a saddle sits on a horse. The concave surface of one bone (i.e., the saddle) straddles the convex surface of another bone (i.e., the horse's back). Generally considered biplanar and biaxial, saddle joints are relatively stable because of the interlocking of the two bones. The carpometacarpal joint of the thumb is an example of a saddle joint. When you twiddle your thumbs, the biplanar nature of this saddle joint becomes obvious.

- *Ball-and-socket joints* are formed when the rounded end of one bone is housed in a depression (socket) of another bone. Ball-and-socket joints, such as the hip and shoulder, are triplanar and triaxial and demonstrate tremendous range of motion.

Go to the web study guide to access tables that detail the articulations of the upper extremities, lower extremities, and the spinal column.

Joint Motion and Movement Description

A full appreciation of human movement requires an understanding of anatomical details and certain movement concepts. Some of these concepts have been outlined in earlier sections, and we build on them here in more depth.

A few key points should be kept in mind when considering joints and motion. Specific movement-related descriptors, such as *flexion* and *extension*, can be used in one of two ways. They may describe movements (e.g., "Maria flexed her elbow") or positions (e.g., "Trevor maintained a flexed elbow while carrying the box"). The context usually makes clear whether the term is being used to describe a movement or a position. Also keep in mind that all movements are described from the mover's perspective. Consider the case of a woman, for example, who curls (flexes) her right elbow in lifting a dumbbell. Viewed face to face, the movement occurs on the observer's left side. Described from the mover's perspective, however, the movement is flexion of the right elbow.

Joint Stability and Mobility

Some joints, such as the hip, are very stable (i.e., they strongly resist dislocation), while others such as the glenohumeral joint are unstable (i.e., they are dislocated fairly readily). From a movement perspective, some joints move freely and have extensive mobility, while others are immovable, or immobile. In general, very stable joints are relatively immobile; conversely, relatively unstable joints typically are very mobile. This concept describes a stability–mobility continuum: Joints built for stability are less mobile, and joints designed for mobility are less stable. The concept of a stability–mobility continuum holds for most joints. Where a given joint falls on this continuum depends on the degree of bony fit, joint capsule tightness, the amount of ligamentous support, and whether the joint contains a stabilizing fibrocartilage structure such as a meniscus. The presence or absence of one or more of these characteristics, however, does not ensure a joint's relative stability or mobility. A strong complement of ligaments and a tight joint capsule, for example, might compensate for poor bony fit and result in a relatively stable joint.

Applying the Concept

Double-Jointed Joints

The term *double jointed* sometimes is used to describe joints that are hypermobile, allowing extreme range of motion. Contrary to what the name might suggest, double-jointed articulations do not contain a second joint but rather have shallow bony articulations, loose joint capsules, or lax ligaments. These structural characteristics allow greater than normal range of motion and possible **subluxation** (i.e., partial dislocation) of the joint.

As with most rules, however, there are exceptions. The hip joint, for example, is both mobile and stable. The hip's mobility is afforded by its ball-and-socket construction, with the head of the femur fitting into the socket (acetabulum) of the pelvis; its stability is provided by the considerable reinforcement of periarticular tissues, especially the large muscle mass surrounding the joint. The depth of the femoral head within the acetabulum contrasts with the large humeral head and shallow glenoid of the unstable glenohumeral (shoulder) joint.

Movement Planes and Joint Motion

If a person moves a body segment away from anatomical position (see figure 1.8), subsequent movements of that segment may occur in a plane different from the one in anatomical position. For example, when in anatomical position, external (or lateral) rotation of the hip joint occurs in the transverse plane. If the hip is flexed (from anatomical position) 90° before it is externally rotated, the rotation now happens in the frontal plane (figure 3.7*a*).

As another example, consider the elbow joint. From anatomical position, elbow flexion and extension occur in the sagittal plane. If the glenohumeral joint is first abducted 90°, subsequent elbow flexion and extension now happen in the transverse plane (figure 3.7*b*).

Movements confined to a single plane often appear restricted and robotic. Full expression of the wide variety and elegance of human movement depends on our ability to move unconstrained in three dimensions. The amount of movement allowed at a particular joint is dictated by its structure and described by the number of planes in which a segment can move or the number of primary rotational axes a joint possesses.

A uniplanar joint (e.g., elbow) has a single axis of rotation. Biplanar and biaxial articulations such as the metacarpophalangeal joints move in two planes. Triplanar joints are free to move in three planes. The hip and glenohumeral joints, for example, move in three planes: flexion and extension in the sagittal plane, abduction and adduction in the frontal plane, and internal and external rotation in the transverse plane.

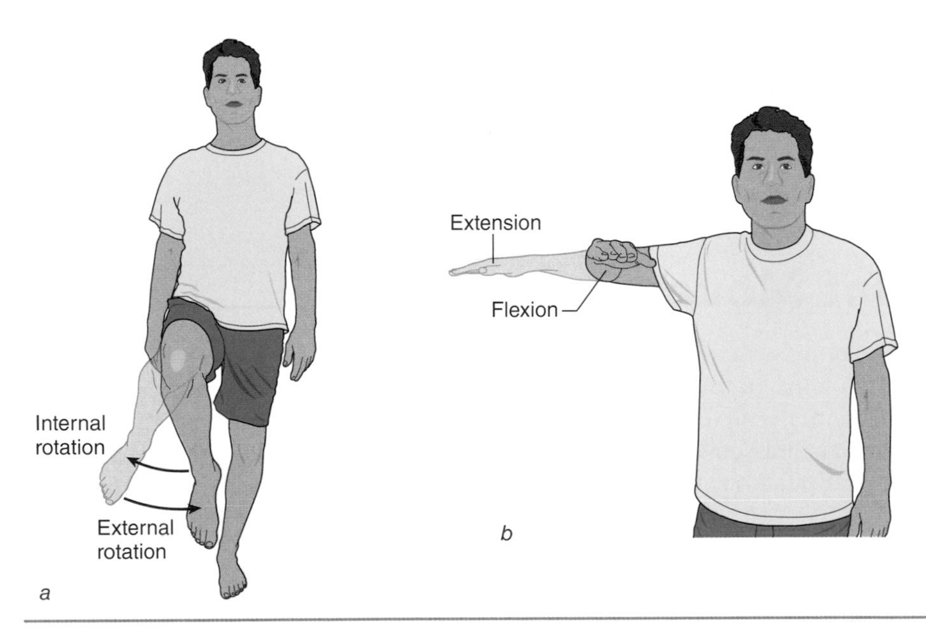

FIGURE 3.7 Movements occurring out of anatomical position: *(a)* After hip flexion, internal and external rotation of the thigh now occurs in the frontal plane, as compared with its movement in the transverse plane when in anatomical position; *(b)* after abduction of the arm, elbow flexion and extension now occur in the transverse plane, as compared with its movement in the sagittal plane when in anatomical position.

Applying the Concept

Contortionism

Contortionism is an art in which performers exhibit extreme flexibility. Contortionists have joint hypermobility, and some can even voluntarily dislocate joints to achieve curious postures. As to the question of whether contortionists are born or made, the answer would appear to be both. Some performers seem genetically predisposed to hypermobility and can achieve contorted positions with little training, while others must spend several hours a day training to maintain their remarkable flexibility and joint laxity.

THIS YOUNG contortionist displays her extreme flexibility.

Types of Joint Movement

As just discussed, the bony structure and periarticular tissues of synovial joints dictate movement potential at each joint. In this section we consider four types of joint movement. Keep in mind that joint movements are described according to the type of movement (e.g., gliding, rotation) and the plane of movement (sagittal, frontal, or transverse) relative to the standard reference of anatomical position.

Gliding and Angular Movements

The simple sliding, or gliding, of two surfaces on one another is considered uniplanar and typically does not involve any rotation. Gliding movements are seen at the intercarpal and intertarsal joints, where the amount of gliding is very limited because of the tightness of the joint capsule and supporting ligaments.

Angular motion occurs when a body segment moves through an angle about an imaginary line called the *axis of rotation*. The axis of rotation is usually located in either the proximal or distal end of the segment. When the forearm is rotated about the elbow during a biceps curl, the axis of rotation is a line through the elbow that is perpendicular to the plane of motion (figure 3.8*a*).

Flexion is angular movement occurring in the sagittal plane (relative to anatomical position) in which the angle between articulating segments decreases. **Extension** is the opposite movement when the relative joint angle increases. For example, the neck flexes when the head nods forward, with the chin moving toward the chest. Neck extension returns the head to its upright position. When a joint angle increases beyond anatomical position, the movement is termed **hyperextension** (figure 3.8*b*), such as when the head is tilted backward from anatomical position. Only certain joints can hyperextend without causing injury. The glenohumeral and hip joints easily swing back posteriorly into hyperextension. The elbow and knee, however, risk serious injury if hyperextended.

In the frontal plane, angular movement that takes a segment away from the body's midline is termed **abduction**. This happens, for example, when the leg is lifted out to the side from anatomical position (figure 3.8*c*). Moving the segment back toward the midline is called

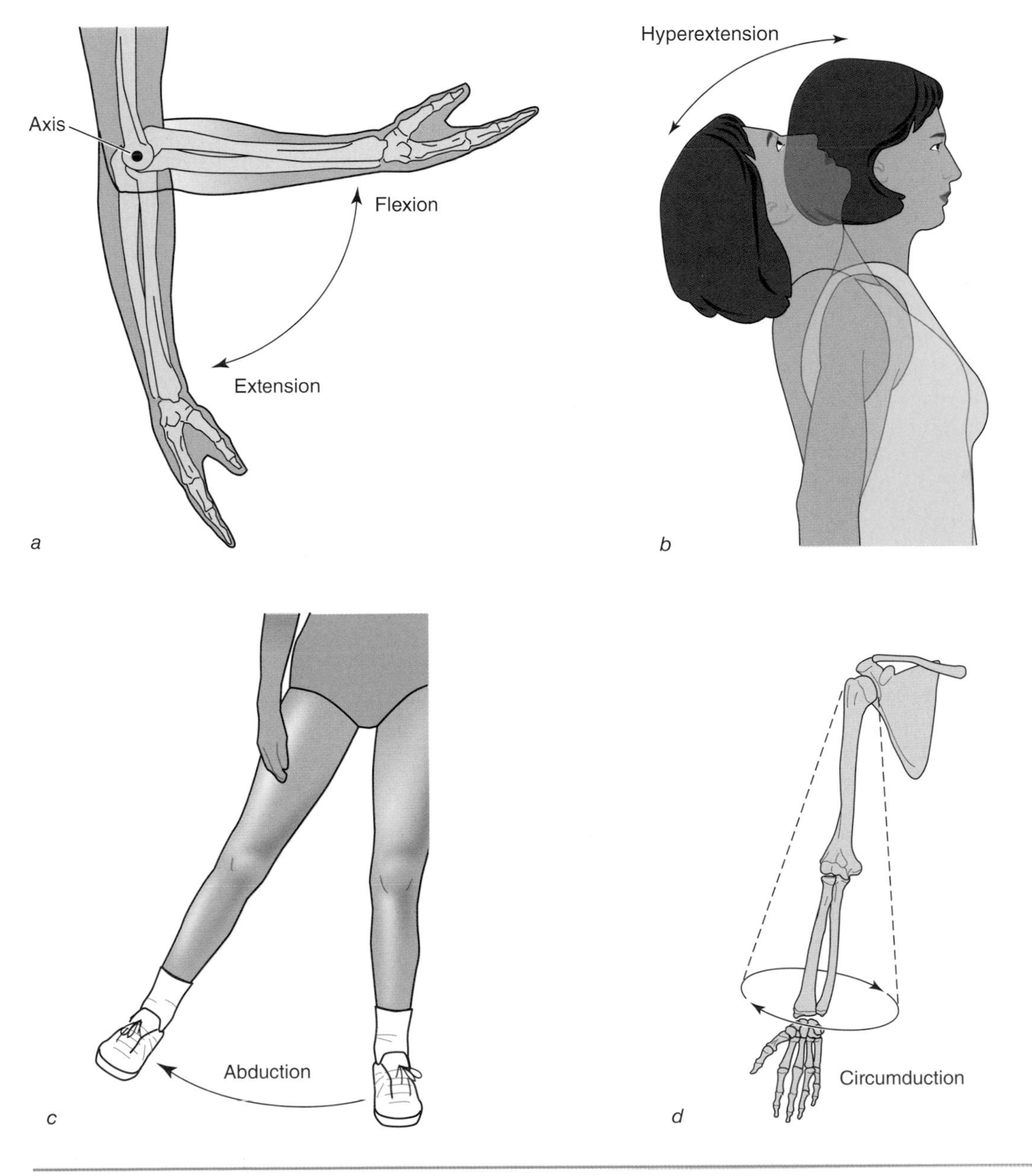

FIGURE 3.8 *(a)* Axis of rotation during flexion and extension at the elbow joint, *(b)* neck hyperextension, *(c)* abduction of the leg from anatomical position, and *(d)* circumduction.

adduction. Abduction and adduction also describe movement of the fingers relative to the midline of the hand. Spreading the fingers apart is abduction; bringing them back together is adduction. The terms *abduction* and *adduction* are used only to describe movements of the appendicular skeleton.

A potentially confusing situation arises when the arm is elevated to the side (abducted) beyond 90°. As the arm moves past 90° to an overhead position, it moves back toward the body's midline, and the movement might be called adduction. While technically correct, we do not describe the motion that way. Frontal plane movement from anatomical position to overhead is called abduction throughout the full range of motion. Similarly, the return movement from an overhead position back to anatomical position is termed adduction.

Circumduction is a special form of angular motion in which the distal end of a limb or segment moves in a circular pattern about a relatively fixed proximal end. In three-dimensional space, circumduction traces out a cone-shaped pattern (figure 3.8*d*). Examples of circumduction include rotating the arm in full circles about the shoulder and moving a finger in a circular pattern about its proximal metacarpophalangeal joint.

Rotational Movements

Rotational movement, or rotation, is distinguished from angular movement by the fact that the axis of rotation is oriented along the long axis of a bone or segment instead of passing through one end of the segment (figure 3.9). From anatomical position, rotation occurs in the transverse plane. (Note that the term *rotation* is technically defined as previously mentioned, but also is used to describe angular movement of a segment about its joint axis [e.g., hip rotation].)

Rotations of the axial skeleton (e.g., when the spine twists from side to side) are simply described as rotation left or rotation right. (Remember, the directions left and right are described relative to the person moving, not from the perspective of someone observing the movement.)

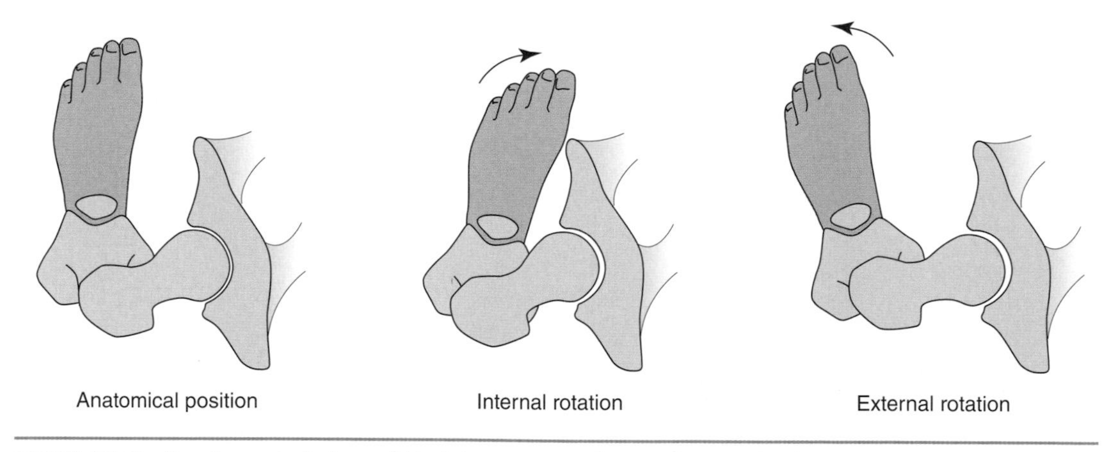

FIGURE 3.9 Top (superior) view of hip joint rotation about a longitudinal axis.

Rotational movements in appendicular segments are referenced by their direction relative to the body's midline. The anterior surface of the segment (in anatomical position) is used as the reference. If movement rotates the segment's anterior surface inward toward the midline, the movement is termed **internal** (or **medial**) **rotation**. Rotational movement away from the midline of the body is called **external** (or **lateral**) **rotation** (see figure 3.10).

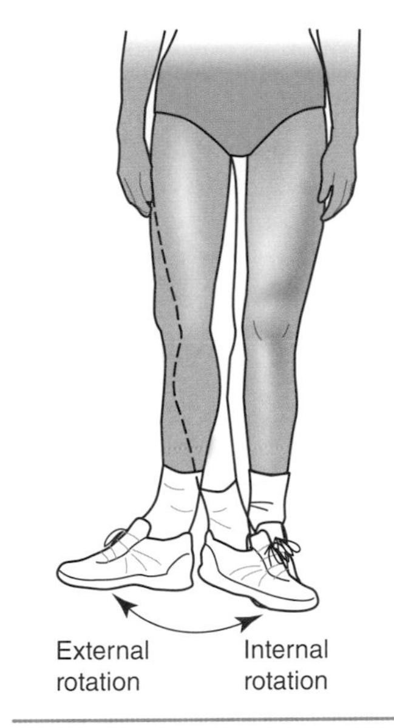

FIGURE 3.10 Hip joint rotation.

Special Movements

The general movement forms just discussed describe most movements. Some movements, however, have special names unique to a particular joint or body region.

▸ Plantar Flexion and Dorsiflexion

At the ankle, movement of the foot away from the lower leg is called **plantar flexion**, as when performing a calf raise exercise or depressing an automobile's gas pedal. The reverse movement of bringing the foot upward toward the lower leg is termed **dorsiflexion** (figure 3.11).

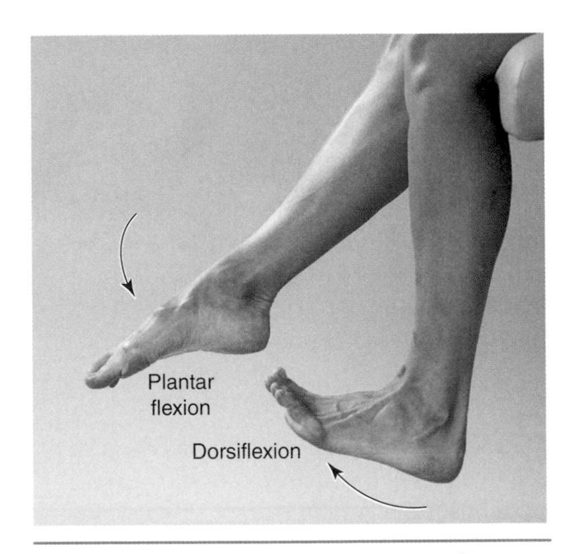

FIGURE 3.11 Plantar flexion and dorsiflexion.

▸ Inversion and Eversion

Inversion is a movement of the intertarsal joints (particularly the talocalcaneal, or subtalar, joint) that results in the sole of the foot being moved inward toward the midline of the body. For **eversion**, the sole of the foot is moved away from the body's midline (figure 3.12). From anatomical position, eversion and inversion happen primarily in the frontal plane.

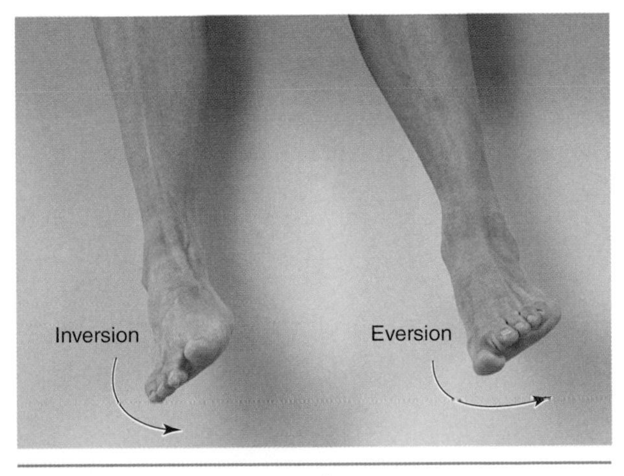

FIGURE 3.12 Inversion and eversion.

▸ Pronation and Supination

Simultaneous rotation of the proximal and distal radioulnar joints (in the forearm) allows the radius (lateral forearm bone) to rotate over the relatively stationary ulna (medial forearm bone). From anatomical position, rotation of the radius over the ulna is termed **pronation**. The return from a pronated position back to anatomical position is called **supination** (figure 3.13). In anatomical position, the radioulnar joints place the forearm in what is termed a *supinated position*. When shaking hands, the forearm is in midposition, halfway between the pronated and supinated positions. (Important note: Many authors, though certainly not all, also use the terms *supination* and *pronation* to describe combined movements of the foot and ankle. Supination of the foot and ankle commonly, though not universally, describes the combination of inversion, plantar flexion, and internal [medial] rotation, or adduction, of the foot. Pronation of the foot and ankle describes the opposite combination of eversion, dorsiflexion, and external [lateral] rotation, or abduction, of the foot.)

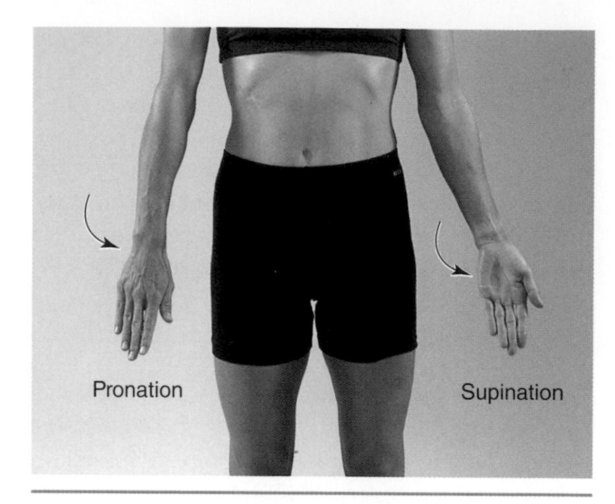

FIGURE 3.13 Pronation and supination.

▸ Radial and Ulnar Deviation

In an earlier section, abduction and adduction were used to describe movements in the frontal plane away from the midline and toward the midline, respectively. Abduction of the wrist and hand, such that the thumb moves closer to the radius, is termed **radial deviation**. Adduction of the wrist and hand, such that the little (fifth) finger moves closer to the ulna, is called **ulnar deviation** (figure 3.14).

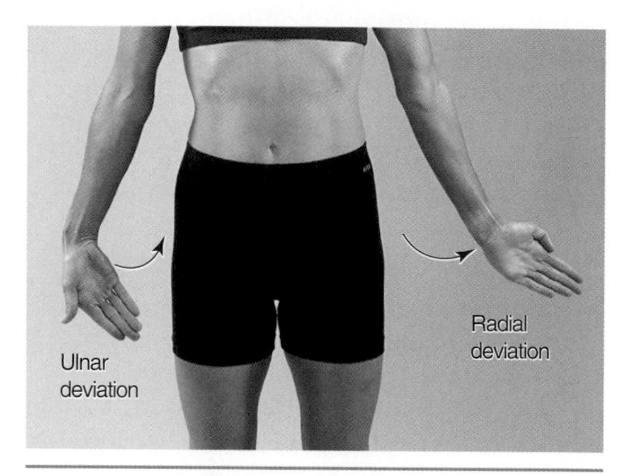

FIGURE 3.14 Radial and ulnar deviation.

▸ Opposition

Opposition refers to the thumb's ability to work with the other four fingers to perform grasping movements (figure 3.15). This ability is of paramount importance in allowing us to manipulate objects (e.g., picking up a pencil, turning a key).

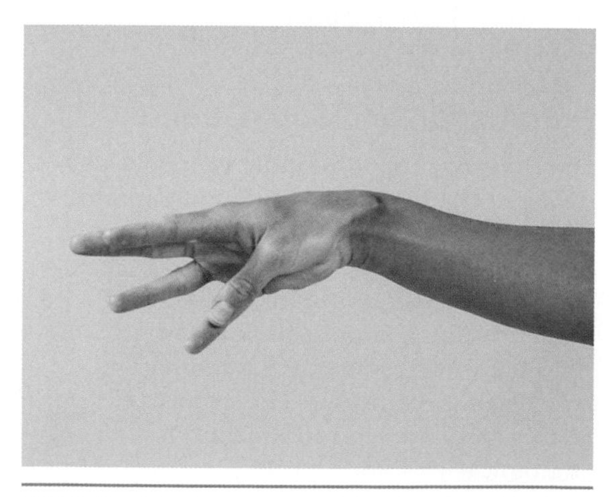

FIGURE 3.15 Opposition of the thumb and fingers.

▸ Elevation and Depression

Elevation refers to a structure moving in a superior, or upward, direction. **Depression** describes an inferior, or downward, movement (figure 3.16). These terms are typically used to describe movements of the jaw (mandible) and scapulae. Elevation is seen when closing the mouth or shrugging the shoulders. Depression results in opening the mouth and dropping the shoulders. These movements usually occur in the frontal plane.

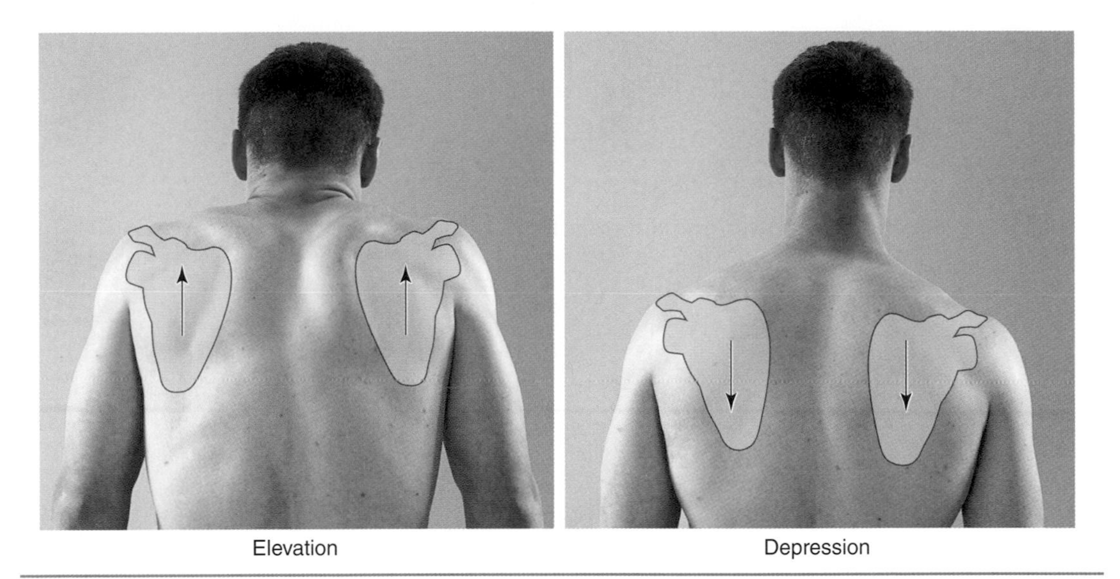

FIGURE 3.16 Elevation and depression of the scapulae.

▸ Protraction and Retraction

Protraction (also abduction) describes movement anteriorly, or toward the front of the body. **Retraction** (also adduction) refers to posteriorly directed movement toward the back. These movements occur in the transverse plane. For example, the clavicles and scapulae protract when one slumps forward with rounded shoulders and crossed arms. These bones retract when the shoulders are pulled back into an upright and correct posture (figure 3.17).

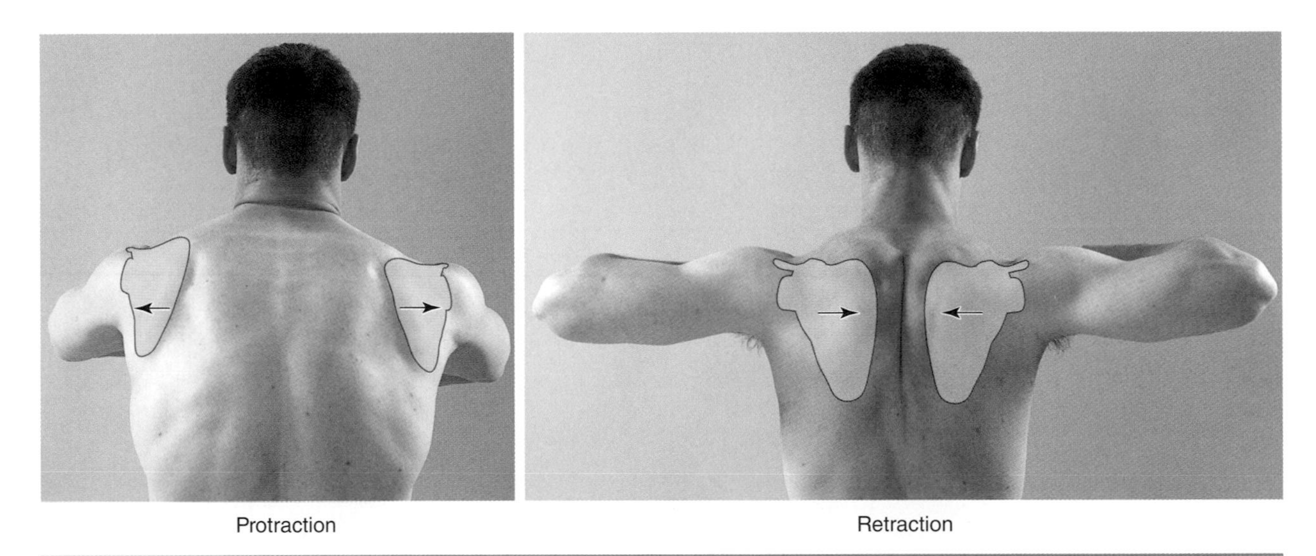

FIGURE 3.17 Protraction and retraction.

▸ Lateral Flexion

Lateral flexion (right and left) refers to sideways bending of the vertebral column in the frontal plane (figure 3.18).

▸ Upward and Downward Rotation

Upward and **downward rotation** refer to rotational movements of the scapulae (figure 3.19).

FIGURE 3.18 Lateral flexion (left).

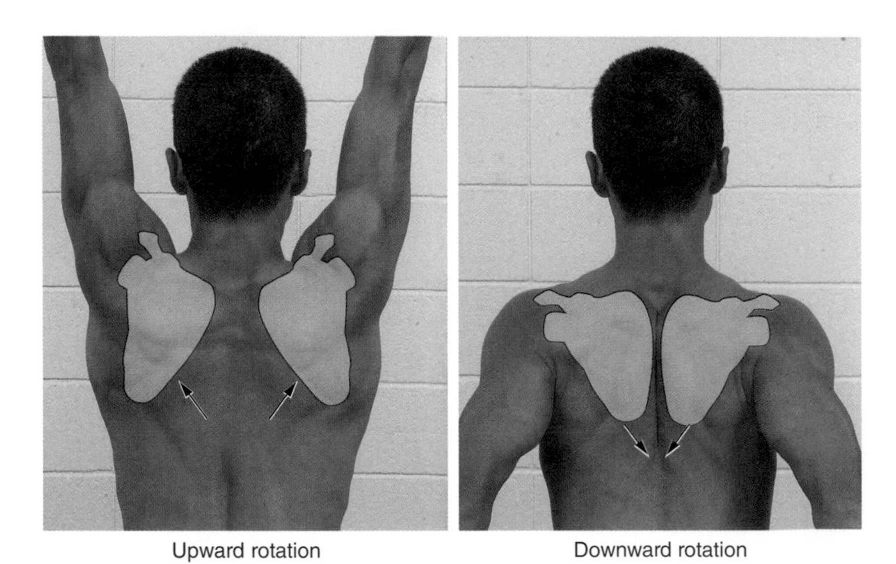

FIGURE 3.19 Upward and downward rotation.

Joint Structure and Movement

Gross movements of the human musculoskeletal system are made primarily by the major synovial joints of the extremities, along with contributions from joints of the head, neck, and spine. This section explores each joint's structure and movements. A summary of joint structure and movements is given in tables 3.1, 3.2, and 3.3. Typical range of motion values are shown in table 3.4.

TABLE 3.1 Summary of Joint Structure and Movement of the Head, Neck, and Trunk

		ALL MOVEMENTS BEGIN FROM ANATOMICAL POSITION		
Joint	**Structural classification**	**Movement**	**Plane**	**Axial/Planarity**
Intercranial	Suture	None		
Temporomandibular	Synovial (condyloid)	Elevation Depression Protraction Retraction	Sagittal Transverse	Biaxial/Biplanar
Atlanto-occipital	Synovial (hinge)	Flexion Extension	Sagittal	Uniaxial/Uniplanar
Vertebral column: atlantoaxial	Synovial (pivot)	Rotation right Rotation left	Transverse	Uniaxial/Uniplanar
C2-L5: (Vertebral bodies: symphysis) (Articular processes: synovial [plane])		Flexion Extension Hyperextension Lateral flexion right Lateral flexion left Rotation right Rotation left	Sagittal Frontal Transverse	Triaxial/Triplanar
Costovertebral	Synovial (plane)	Gliding		Nonaxial/Nonplanar
Sternomanubrial	Symphysis	Sternal angle increase Sternal angle decrease		Nonaxial/Nonplanar

TABLE 3.2 Summary of Joint Structure and Movement of the Upper Extremity

		ALL MOVEMENTS BEGIN FROM ANATOMICAL POSITION		
Joint	**Structural classification**	**Movement**	**Plane**	**Axial/Planarity**
Sternoclavicular (shoulder girdle)	Synovial (ball and socket)	Anterior rotation Posterior rotation Upward rotation Downward rotation Abduction Adduction	Sagittal Frontal Transverse	Triaxial/Triplanar
Acromioclavicular	Synovial (plane)	Gliding		Nonaxial/Nonplanar
Glenohumeral (shoulder)	Synovial (ball and socket) [starting with shoulder flexed 90°]	Flexion Extension Hyperextension Abduction Adduction Internal (medial) rotation External (lateral) rotation Horizontal abduction (horizontal extension) Horizontal adduction (horizontal flexion)	Sagittal Frontal Transverse Transverse	Triaxial/Triplanar
Elbow	Synovial (hinge)	Flexion Extension	Sagittal	Uniaxial/Uniplanar
Radioulnar	Proximal: synovial (pivot) Middle: syndesmosis Distal: synovial (pivot)	Pronation Supination	Transverse	Uniaxial/Uniplanar
Radiocarpal (wrist)	Synovial (condyloid)	Flexion Extension Hyperextension Radial deviation (abduction) Ulnar deviation (adduction)	Sagittal Frontal	Biaxial/Biplanar
Intercarpal	Synovial (plane)	Gliding		Nonaxial/Nonplanar
Carpometacarpal	Synovial (plane)	Gliding		Nonaxial/Nonplanar
Metacarpophalangeal	Thumb: synovial (saddle) (2-5): synovial (condyloid)	Flexion Extension Hyperextension Abduction Adduction	Frontal (2-5) Sagittal Sagittal (2-5) Frontal	Biaxial/Biplanar
Interphalangeal	Synovial (hinge)	Flexion Extension	Sagittal	Uniaxial/Uniplanar

TABLE 3.3 Summary of Joint Structure and Movement of the Pelvis and Lower Extremity

		ALL MOVEMENTS BEGIN FROM ANATOMICAL POSITION		
Joint	**Structural classification**	**Movement**	**Plane**	**Axial/Planarity**
Sacroiliac	Synovial (plane)	Gliding		Nonaxial/ Nonplanar
Pubic symphysis	Symphysis	Distraction/Separation during childbirth		
Pelvic girdle (movement of pelvis relative to femur)	Synovial (ball and socket)	Anterior tilt Posterior tilt Lateral tilt right Lateral tilt left Rotation right Rotation left	Sagittal Frontal Transverse	Triaxial/Triplanar
Hip (movement of femur relative to pelvis)	Synovial (ball and socket)	Flexion Extension Hyperextension Abduction Adduction Internal (medial) rotation External (lateral) rotation	Sagittal Frontal Transverse	Triaxial/Triplanar
	[starting with hip flexed 90°]	Horizontal abduction (horizontal extension) Horizontal adduction (horizontal flexion)	Transverse	
Patellofemoral	Synovial (plane)	Gliding		Nonaxial/ Nonplanar
Tibiofemoral (knee)	Synovial (bicondyloid)	Flexion Extension Internal (medial) rotation External (lateral) rotation [with knee flexed]	Sagittal	Biaxial/Biplanar
Ankle	Synovial (hinge)	Dorsiflexion Plantar flexion	Sagittal	Uniaxial/Uniplanar
Subtalar	Synovial (plane)	Inversion Eversion	Frontal	Uniaxial/Uniplanar
Intertarsal	Synovial (plane)	Gliding		Uniaxial/Uniplanar
Tarsometatarsal	Synovial (plane)	Gliding		Nonaxial/ Nonplanar
Metatarsophalangeal	Synovial (condyloid)	Flexion Extension Hyperextension Abduction Adduction	Sagittal Transverse	Biaxial/Biplanar
Interphalangeal	Synovial (hinge)	Flexion Extension	Sagittal	Uniaxial/Uniplanar

TABLE 3.4 Average Ranges of Joint Motion*

Joint	Joint motion	ROM (degrees)
Hip	Flexion	90-125
	Hyperextension	10-30
	Abduction	40-45
	Adduction	10-30
	Internal rotation	35-45
	External rotation	45-50
Knee	Flexion	120-150
	Rotation (when flexed)	40-50
Ankle	Plantar flexion	20-45
	Dorsiflexion	15-30
Shoulder	Flexion	130-180
	Hyperextension	30-80
	Abduction	170-180
	Adduction	50
	Internal rotation**	60-90
	External rotation**	70-90
	Horizontal flexion**	135
	Horizontal extension**	45
Elbow	Flexion	140-160
Radioulnar	Forearm pronation (from midposition)	80-90
	Forearm supination (from midposition)	80-90
Cervical spine	Flexion	40-60
	Hyperextension	40-75
	Lateral flexion	40-45
	Rotation	50-80
Thoracolumbar spine	Flexion	45-75
	Hyperextension	20-35
	Lateral flexion	25-35
	Rotation	30-45

*Range of motion (ROM) for movements made from anatomical position (unless otherwise noted). Averages reported in the literature vary, sometimes considerably, depending on method of measurement and population measured. Values here are representative of the ranges of reported maximum ROM.

**Movement from abducted position.

Movements of the Head, Neck, and Spine

The head is made up of the skull, the brain, and its associated structural components. Structures in the head are protected by an intricate collection of 22 bones (figure 3.20). The brain and its protective covering are contained in the cranium, composed of 8 bones: frontal, occipital, ethmoid, and sphenoid bones, along with paired temporal and parietal bones. The anterior and anterolateral aspects of the head are formed by 14 facial bones.

The cranial bones are connected by tight suture joints that are immovable, in the same way as interlocking pieces of a jigsaw puzzle. Normally, the only movable joint in the head is the mandible (jawbone) at the temporomandibular (TMJ) joint formed by the articulation of the mandibular ramus with the temporal bone. The TMJ is a synovial condyloid joint, with joint surfaces separated by an articular fibrocartilage disc.

The head rests on the shoulders atop the most superior portion of the spine. The vertebral column (spine) is a group of 26 vertebrae extending from the base of the skull to its inferior termination at the coccyx (tailbone). The spine is divided into five regions (see figure 2.5): cervical (7 vertebrae), thoracic (12), lumbar (5), sacral (1), and coccygeal (1). The sacral and coccygeal vertebrae, while each considered as a single bone, are formed by 5 and 4 fused vertebrae, respectively. Vertebrae in the cervical, thoracic, and lumbar regions are separated by **intervertebral** (IV) **discs** composed of a gelatin-like inner mass (**nucleus pulposus**) surrounded by a layered fibrocartilage network (**annulus fibrosus**) as shown in figure 3.21*a*. Adjacent vertebrae and the intervening IV disc form a **motion segment** (figure 3.21*b*). These articulations form symphysis joints and do not have a synovial joint capsule.

At the apex of the spine, the occipital bone of the cranium articulates with the first cervical (C1) vertebra (atlas). These atlanto-occipital joints are formed between the concave superior surfaces of the atlas and the convex occipital condyles of the skull.

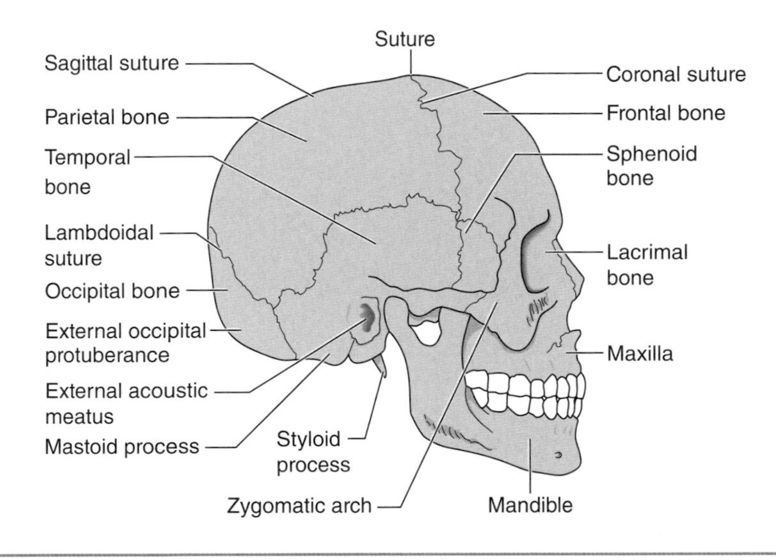

FIGURE 3.20 Cranial bones.

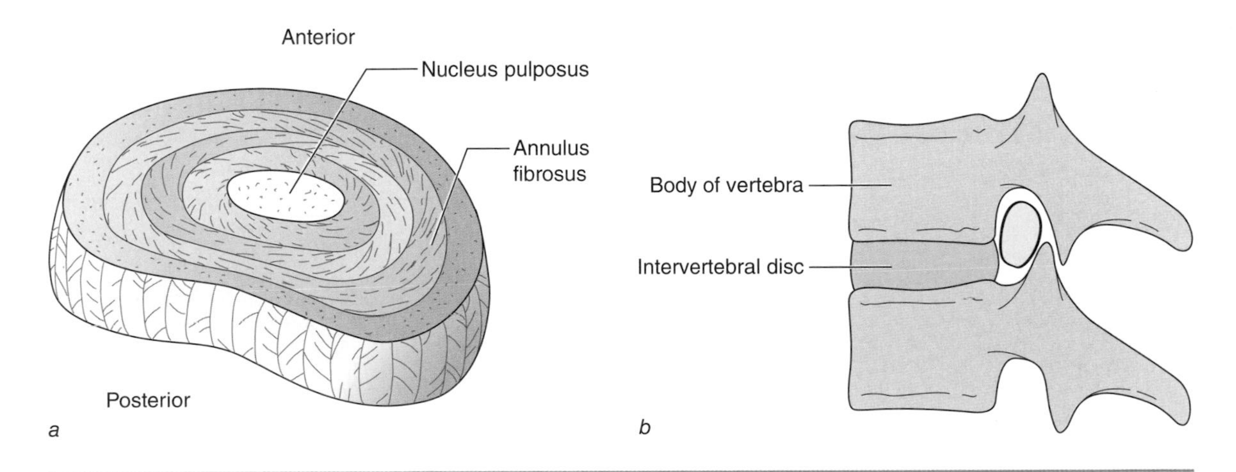

FIGURE 3.21 *(a)* Intervertebral (IV) disc; *(b)* motion segment, consisting of two adjacent vertebrae and the intervening IV disc.

The joint between the atlas (C1) and the axis (C2) has a unique structure, as shown in figure 3.22. A toothlike process (dens) projects superiorly from the axis to articulate with the anterior arch of the atlas to form the atlantoaxial joint. Synovial joints are found between the articular processes (**zygapophyses**) of adjacent vertebrae. The flatness of adjoining surfaces at these joints permits limited gliding between the segments.

The atlanto-occipital joint allows for about 60% of flexion and extension of the head. Head rotation occurs mostly at the atlantoaxial joint, where the unique articulation between the dens of the axis and the atlas allows considerable turning of the head from side to side (figure 3.23).

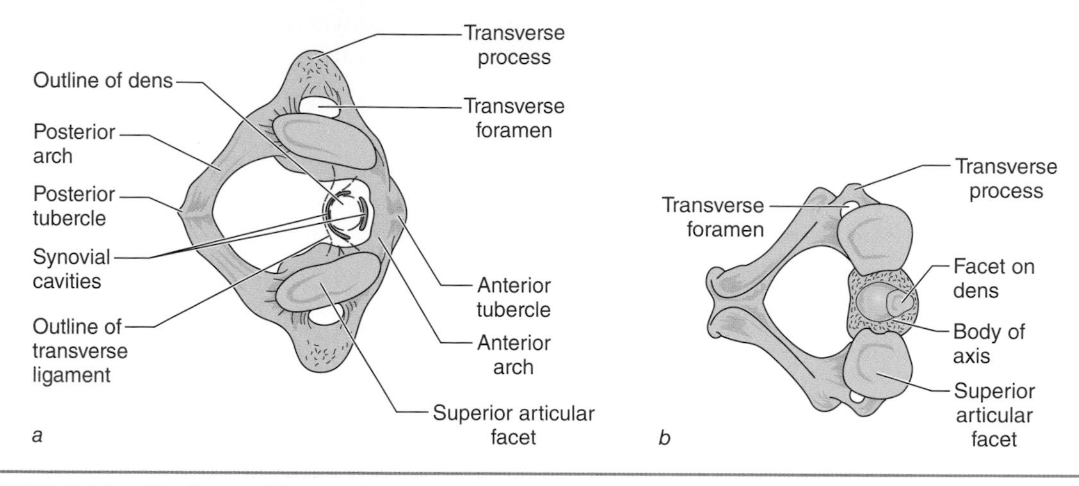

FIGURE 3.22 *(a)* Atlas and *(b)* axis forming the atlantoaxial joint.

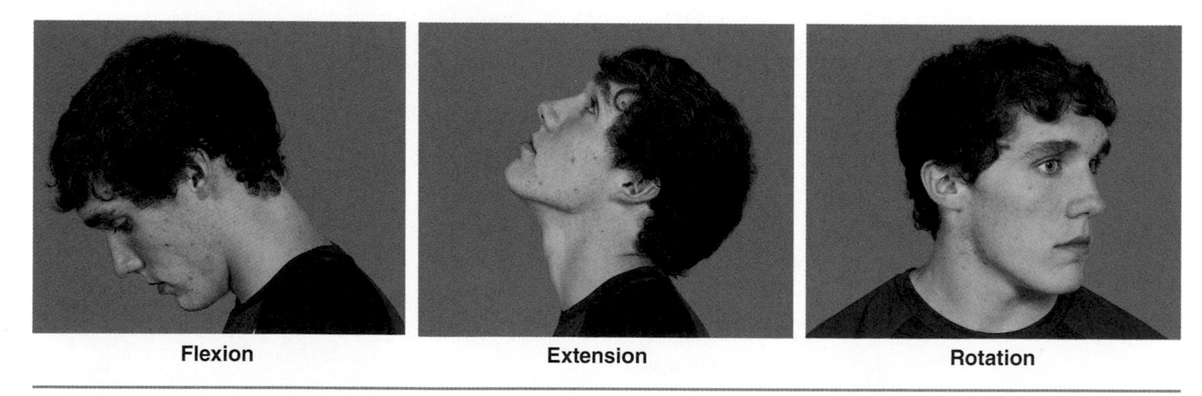

FIGURE 3.23 Head movements.

As triplanar articulations, the IV joints can move (figure 3.24) in flexion and extension (sagittal plane), lateral flexion (frontal plane), and rotation (transverse plane). In general, each motion segment has a relatively limited range of motion, depending on the spinal region. Overall spinal movement represents the sum of all motion segment movements.

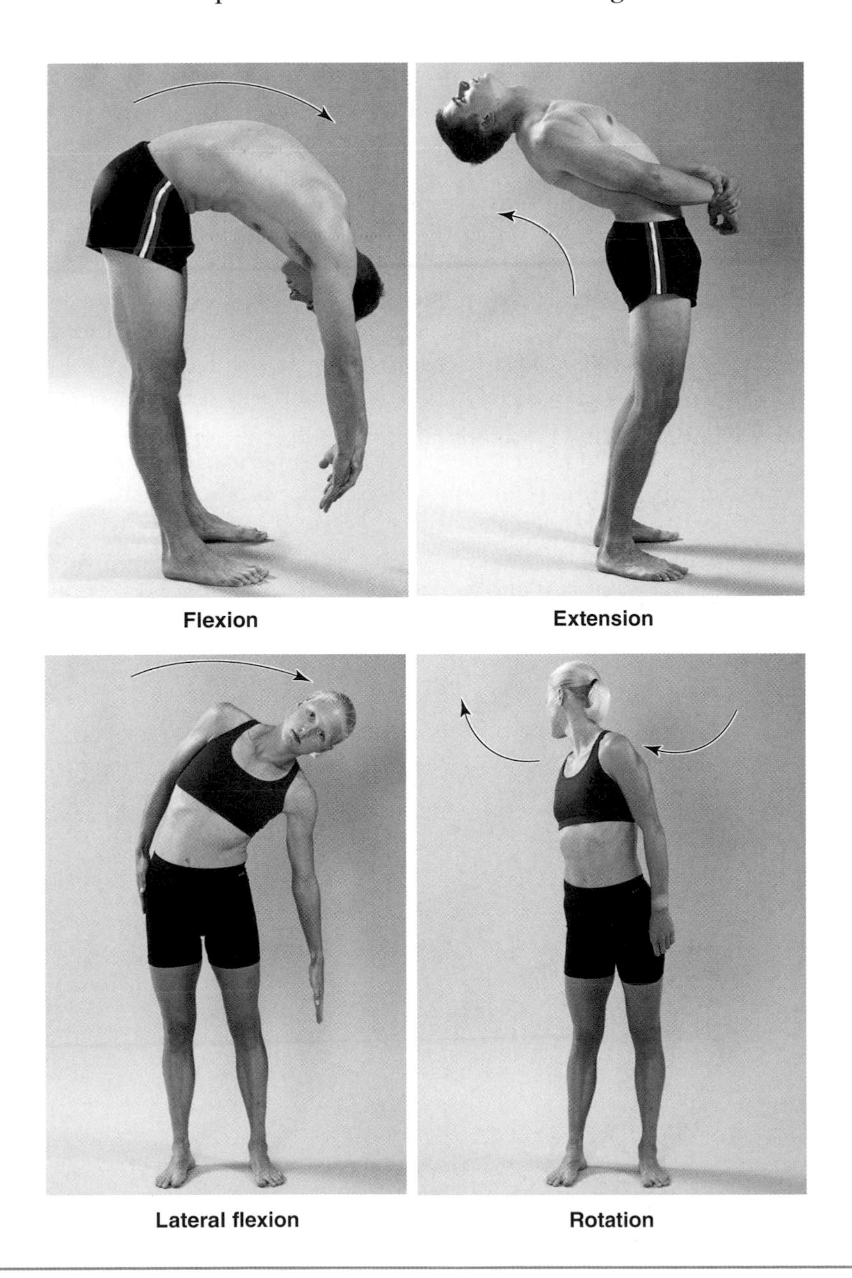

FIGURE 3.24 Spine movements.

Movements of the Shoulder

The shoulder complex includes joints involving the scapula, sternum, clavicle, and humerus (figure 3.25). The clavicle attaches medially to the sternal manubrium at the sternoclavicular joint, a synovial joint with a fibrocartilage disc separating the bony surfaces. At its lateral end, the clavicle articulates with the acromion process of the scapula at the acromioclavicular (AC) joint. The AC joint is a gliding synovial joint with articular surfaces separated by an articular disc.

The humerus of the upper arm articulates with the scapula at the glenohumeral (GH) joint (also shoulder joint). The GH joint is the body's most mobile joint, where the humeral head fits loosely into the shallow glenoid fossa of the scapula. The shoulder's loose joint capsule provides little stability to the joint, which explains why the glenohumeral articulation is a frequent site of joint dislocation. A fibrocartilage glenoid labrum attaches to the rim of the glenoid fossa and improves the joint's bony fit.

The glenohumeral joint is a triplanar articulation whose ball-and-socket structure allows it the greatest mobility of any joint in the body. In the sagittal plane, the arm flexes and extends (figure 3.26). Abduction and adduction occur in the frontal plane, with internal (medial) rotation and external (lateral) rotation happening in the transverse plane. When the upper arm is medially rotated, the range of abduction is limited to about 60° because of pinching (impingement) of the greater tubercle of the humerus on the acromion process of the scapula. Abduction beyond this point requires external rotation of the upper arm, which frees the tubercle from its acromial restriction and permits further movement. In addition to these movements in the primary planes, the arm at the GH joint can move in horizontal abduction (also horizontal extension) and horizontal adduction (also horizontal flexion). Also, in the same manner as the leg acting at the hip, the arm can move in a circular, or conical, pattern known as circumduction.

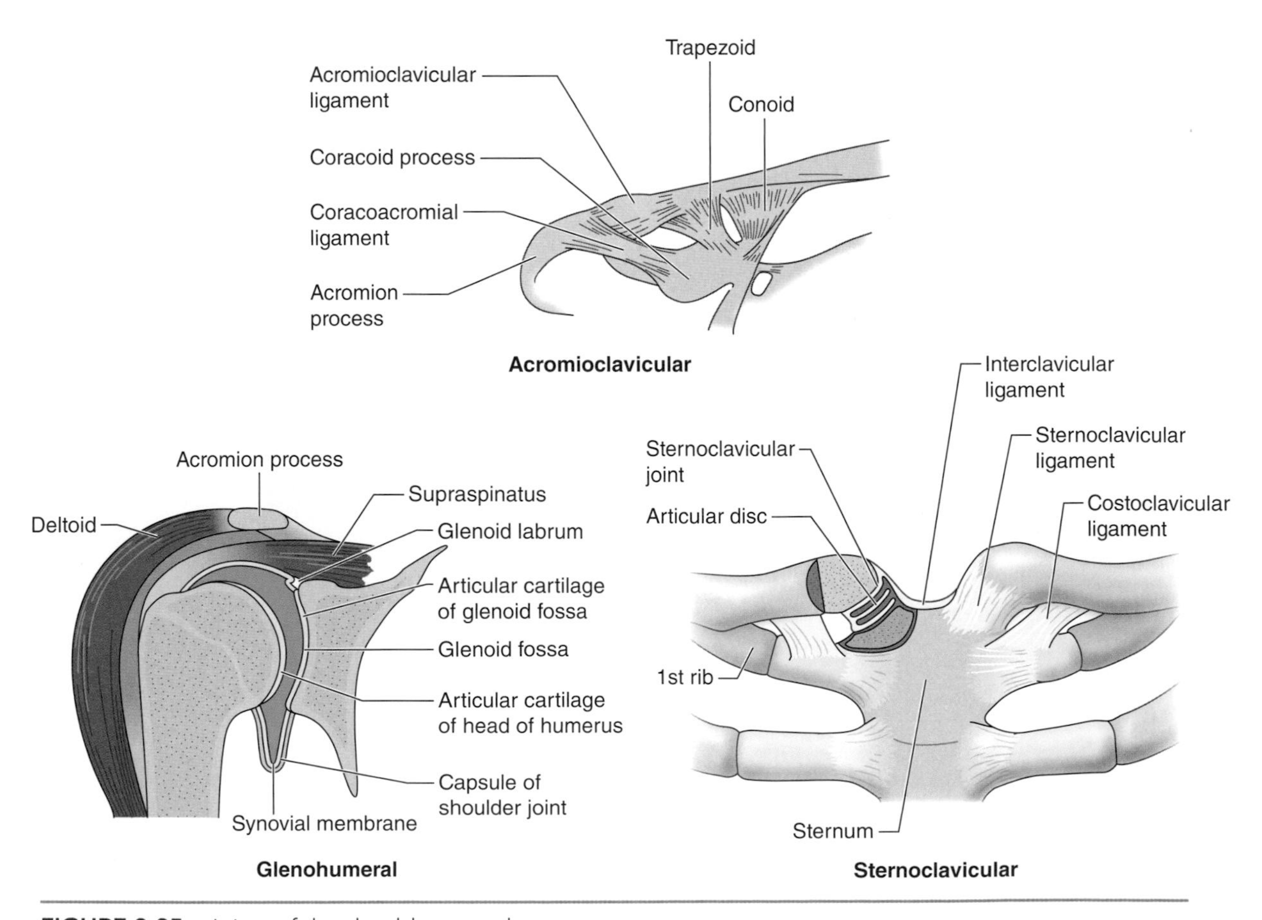

FIGURE 3.25 Joints of the shoulder complex.

The scapula is primarily anchored by muscles and thus can be described as being in muscular suspension. Its only bony articulations are with the humerus and clavicle. The many muscles attaching to the scapula dictate its movement. These movements include elevation and depression, retraction (also adduction) and protraction (also abduction), and upward and downward rotation (figure 3.27).

Glenohumeral joint movement coordinates with scapular motion. In many movements involving the shoulder complex, the humerus and scapula work in concert with one another. For example, after approximately the first 30° of abduction, every 2° of further humeral abduction is accompanied by 1° of scapular rotation. This coordinated action is termed **scapulohumeral rhythm**.

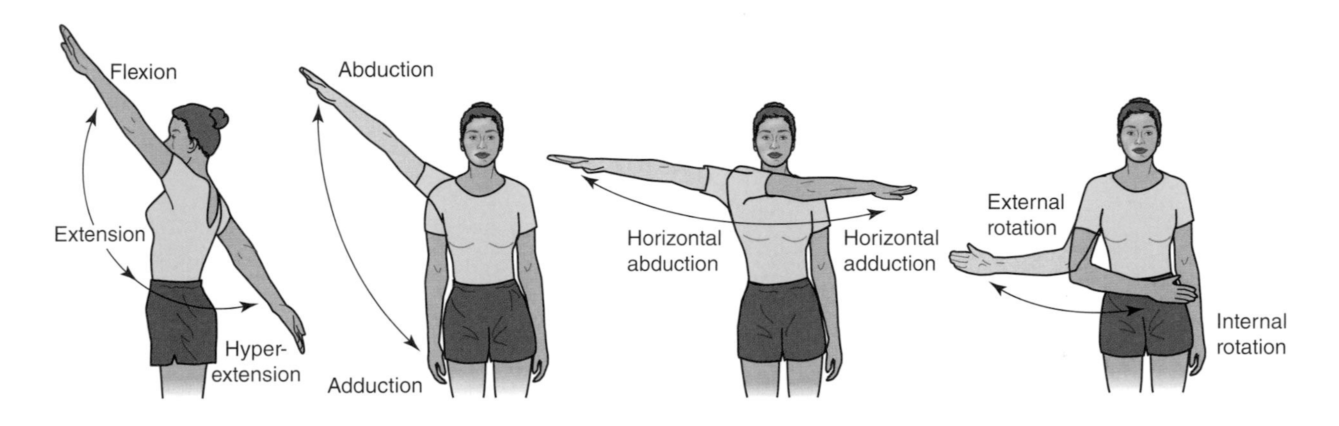

FIGURE 3.26 Glenohumeral (shoulder) movements.

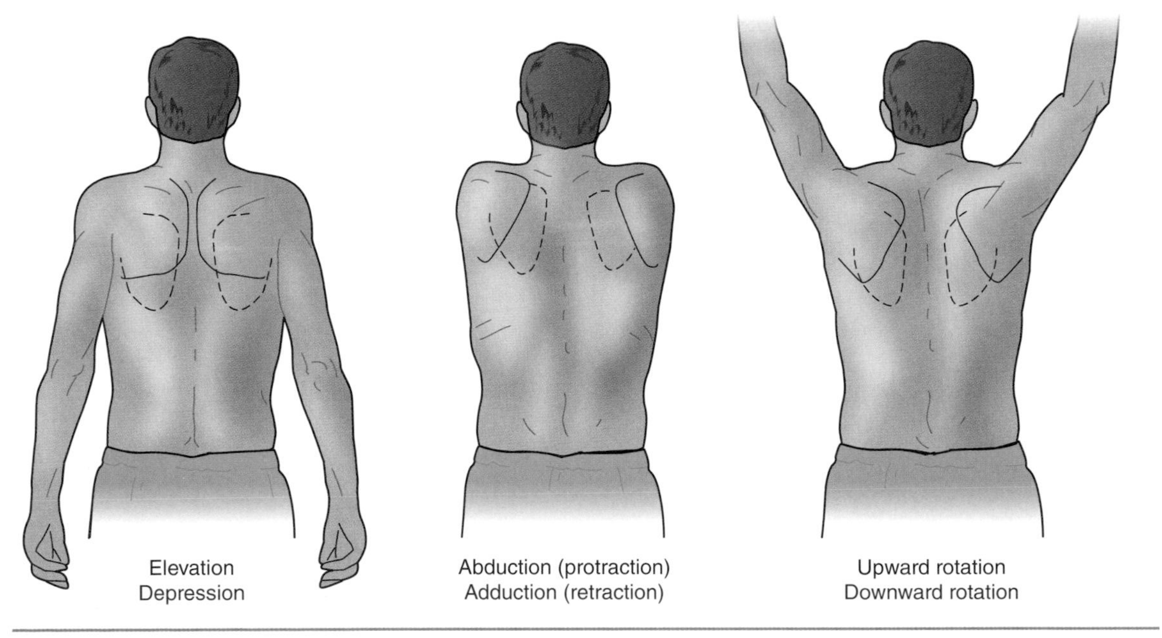

FIGURE 3.27 Scapula movements.

Movements of the Elbow and Forearm

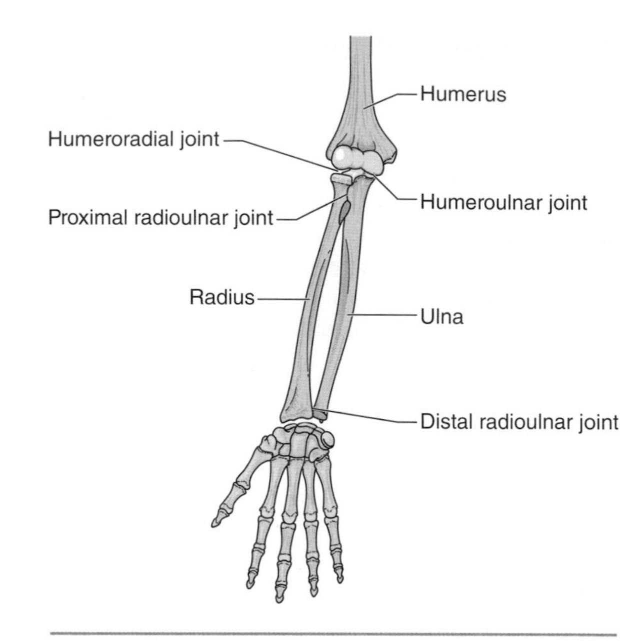

FIGURE 3.28 Elbow and forearm joints.

The elbow complex consists of three joints: humeroulnar, humeroradial, and proximal radioulnar (figure 3.28). The elbow joint proper is formed by the humeroulnar and humeroradial joints. The proximal radioulnar joint works in concert with the distal radioulnar joint to produce movements of the forearm. As a synovial joint, the elbow is surrounded by a thin fibrous capsule that extends continuously from its proximal humeral attachment to the capsule of the proximal radioulnar joint.

The humeroulnar joint is formed by the articulation of the trochlea of the humerus with the trochlear notch of the ulna. The humeroradial joint forms at the junction of the capitulum of the humerus with a shallow depression on the head of the radius.

The proximal radioulnar joint is the articulation between the head of the radius and the ulnar radial notch. At the distal end of the forearm, the ulnar head joins with the ulnar notch of the radius. (Note: The radial head is at the proximal end of the radius, while the ulnar head is at the distal end of the ulna.)

Normal elbow movement is confined to uniplanar flexion and extension at the humeroulnar and humeroradial joints (figure 3.29*a*). Forearm movements of supination and pronation are produced by conjoint rotations at the proximal and distal radioulnar joints. In anatomical position, the forearm is in a supinated position. From this position, pronation causes the radius to roll over a relatively fixed ulna (figure 3.29*b*). The reverse occurs in supination when the radius returns to its anatomical position. When the forearm is in a position halfway between full pronation and supination, as when you shake hands, the forearm is in midposition.

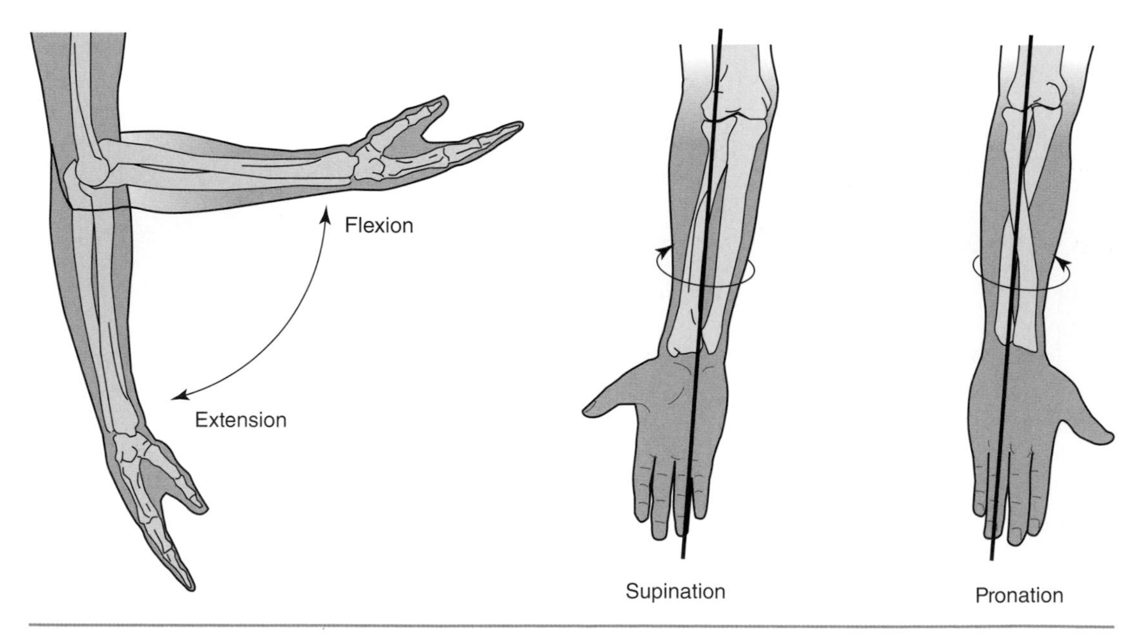

FIGURE 3.29 Elbow and radioulnar joint movements.

Movements of the Wrist and Hand

The wrist (carpus) is not a single joint but rather a group of articulations involving the distal ends of the radius and ulna, along with the carpal bones (figure 3.30). The wrist complex includes the distal radioulnar, radiocarpal, and intercarpal joints. The hand contains numerous articulations, namely the carpometacarpal (CM), metacarpophalangeal (MP), and interphalangeal (IP) joints. All of these are synovial joints. Structurally, the MP joints are condyloid, while the IP articulations are hinge joints.

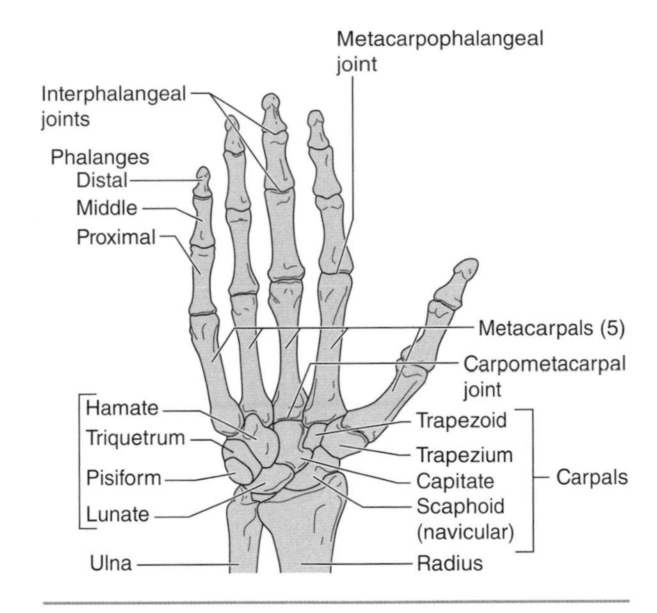

FIGURE 3.30 Bones of the wrist and hand.

The largest joint of the wrist complex is the radiocarpal joint (between the radius and carpals), which allows for flexion and extension (figure 3.31). The wrist also moves in radial deviation (abduction) and ulnar deviation (adduction).

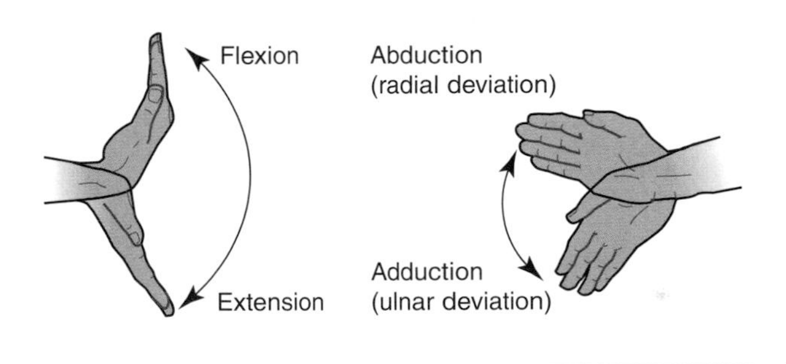

FIGURE 3.31 Wrist movements.

The carpometacarpal (CM) joint of the thumb is a saddle joint that accounts for much of thumb movement. Its structure allows for biplanar movements of flexion, extension, abduction, and adduction (figure 3.32). The second and third CM joints are relatively immovable. The fourth CM joint has limited movement, while the fifth CM joint shows somewhat greater mobility.

Metacarpophalangeal (MP) joints of fingers two through five are synovial condyloid joints that allow the fingers to move in flexion, extension, abduction, and adduction. The thumb's MP joint is a condyloid articulation that permits flexion and limited extension, abduction, and adduction.

Each finger has two interphalangeal (IP) joints (proximal and distal) whose hinge structure permits flexion and extension. The thumb has a single IP joint between its two phalanges.

Movements of the Hip and Pelvis

The hip joint is formed by articulation of the head of the femur with the acetabulum of the pelvis (figure 3.33). Also known as the coxofemoral (or coxal) joint, the hip is a synovial ball-and-socket joint capable of movement in all three primary planes (i.e., triplanar). As described earlier in the chapter, the hip joint is relatively stable and resists dislocation, largely due to the good fit of the femoral head in the acetabulum and the large muscle mass surrounding the joint. The joint's stability is enhanced by its **labrum**, a U-shaped ring of fibrocartilage around the rim of the acetabulum. The labrum deepens the socket (acetabulum) and thereby improves the joint's bony fit.

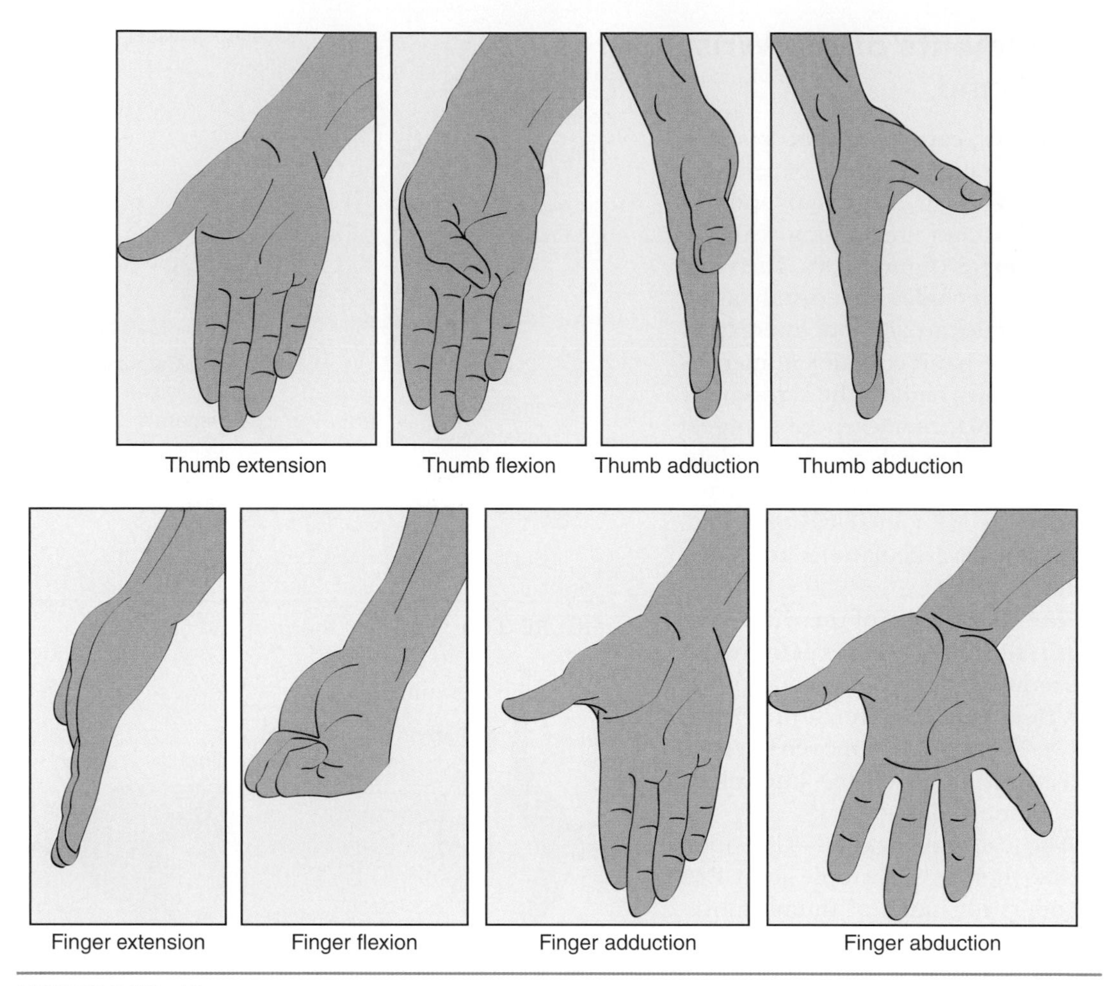

FIGURE 3.32 Finger movements.

Relative movement between the pelvis and femur can be viewed in two ways: (1) With the pelvis fixed, the femur is free to move through all three planes, and (2) with the femur fixed, the pelvis can perform triplanar movement.

In the first case, with a fixed pelvis, the femur can move (figure 3.34) in flexion and extension (sagittal plane), abduction and adduction (frontal plane), and internal (medial) and external (lateral) rotation (transverse plane). A circular, or cone-shaped, movement pattern known as circumduction combines the movements of flexion, extension, abduction, and adduction.

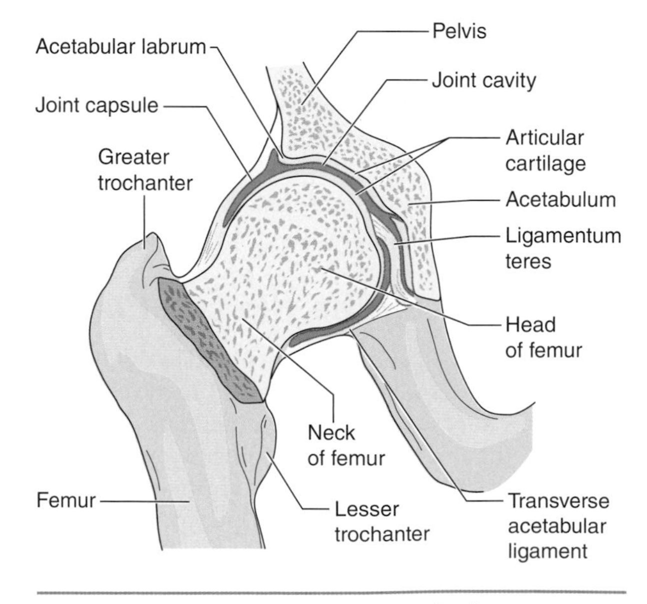

FIGURE 3.33 The hip joint (longitudinal section).

When the femur is fixed, as in standing, and the pelvis is allowed to move (figure 3.35), possible movements include anterior and posterior tilt in the sagittal plane, lateral tilt left and right in the frontal plane (abduction and adduction), and rotation left and right in the transverse plane (internal and external rotation).

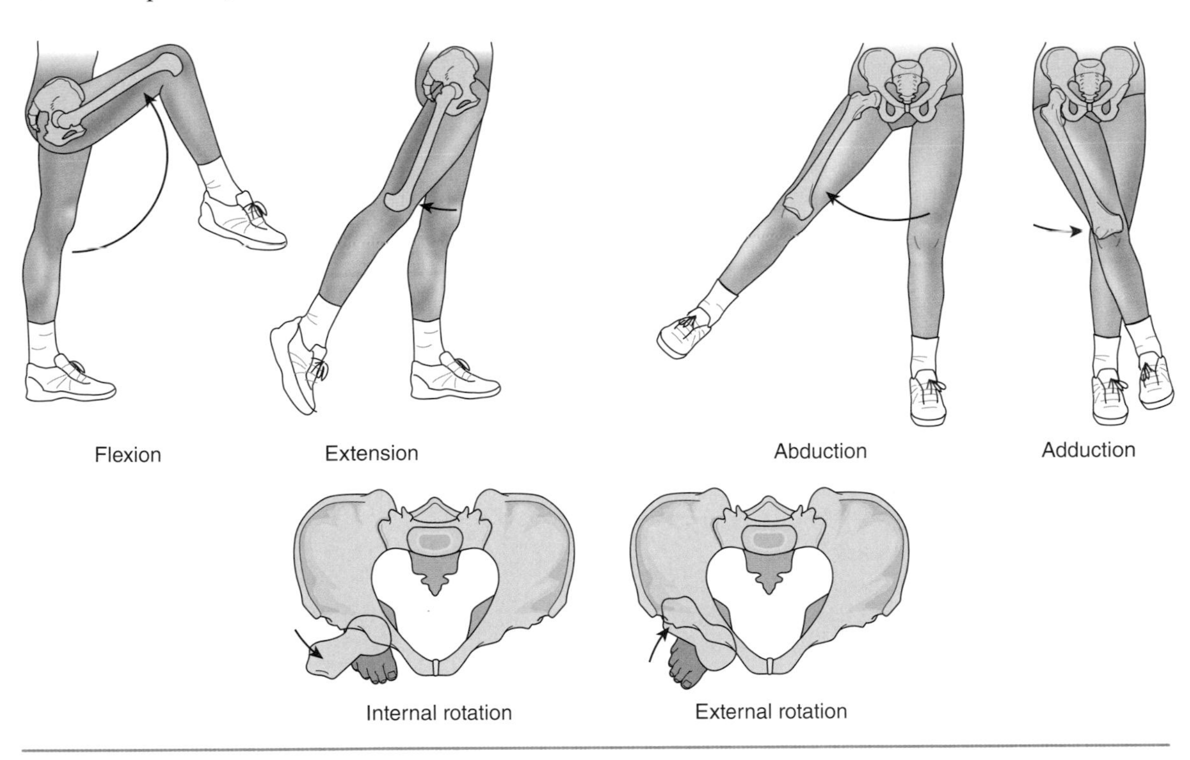

FIGURE 3.34 Hip movement about a fixed pelvis.

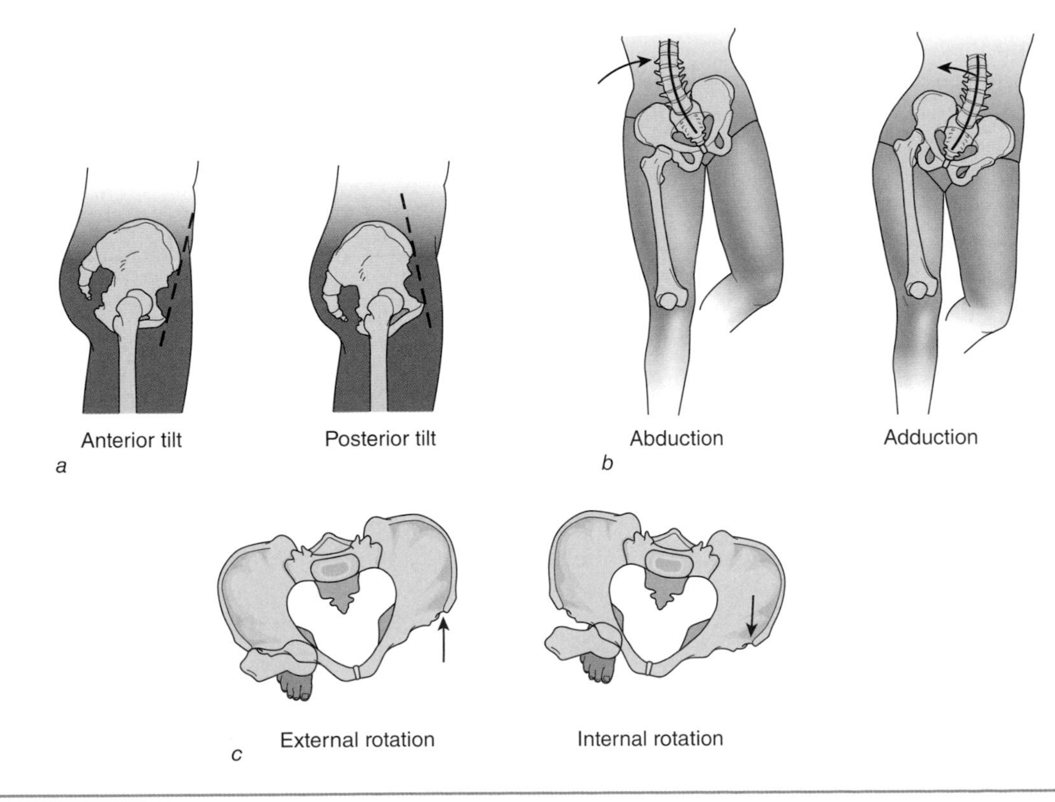

FIGURE 3.35 Pelvic movement about a fixed femur.

Movements of the Knee

The knee (tibiofemoral) joint, the largest joint in the body, is formed by articulation of the medial and lateral condyles at the distal end of the femur with matching condylar surfaces at the proximal end of the tibia (figure 3.36). The knee joint is the most complex of the major synovial joints. Although it functions as a modified hinge joint, the knee joint is more commonly referred to as a *bicondyloid* (or double condyloid) *joint*. The articulation of the femoral condyles with the tibial surfaces is enhanced by the presence of two menisci (medial and lateral). The medial and lateral menisci are semicircular rings of fibrocartilage that deepen the articular surface, improve the bony fit, and stabilize the knee joint.

The anterior surface of the distal femur also articulates with the patella (kneecap) to form the patellofemoral joint. The patella improves the leverage of the muscles (quadriceps group) responsible for extending the knee by moving the muscles' line of action away from the knee joint's axis of rotation.

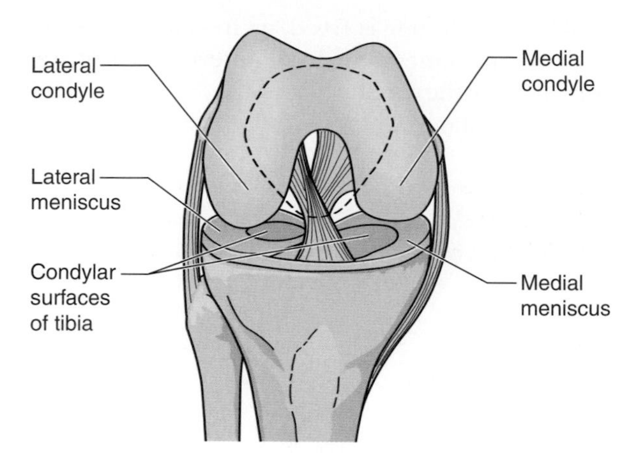

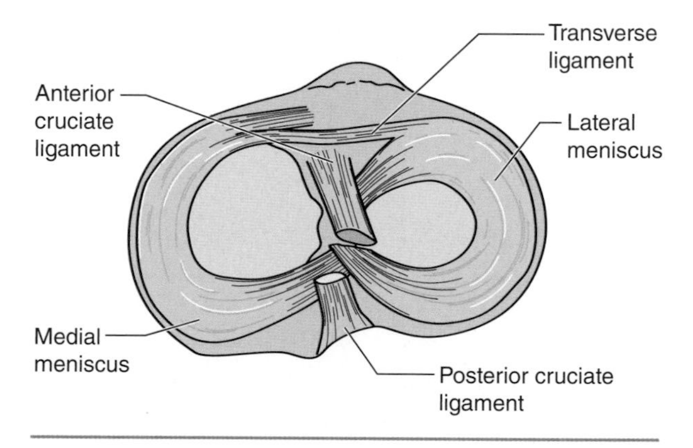

FIGURE 3.36 Knee bones, ligaments, and menisci.

The knee acts primarily as a modified hinge joint, with flexion and extension happening in the sagittal plane (figure 3.37). As the knee flexes, however, some rotation is possible between the lower leg (tibia) and the thigh (femur). During the final few degrees of extension (as the knee approaches full extension), the tibia and femur rotate relative to one another to prevent rotation of the lower leg. This tibiofemoral rotation, termed the **screw-home mechanism**, stabilizes the knee at full extension for the initial stance phase of walking and running.

At the patellofemoral joint, the patella slides into the intercondylar groove of the femur as the knee moves from full extension into flexion. This movement is termed **patellar tracking**. In a well-functioning knee, the patella tracks in the middle of the groove. In some cases, however, conditions such as injury, muscle weakness, or paralysis may cause the patella to track to one side of the groove, usually to the lateral side. This maltracking can be painful and limit performance of tasks involving the knee (e.g., running and jumping).

Movements of the Ankle and Foot

The numerous bones, ligaments, and articulations in the ankle and foot region make this one of the body's most structurally complex areas. The ankle (talocrural) joint is a synovial joint formed by articulation of the distal ends of the tibia and fibula with the superior surface of the talus. The talus fits into a deep socket, or mortise, formed by the tibia and fibula. The ankle functions as a hinge joint, with the talus rotating between the tibial and fibular malleoli. In a dorsiflexed position, the talus fits snugly within the mortise and the ankle joint is very stable. As the ankle plantar flexes, a narrower portion of the talus rotates into the mortise. This results in a looser bony fit, compromising joint stability; the ankle therefore is relatively unstable in the plantar flexed position.

As described in chapter 2, each foot contains 26 bones, including 7 tarsals, 5 metatarsals, and 14 phalanges (see figure 2.13). Each bone articulates with one or more adjacent bones. Joints between neighboring tarsal bones are termed intertarsal joints, in general, with specific joints identified by the involved bones (e.g., cuboideonavicular joint).

One of the most important intertarsal joints is the articulation of the superior surface of the calcaneus with the inferior surface of the talus. This talocalcaneal, or subtalar, joint plays an essential role in proper functioning of the foot and ankle complex during load-bearing activities such as walking and running. The subtalar joint axis runs obliquely, as shown in figure 3.38.

The distal tarsal bones (cuboid and cuneiforms) join with the proximal ends of the metatarsals to form five tarsometatarsal joints. Similarly, the distal ends of the metatarsals articulate with heads of the proximal phalanges to create five metatarsophalangeal (MP) joints. Finally, interphalangeal (IP) joints are hinge joints formed between adjacent phalanges in each of the toes.

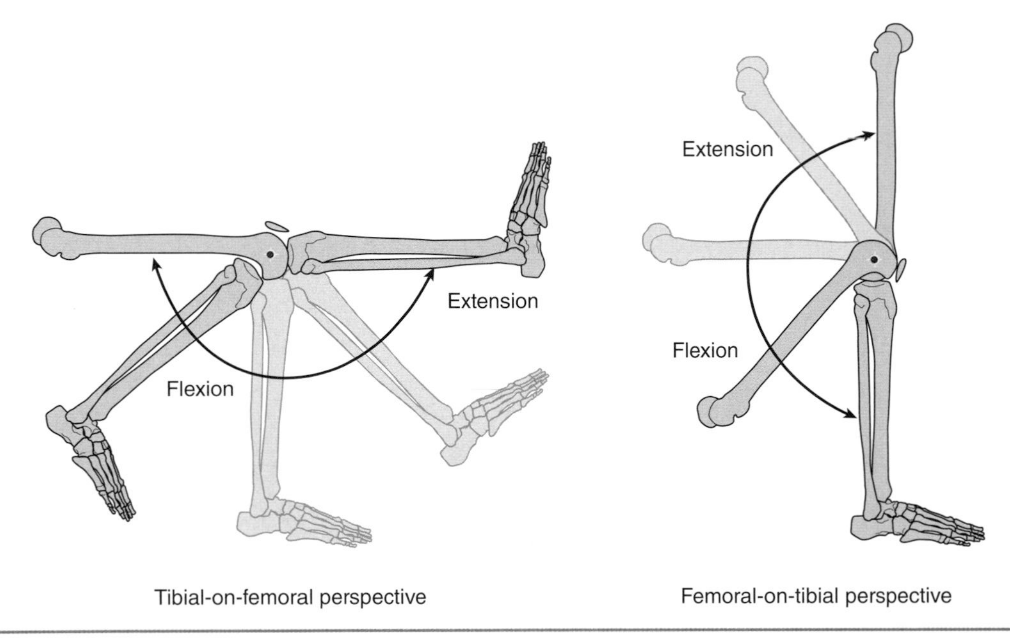

FIGURE 3.37 Knee joint movements in the sagittal plane: tibial motion about a fixed femur and femoral movement about a fixed tibia.

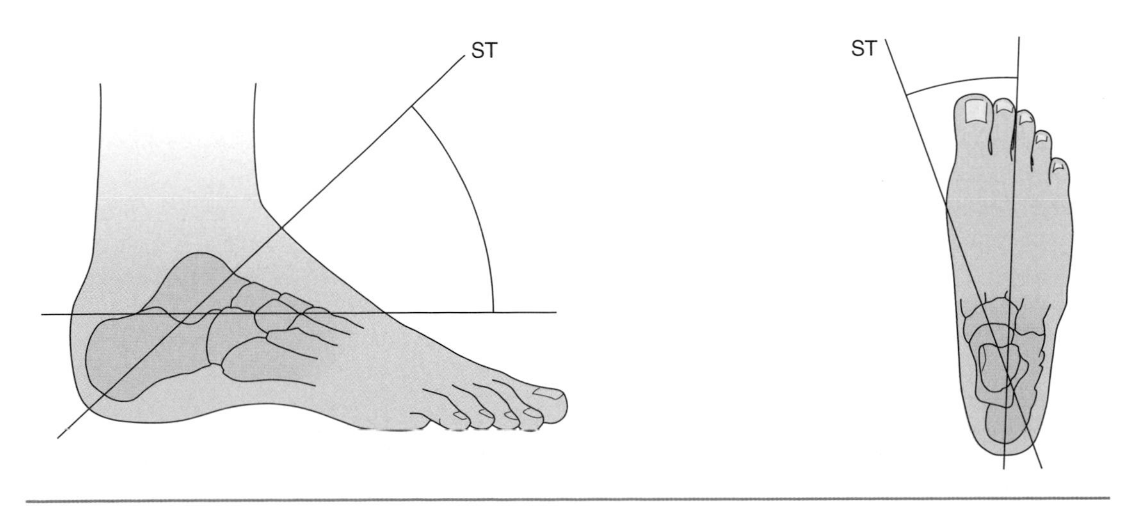

FIGURE 3.38 Subtalar joint axis, inclined superiorly and medially. ST = subtalar joint.

Applying the Concept

Spinal Deformities

As described earlier, the spine forms normal curvatures that help the body accept compressive loads. Injury, disease, and congenital predisposition can cause deformities of the spinal column, leading to abnormal structural alignment or alteration of spinal curvatures. These deformities often result in altered force distribution patterns and pathological tissue adaptations that may lead to or exacerbate other musculoskeletal injuries. Three primary types of spinal deformity exist: scoliosis, kyphosis, and lordosis (figure 3.39). These deformities are classified by their magnitude, location, direction, and cause, and can occur in isolation or in combination.

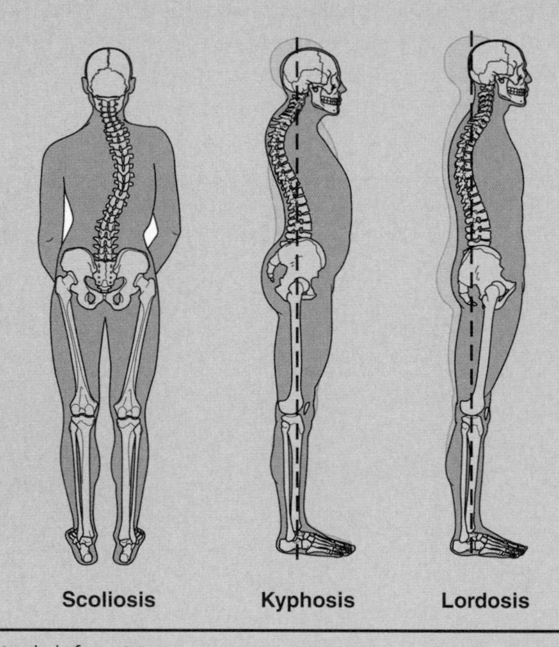

FIGURE 3.39 Spinal deformities.

Scoliosis is a lateral (frontal plane) curvature of the spine, which is also usually associated with some twisting of the spine. Mild spinal deviations are well tolerated and usually asymptomatic (i.e., present no symptoms). Severe deformities, in contrast, can markedly compromise cardiopulmonary function and upper and lower limb mechanics.

Kyphosis is a sagittal-plane spinal deformity, usually in the thoracic region, characterized by excessive flexion that produces a hunchback posture. Kyphosis is more severe in women than in men and is more prevalent with advancing age in both genders. Elderly postmenopausal women are at particular risk, largely because of the strong association between kyphosis and osteoporosis. (Note: In general, *kyphosis* refers to a forward curvature that is concave anteriorly. The thoracic and sacral regions have a natural kyphosis. Thus, hunchback cases actually involve a *hyper*kyphosis, or exaggerated kyphotic curvature. Clinically, the term *kyphosis* often is used, as here, to describe this hyperkyphotic condition.)

Clinically, **lordosis** is an abnormal extension deformity, usually in the lumbar region, that produces a hollow or swayback posture. (Note: In general, *lordosis* refers to a backward curvature that is concave posteriorly. The cervical and lumbar regions have a natural lordosis. Thus, swayback cases actually involve a *hyper*lordosis, or exaggerated lordotic curvature. Clinically, the term *lordosis* often is used, as here, to describe this hyperlordotic condition.)

The primary ankle movements are dorsiflexion and plantar flexion. Relative to anatomical position, both movements occur in the sagittal plane. Dorsiflexion describes movement of the dorsal (top) surface of the foot toward the lower leg (see figure 3.9). The opposite movement, plantar flexion, involves movement of the plantar (bottom) surface of the foot away from the lower leg. Standing on one's tiptoes and pushing down on the accelerator pedal in a car involve ankle plantar flexion.

The subtalar joint is primarily responsible for inversion and eversion. Inversion involves a tilting of the foot in the frontal plane so that the bottom of the foot faces medially (see figure 3.10). In eversion, the foot tilts so that the sole of the foot faces laterally. Inversion and eversion both happen primarily at the subtalar joint.

As noted earlier, the terms *supination* and *pronation* are commonly (though not universally) used to describe combined movements of the ankle and foot. Supination of the foot and ankle describes the combination of ankle plantar flexion, subtalar inversion, and internal (medial) rotation of the foot. Pronation of the ankle and foot describes the opposite combination of ankle dorsiflexion, subtalar eversion, and external (lateral) rotation of the foot.

In addition to the subtalar joint, other intertarsal articulations contribute to a lesser degree to inversion and eversion. These joints include the calcaneocuboid joint and the talonavicular portion of the talocalcaneonavicular joint. In general, the tight fit between the intertarsal bones largely restricts them to slight gliding relative to adjacent bones.

The tarsometatarsal articulations are synovial joints limited to slight gliding movements. The metatarsophalangeal joints, as condyloid synovial joints, allow flexion, extension, and hyperextension in the sagittal plane, along with abduction and adduction in the transverse plane. The interphalangeal joints of the toes act as hinge joints allowing flexion and extension in the sagittal plane.

Concluding Comments

Each joint in the body has its own movement potential. Some articulations, such as the glenohumeral joint, possess tremendous freedom and range of movement. Others, such as the tibiofibular joints, have limited movement potential. The body's neuromuscular system considers the movement potential at each joint under its control, coordinates the action of joints, and thereby determines our ability to execute purposeful movements, ranging from simple tasks of daily living to the intricate movements of skilled performers.

Go to the web study guide to access critical thinking questions for this chapter.

Suggested Readings

Alexander, R.M. (2000). *Bones: The unity of form and function.* New York: Basic Books.

Jenkins, D.B. (2008). *Hollinshead's functional anatomy of the limbs and back* (9th ed.). Philadelphia: Saunders.

Levangie, P.K., & Norkin, C.C. (2011). *Joint structure and function: A comprehensive analysis* (5th ed.). Philadelphia: Davis.

MacKinnon, P., & Morris, J. (2005). *Oxford textbook of functional anatomy* (2nd ed., Vols. 1-3). Oxford: Oxford University Press.

Neumann, D.A. (2016). *Kinesiology of the musculoskeletal system.* (3rd ed.). St. Louis: Mosby.

Nordin, M., & Frankel, V.H. (2001). *Basic biomechanics of the musculoskeletal system* (3rd ed.). Philadelphia: Lippincott Williams & Wilkins.

4

Skeletal Muscle

Objectives

After studying this chapter, you will be able to do the following:

- Describe the structure and functions of skeletal muscle
- Describe the types of muscle action
- Explain the steps in muscle action
- Identify muscle fiber types and muscle fiber arrangement
- Explain the length–tension and force–velocity relationships of muscle
- Describe the stretch–shortening cycle and its functional implications
- Explain the process of muscle hypertrophy
- Explain how muscles are named
- Determine the functional actions of muscle
- Identify the muscles acting at the major joints of the body
- Describe muscle injury and its consequences

Of all the body's tissues, muscle is unique in its ability to generate force. To understand muscle's role in movement, we first need to understand the fundamentals of muscle structure and function. This chapter focuses on the structure and function of skeletal muscle, including the physiology and mechanics of muscle action, factors affecting muscle force production, neural control of muscle action, and muscle adaptations.

Properties of Skeletal Muscle

Skeletal muscle, which accounts for 40% to 45% of our body weight, plays several important roles in the overall functioning of the human body. In the current context, muscle's most essential role is to produce the forces necessary for human movement, from the basic activities of daily living to the extremes of athletic performance. Muscular actions are largely under voluntary control, but may also be elicited through reflex action (e.g., pulling away rapidly from a painful stimulus) or seen in nonreflex movements such as walking that require little, if any, voluntary control. Movements we are so accustomed to performing that they become automatic (e.g., walking, breathing, chewing, coughing, reaching for objects) are sometimes referred to as *stereotypical movements*.

All muscle tissue is distinguished by four functional properties, or characteristics:

- **Excitability** (also **irritability**), describes the ability of muscle to respond to a stimulus. The stimulus for skeletal muscles typically comes from the nervous system. A muscle's fibers are stimulated by a wave of stimulation conducted along the muscle's length. This conductivity is an important feature of muscle's force-production capability.
- **Contractility**, also known as *activity* or *action,* refers to muscle's ability to generate a pulling (or tension, tensile) force. (Note: *Contraction* is used by some authors to refer to a muscle's ability to shorten, or change its length.)
- **Extensibility** describes the ability of muscle to lengthen, or stretch, and as a consequence to generate force over a range of lengths. For example, when you shorten the biceps brachii, thereby flexing your elbow, the muscle's extensibility allows it to subsequently lengthen as the elbow extends.
- **Elasticity** refers to a tissue's ability to return to its original length and shape after an applied force is removed. When a muscle and its associated connective tissues are stretched by an external pulling force, for example, its elastic properties allow it to return to its unloaded length once the force is removed.

Absence or compromise of any (or all) of these properties limits a muscle's ability to function effectively.

Structure and Function of Skeletal Muscle

Skeletal muscles are composed of structures that can produce force (**contractile component**) and connective tissues that cannot produce force (**noncontractile component**) but are nonetheless important for the muscle's physiological and mechanical performance. Since muscle cells (fibers) are delicate and easily damaged, the collagenous connective tissues around and within the muscle protect the cells. From a mechanical perspective, the connective tissues play a role in accepting and transmitting forces within the muscle and to connected structures.

Muscle's structural hierarchy proceeds from the whole skeletal muscle down to the myofilament level, where active force production actually occurs. We move through these structural levels by first exploring macroscopic (gross) muscle structure and then proceeding to microscopic structures.

Skeletal Muscle Anatomy and Muscle Action

Skeletal muscles contain three connective tissue layers that protect and support the muscle, help give it its shape, and contribute collagen fibers to form the tendons at both ends of the muscle. The collagen fibers of the tendons then pass into the bone matrix, enabling the contracting muscle to produce joint movement. The outer connective tissue layer, which surrounds the whole muscle and separates the muscle from surrounding tissues, is the **epimysium**. Within each muscle, bundles of muscle fibers called **fascicles** are separated from one another by **perimysium**. Continuous with the perimysium, and forming the innermost layer of connective tissue, is the **endomysium** that surrounds each muscle fiber (figure 4.1).

A plasma membrane, called the **sarcolemma**, surrounds each muscle fiber and forms narrow tubes, called **transverse tubules**, or **T-tubules**, that pass through the muscle fiber (figure 4.2). These T-tubules are critical because they form channels that allow the electrical signal, needed to stimulate muscle contraction, to pass through the fiber.

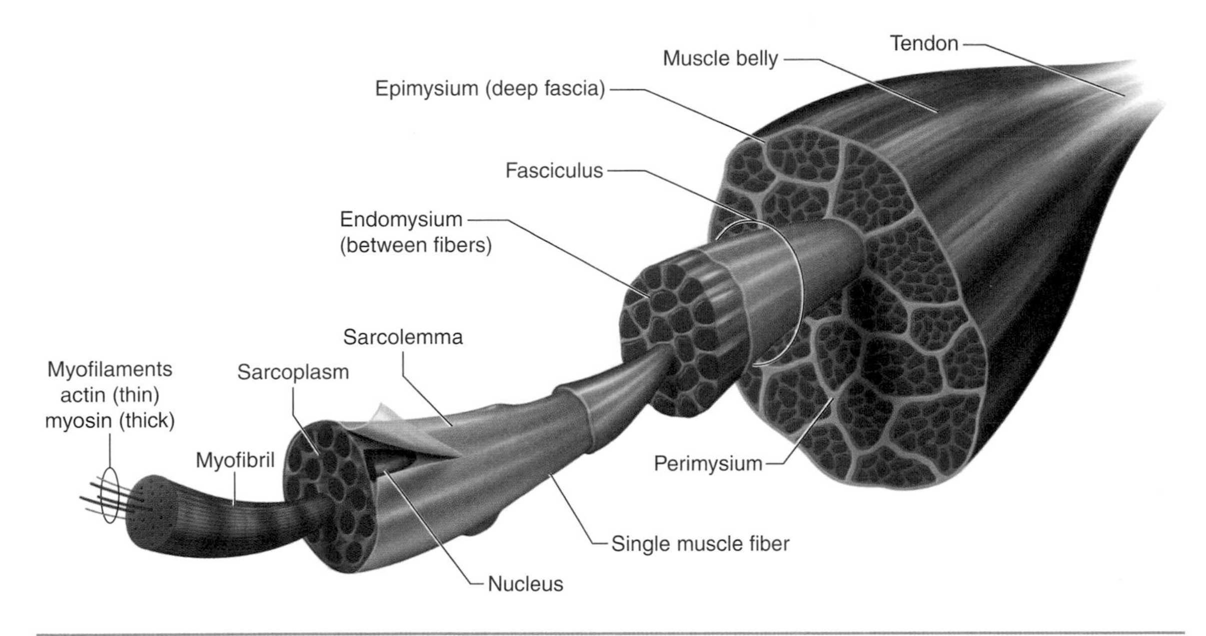

FIGURE 4.1 Structure of skeletal muscle.

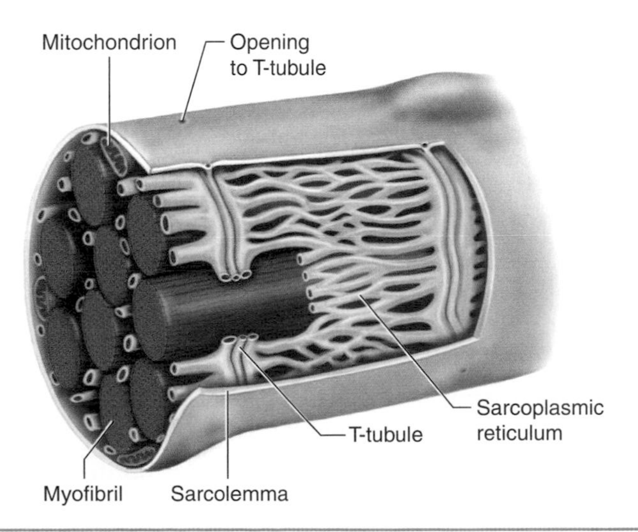

FIGURE 4.2 Structure of a muscle fiber.

At a microscopic level, each muscle fiber consists of myofibrils, and each myofibril is composed of structural, regulatory, and contractile proteins organized into discrete units called *sarcomeres*. Each myofibril is surrounded by a fluid-filled system of membranous sacs called the **sarcoplasmic reticulum**. On either side of the T-tubules, expanded ends of the sarcoplasmic reticulum, called *terminal cisternae*, store the calcium needed for muscle contraction. The **sarcomere** makes up the functional contractile unit of each myofibril (figure 4.3). Each sarcomere is bounded by Z discs (Z lines) and contains actin and myosin myofilaments. How small is a sarcomere? A myofibril 10 mm long contains approximately 4,000 sarcomeres connected end to end. Because of the parallel arrangement of the actin and myosin filaments and the alignment of myofibrils and their respective sarcomeres, skeletal muscles have a striped, or striated, appearance.

Sarcomeres contain three kinds of proteins: (1) contractile proteins (actin and myosin) that produce force during contraction, (2) regulatory proteins (tropomyosin and troponin) that help turn the contractile process on and off, and (3) structural proteins (e.g., titin, myomesin) that maintain alignment of the actin and myosin filaments.

The contractile proteins consist of **myosin** and **actin** filaments, otherwise known as thick and thin filaments, respectively. The tail regions of all the myosin molecules connect at the M line, whereas the myosin heads are all located closer to the Z lines and between the actin filaments (figure 4.3).

Every actin filament is anchored to a Z line and extends toward the middle, or M line, of the sarcomere. The actin filaments also contain two regulatory proteins known as **tropomyosin** and **troponin**. Tropomyosin molecules bonded together form continuous strands that extend the length of the actin filament, while troponin molecules are located at the junction of each successive pair of tropomyosin molecules. When the muscle is relaxed, these tropomyosin strands cover the myosin-head binding sites on the actin filament.

In addition to the contractile and regulatory proteins just discussed, each sarcomere contains structural proteins that contribute to the alignment, stability, elasticity, and extensibility of the myofibril. Two prominent structural proteins are titin and myomesin. **Titin** stabilizes the alignment of the thick filaments by connecting each myosin filament to both the Z disc and the M line. The titin that extends from the Z disc to the beginning of the thick filament is believed to help keep the myosin filaments centered between the Z discs, and it also accounts for much of the elasticity and extensibility of myofibrils. Titin therefore may help the sarcomere return to its resting length after the muscle has contracted or been stretched. **Myomesin** molecules form the M line that anchors the myosin filaments, and the titin filaments, in the center of the sarcomere.

Types of Muscle Action

Muscle **action** (also **contraction**), is the internal state in which a muscle actively exerts a force, regardless of whether it shortens, lengthens, or stays the same length. There are three types of muscle action:

1. **Concentric** (shortening): The turning, or rotational, effect (torque) produced by the muscle at a given joint is greater than the external torque (created by an external force such as a held weight); therefore, the muscle is able to shorten while overcoming the external load. For example, concentric muscle action is used to flex the elbow from a fully extended position.
2. **Isometric** (*iso* = same; *metric* = length): The torque produced by the muscle is equal and opposite to the external torque, therefore, there is no limb movement (see sidebar). For example, isometric muscle action is used to hold the elbow at 90° of flexion.
3. **Eccentric** (lengthening): The torque produced by the muscle is less than the external torque, but the torque produced by the muscle causes the joint movement to occur more slowly than the external torque, acting by itself, would tend to make the limb move. For example, eccentric muscle action is used to slowly extend the elbow from a flexed position.

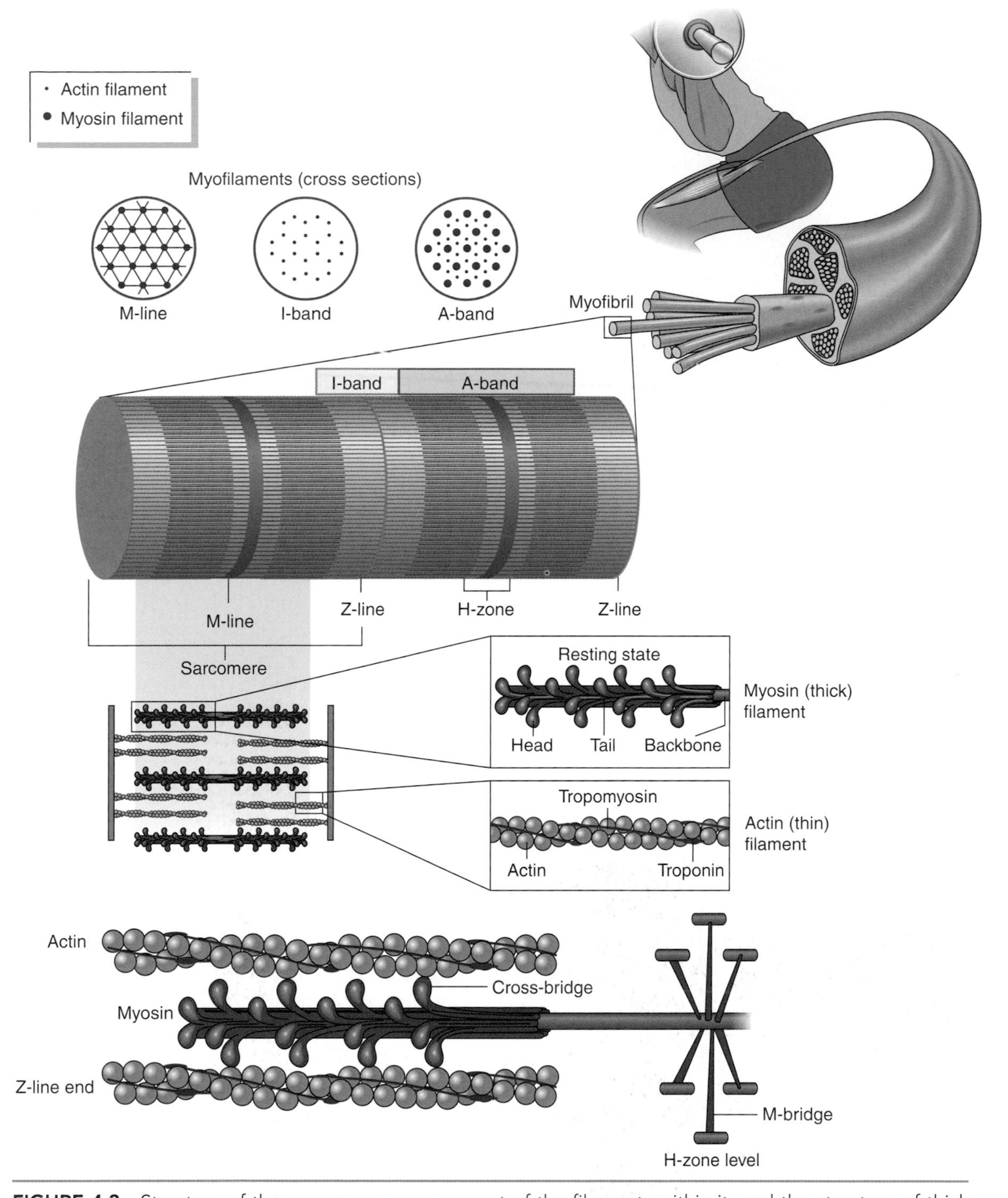

FIGURE 4.3 Structure of the sarcomere, arrangement of the filaments within it, and the structure of thick (myosin) and thin (actin) filaments.

How do you determine which type of muscle action is occurring during exercise? Perhaps the easiest application is weightlifting. For any exercise (e.g., bench press, squat, shoulder press, elbow curl, seated row), determine when during the lift you are overcoming the weight (i.e., overcoming gravity). That is the concentric phase. In contrast, when you let the weight overcome the force your muscles are producing (i.e., when the weight is moving with gravity), you are performing the eccentric phase. If at any time during the movement you stop and hold the weight in a fixed position, you are performing an isometric phase. For example, in the flat bench press, lowering the weight to your chest is the eccentric phase, and press-

Applying the Concept

Isometric: Is It Really?

Derived from *iso*, meaning same, and *-metric*, meaning measure or length, the term *isometric* typically is used to describe tasks in which muscle force is produced with no resulting movement. Holding a weight in a fixed position, for example, requires isometric action. It is often inferred that the absence of joint movement means the muscle is also acting isometrically, or not changing its length. In actuality, when a joint is held motionless while muscle force is generated, the entire musculotendinous unit is isometric. Within the musculotendinous unit, however, the active muscle shortens slightly and pulls to lengthen the tendon. Slight shortening of the muscle combined with a bit of tendon lengthening results in no net length change (isometric) in the musculotendinous unit. The length changes in the muscle and tendon are minimal, of course, but to say that constant, or unchanging, joint angles are associated with true isometric muscle action is technically incorrect.

ing the weight off the chest is the concentric phase. If at any time during the bench press you stop and hold the weight, then you are performing an isometric phase. Muscle control of movements is discussed in detail in chapter 6 (Muscular Control of Movement and Movement Assessment).

In addition to the three action types just listed, two important terms are often used to define properties of a muscle action. The term **isokinetic** describes constant angular velocity about a joint. It is possible, therefore, to produce an isokinetic concentric or an isokinetic eccentric action. Isokinetic actions are commonly used in experiments that test the relative effectiveness of different exercise devices for recruiting specific muscle groups. By using isokinetic actions, the experimenters can control for the effects of contraction velocity on muscle force production.

Isotonic literally means constant tension. This condition does not occur in intact human subjects (i.e., *in vivo*) because the level of muscle force varies continuously and rarely, if ever, is constant throughout a movement. Isotonic conditions are practical only in isolated muscle preparations in laboratory experiments. A more accurate term for human actions is **isoinertial**, which means constant resistance. For example, when performing an elbow curl with a 25 lb (11.4 kg) dumbbell, you are performing an isoinertial action throughout the movement because the external resistance remains constant.

Excitation–Contraction Coupling

The physiological steps that produce a muscle contraction involve passage of an electrical signal (**action potential**) through the sarcolemma and the eventual swiveling (pivoting) of the myosin heads to produce force. This process, called **excitation–contraction coupling**, is outlined as follows (figure 4.4):

1. The electrical signal (action potential) passes along the axon of the **lower motor neuron** (LMN) to the presynaptic membrane of the terminal ending, causing calcium channels to open.
2. Calcium (Ca^{2+}) from the surrounding fluid flows into the terminal ending and facilitates the movement of the synaptic vesicles to the presynaptic membrane. Each vesicle contains the neurotransmitter **acetylcholine** (ACh).
3. The synaptic vesicles bind to the presynaptic membrane of the terminal ending and release acetylcholine into the synaptic cleft (figure 4.4*a*). This process is called **exocytosis**.

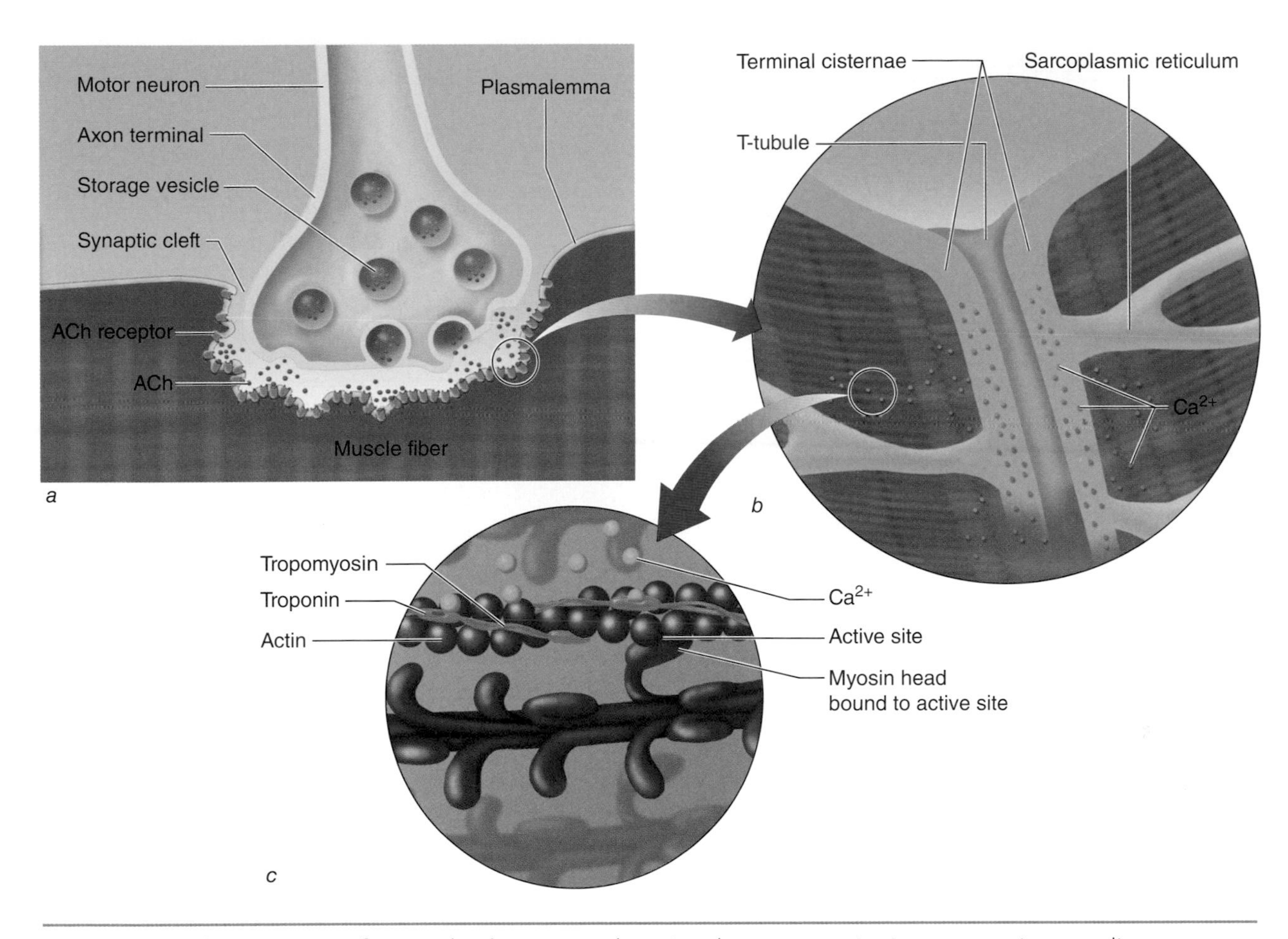

FIGURE 4.4 The sequence of events leading to muscle action, known as excitation–contraction coupling.

4. Acetylcholine crosses the synaptic cleft and binds to ACh receptors on the sarcolemma. This binding causes the sodium channels to open, and if enough sodium rushes into the muscle fiber, an action potential will pass across the sarcolemma in all directions.
5. The action potential then passes along the sarcolemma and into the fiber through the T-tubules. The T-tubules pass through the muscle fiber and around all of the myofibrils.
6. As the electrical signal passes through the T-tubules, it causes the release of calcium from the sarcoplasmic reticulum (figure 4.4*b*).
7. The calcium then binds to troponin. This interaction between the troponin and calcium is believed to cause a shift of the tropomyosin strands off the myosin-head binding sites on the actin filament (figure 4.4*c*).
8. The myosin heads bind to the actin filament and swivel (pivot) toward the M line. A new energy molecule, **adenosine triphosphate** (ATP), must be added to the myosin head for it to detach from the actin. This cycling of the myosin heads continues as long as the neural signal is maintained, calcium stays bound to troponin, and the ATP needed to supply the energy for contraction is replenished.
9. The muscle will stop contracting, however, when the stimulation stops and acetylcholine is no longer released from the terminal ending of the LMN.
10. If the signal stops, calcium will then be actively pumped back into the sarcoplasmic reticulum, and the troponin–tropomyosin complex will settle back into its original position on the actin filaments, blocking the myosin-head binding sites. With no more interaction between the actin and myosin filaments, the muscle relaxes and returns to its resting, or starting, length.

Applying the Concept

Negative Training

When performing overload training with a weight that exceeds your concentric limit, the muscle has no choice but to perform an eccentric contraction. Eccentric actions, particularly at this intensity, produce the most muscle damage and therefore stimulate the most regeneration and growth. This explains why high-intensity eccentric contractions, referred to as *negative training*, or *forced negatives*, are popular with bodybuilders seeking to maximize muscle size. For example, consider a person whose maximum in the bench press is 180 lb (81.8 kg). If the weight is increased to 190 lb (86.4 kg), the lifter will be unable to raise the bar from his chest using concentric muscle action. If, however, the lifter is handed the 190 lb bar with his arms fully extended, he could control the heavier weight down to his chest using eccentric muscle action. Controlling the bar's descent using eccentric action is termed a *negative* since the muscle is doing negative (eccentric) work.

The mechanical sum of the 10 steps just described makes up the **sliding filament model** of muscle contraction. The sliding filament model is generally accepted as the best description of the steps that produce muscle contraction.

How many times must the myosin heads swivel to produce a muscle contraction? The range of joint motion determines how many cycles each myosin head needs to perform. In other words, during shortening or lengthening muscle actions, each myosin head undergoes many repeated but independent cycles of asynchronous movement. During an isometric contraction, the myosin heads still bind to the actin and swivel to produce force, but unlike a concentric contraction, this cycle is not repeated to shorten the muscle. During an eccentric contraction, the external load, or body weight, overcomes the normal tendency for the myosin heads to swivel toward the M line. The actin–myosin coupling is still forming, but the bond between the proteins is broken as the myosin heads are forcefully pulled toward the Z line.

Muscle Fiber Types

Human skeletal muscles are a composite of different fiber types, with the percentage of fiber type varying from muscle to muscle and person to person. Based on their mechanical and contractile properties, skeletal muscles are divided into two distinct fiber types: type I (slow twitch) and type II (fast twitch). Type II fibers are subdivided into type IIa and type IIx.

Metabolically, muscle fibers can be classified as **oxidative** or **glycolytic**. Because of their different metabolic profiles, type I fibers are classified as slow oxidative (SO) fibers, and type II fast-twitch fibers are further divided into two principal subcategories: fast oxidative glycolytic (FOG) fibers, or type IIa, and fast glycolytic (FG) fibers, or type IIx. SO fibers have the highest number of mitochondria, oxidative enzymes, myoglobin, and capillaries, and therefore generate ATP primarily through aerobic (i.e., using oxygen) pathways.

In contrast, FG fibers have relatively few mitochondria, oxidative enzymes, and capillaries and a low myoglobin content. They do store large amounts of glycogen, however, and generate ATP mainly through anaerobic glycolysis (the metabolism of sugar without using oxygen). Because FG fibers contain the most myofibrils and use ATP quickly, they are the strongest and fastest fibers. In other words, FG fibers are designed for intense, short-duration anaerobic movements. As indicated in table 4.1, type IIa fibers are often referred to as *intermediate fibers* because they share characteristics of both type I and type IIx. Some of the contractile, anatomical, and metabolic characteristics that distinguish the different fiber types are listed in table 4.1.

TABLE 4.1 Names and Properties of Skeletal Muscle Fiber Types

Property	Type I (SO)	Type IIa (FOG)	Type IIx (FG)
Color	Red	White/red	White
Myoglobin content	High	Intermediate	Low
Capillary density	High	Intermediate	Low
Oxidative enzyme content	High	Intermediate/high	Low
Mitochondrial density	High	Intermediate	Low
Stored lipids	High	Intermediate/high	Low
Fatigue resistance	High	Intermediate	Low
Glycogen content	Low	Intermediate	High
Glycolytic enzyme content	Low	Intermediate/high	High
Myosin ATPase activity	Low	Intermediate	High
Fiber diameter	Small	Intermediate	Large

Alternative names of fibers:

Slow twitch = type I = slow oxidative (SO)

Fast twitch = type IIa = fast fatigue resistant (FR) = fast oxidative glycolytic (FOG)

Fast twitch = type IIx = fast fatigable (FF) = fast glycolytic (FG)

Is it possible to change fiber types through training? Training may not change the neural control (i.e., the specific neuron and therefore its size, threshold, and speed of conduction velocity), but it can change the metabolic profile of the fibers. For example, endurance training can increase the oxidative capacity of all three fiber types, so with intense endurance training, what started off as an FG fiber metabolically may take on the metabolic profile of an FOG fiber. Although neurologically FG fibers are still the fastest and have the highest myosin ATPase activity (the enzyme used to break down ATP on the myosin molecule), a great deal of movement speed comes from the proper training of the nervous system. A well-trained primarily slow-twitch athlete can still be fast. The genetic advantage in speed and power, however, still goes to the predominantly fast-twitch individual. Likewise, the genetic advantage for endurance still goes to the individual with a predominance of SO fibers.

Muscle Fiber Recruitment

The neuromuscular junction represents the synaptic connection between the lower motor neuron (LMN) and the sarcolemma. A **motor unit** is a single LMN plus all the muscle fibers it innervates or has **innervation** with. Because a lower motor neuron innervates only one muscle fiber type, the SO, FOG, and FG classification applies not only to muscle fibers but also to motor units. Muscle contraction is the result of many motor units firing asynchronously and repeatedly. When exercise progresses from low to high intensity, type I fibers are recruited first, followed by type IIa and then type IIx fibers (figure 4.5). The type I fibers produce the least amount of force because

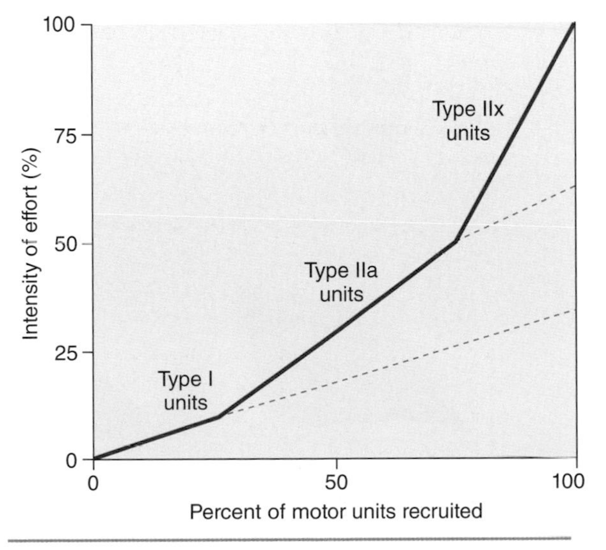

FIGURE 4.5 Recruitment order of type I, type IIa, and type IIx.

they contain the fewest actin and myosin filaments. On the other end of the spectrum are the type IIx fibers, with the highest number of actin and myosin filaments and the highest myosin ATPase activity.

When type IIx fibers are stimulated, therefore, they produce the most force and do so the most quickly. When you progress from low- to high-intensity exercise, your nervous system recruits your muscle fibers in order from the most aerobic (the most resistant to fatigue) to the least aerobic (the easiest to fatigue). This orderly progression of recruitment is known as **Henneman's size principle**. However, if you choose to quickly or explosively produce a great deal of force or power, your nervous system will override this normal aerobically energy-efficient hierarchy by firing all three fiber types at the same time. In this case, there would actually be a preferential recruitment of fast-twitch fibers.

Muscle Fiber Arrangement

Muscle fiber arrangement within our skeletal muscles falls into one of several different categories (figure 4.6):

Fusiform, or longitudinal

Unipennate

Bipennate

Multipennate

Triangular, or radiate

To determine the fiber arrangement of a muscle, on a sketch of the muscle simply draw a line from the muscle's origin through its insertion. If the muscle fibers run parallel to that line, the muscle is fusiform (e.g., biceps brachii, semitendinosus), longitudinal, or quadrate. In pennate muscles, fibers are oriented at oblique angles (normally <30°) to the tendon's line of pull. A unipennate muscle has one set of fibers, all with the same line of pull (e.g., semimembranosus), a bipennate muscle has two sets of fibers with different angles (e.g., rectus femoris), and a multipennate muscle has many sets of fibers at a variety of angles (e.g., deltoid).

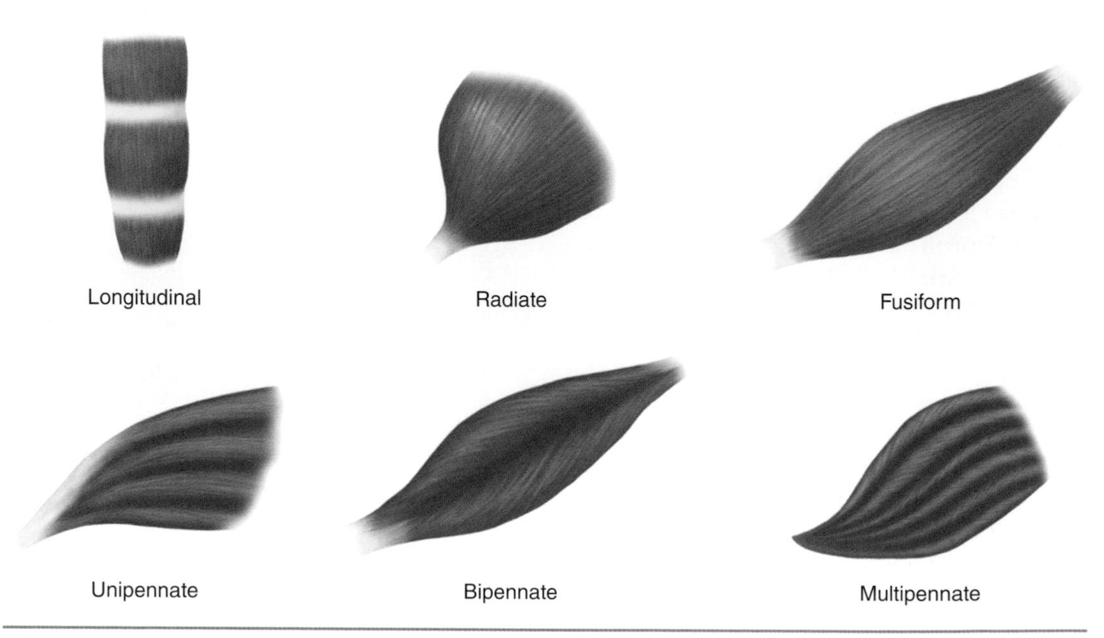

FIGURE 4.6 Muscle fiber arrangements.

The advantage of **pennation** is that for any given volume, more muscle fibers function in parallel, giving the muscle a greater functional cross-sectional area. In other words, pennation is the body's way of packing a greater number of muscle fibers into a smaller volume, therefore producing a muscle designed more for force production than for speed of contraction.

Although fusiform muscles can produce considerable force, they are designed to maximize speed of contraction. For example, given a pennate and fusiform muscle of equal length and volume, the fusiform muscle will have more sarcomeres in series because the muscle fibers are in line with the tendon. If both muscles are stimulated and all their sarcomeres shorten the same distance, then the muscle with the greater number of sarcomeres in series will shorten through a greater distance in the same period of time.

In muscles classified as triangular (also called radiate or fan shaped), fibers fan out from a relatively large origin to a relatively small insertion (e.g., pectoralis major, latissimus dorsi). This design combines high force production with high speed of contraction by packing numerous sarcomeres both in parallel and in series.

Although training cannot change a muscle's architecture, understanding design differences can help us recognize a muscle's potential for injury. For example, the quadriceps are designed for force production, whereas the hamstrings are designed for rapid shortening. Because of these design differences and the fact that the hamstring muscles (semitendinosus, semimembranosus, and biceps femoris long head) cross both the hip and the knee, these muscles are more susceptible to tearing than are the quadriceps in explosive, high-power events such as sprinting.

Length–Tension Relationship

We have known since the late 1800s that the length of a muscle affects its isometric force–production capability (Lieber, 2009). Earlier in this chapter we discuss how the nervous system stimulates muscle fibers to produce force. A muscle's force production also is affected by its passive elements (e.g., titin strands, tendons, and connective tissue sheaths) that are not under neural control.

A muscle's force-production capability as a function of its length is shown in figure 4.7. The active component represents the force-generating capacity of the sarcomeres and, therefore, a myofibril. If a sarcomere is too short, the actin filaments will completely overlap, the myosin filaments will press against the Z lines, the myosin heads will be unable to bind to the actin, and the sarcomere will be incapable of producing force. As the sarcomere is lengthened, it will reach its optimum force-production capability when there is maximum overlap (and binding capability) between the myosin heads and the actin molecules. If the sarcomere is stretched too much, however, there won't be any actin and myosin overlap and force production will drop to zero.

As the sarcomere is lengthened beyond its optimal force potential, the passive elements begin to lengthen and mechanically want to recoil. As the **length–tension relationship** in figure 4.7 shows, the passive tension continues to increase as the muscle is lengthened and helps compensate for the loss in force production by the sarcomere. The muscle's total force-production capability, therefore, is calculated by summing the active and passive components.

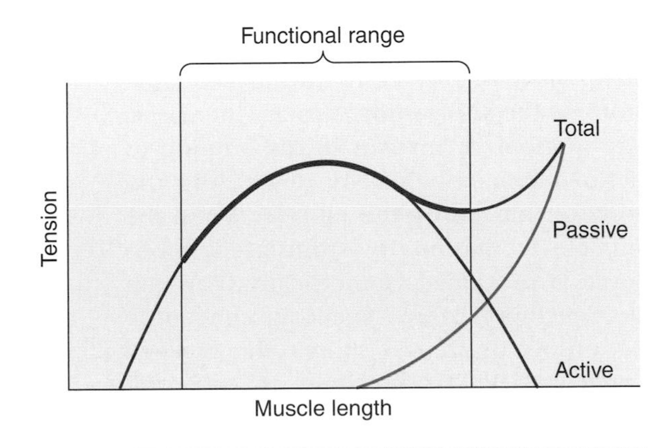

FIGURE 4.7 Length–tension relationship of skeletal muscle.

The data used to generate the length–tension curve (figure 4.7) were originally collected from isolated animal muscle using isometric contractions. How then does the length–tension relationship apply to the muscles in our body? The vertical lines added to figure 4.7 represent the fact that *in vivo* our joints do not allow us to shorten or lengthen a muscle to such extremes that all our sarcomeres would be incapable of producing force. Please note, the exact location of these vertical lines are only approximations and will vary depending on the muscle being analyzed. Figure 4.7 shows that when a muscle is maximally shortened, its force production capability is due solely to its active component, but when the same muscle is maximally lengthened, its force production now depends on both its active and passive components.

The length–tension relationship of muscle is easier to understand in application. For example, can you lift more weight in a standing or a seated calf raise? Can you leg curl more weight when your pelvis is anteriorly tilted than when it is flat on the bench? The answer to both questions is yes. Why? In each case you are changing a joint position to prelengthen one or several of the principal muscles being trained. Because the gastrocnemius crosses both the ankle and the knee, the extended knee lengthens the muscle and increases its force production capability when performing a standing calf raise. In a seated calf raise, the gastrocnemius is too short to maximize force production, making this an excellent exercise to target the soleus, which crosses only the ankle joint and whose length is unaffected by knee flexion and extension.

Anteriorly tilting the pelvis, or flexing the hip, in a leg curl exercise lengthens the long head of the biceps femoris, semitendinosus, and semimembranosus, and therefore increases their force-production capability during the lift. During the leg curl, the ankle is usually in anatomical position or slightly dorsiflexed. Because the gastrocnemius both flexes the knee and plantar flexes the ankle, dorsiflexion lengthens the muscle and increases its force-production capability. A similar action happens in cycling, where anterior pelvic tilt is used not only to improve aerodynamics, but also to increase the force output of the hamstrings and gluteus maximus by increasing their length.

Force–Velocity Relationship

The force-production capability of skeletal muscle depends, in part, on the contraction velocity. This dependency is known as the **force–velocity relationship**. Each point on the force–velocity diagram (figure 4.8) represents the maximum force for any given contraction velocity when the muscle is maximally stimulated. According to the diagram, the faster a muscle contracts concentrically, the less force it can produce. To understand why this is true, we need to consider what happens to the myofibril during a contraction. As the shortening velocity increases, the number of myosin heads binding to actin decreases, thereby decreasing the force produced by the muscle. When the muscle is maximally stimulated, more force is produced isometrically than can be developed at any speed of concentric action. As figure 4.8 shows, the greatest force can be produced in an eccentric action. Because the curve on the force–velocity diagram represents maximal stimulation of the muscle, the muscle functions submaximally underneath the curve in most everyday tasks.

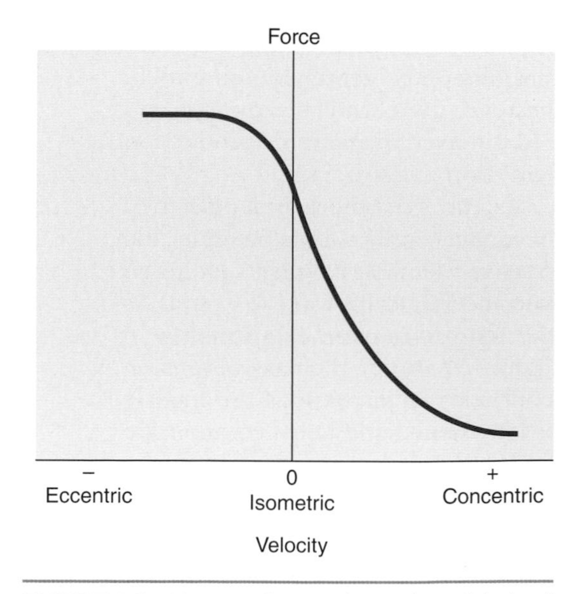

FIGURE 4.8 Force–velocity relationship of skeletal muscle.

To experience the force–velocity relationship, try the following experiment. Perform a series of dumbbell elbow curls. Execute each repetition with maximum effort (speed) using the same range of motion and proper form. Starting with no dumbbell, then using dumbbells that increase in 5 lb (2.3 kg) increments, perform one maximum-speed curl for each weight, allowing sufficient rest between lifts to reduce fatigue effects. Rate the speed of contraction for each lift using subjective terms such as *very fast*, *fast*, *moderate*, *slow*, and so on. What happens to movement speed as you increase the weight? It slows down. Once you reach a weight you can't lift through the full range of motion, you have reached or are very close to your isometric limit. The weight you are holding in that static (stationary) position is heavier than any weight you can lift concentrically.

For illustration purposes, assume your isometric limit for the dumbbell elbow curl is 50 lb (22.7 kg). There would be no point in using a 55 lb (25 kg) dumbbell from the starting position because this weight exceeds your concentric limit. But if a spotter hands you the 55 lb (25 kg) dumbbell at the top position (i.e., the most flexed elbow position), the descent of the weight can be controlled using eccentric muscle action. Notice the movement occurs in the same direction as gravity, but happens more slowly than gravity wants the dumbbell to move. With each additional 5 lb (2.3 kg) increase in dumbbell weight, the movement will not only speed up but also get harder to control. Soon you will reach a weight that is too heavy to resist, or, if you do, you will risk injury. You have just proved that when a muscle is maximally stimulated, it can handle a greater load during eccentric actions than during isometric or concentric actions and that the faster the muscle contracts concentrically, the less force it can produce.

Stretch–Shortening Cycle

The **stretch–shortening** (also **stretch-shorten**) **cycle** is an eccentric action immediately followed by a concentric action. The benefit of this cycle is that the normal concentric force-production capability of the muscle is enhanced. Four mechanisms have been proposed to explain this enhancement: time for force development, elastic energy, force potentiation, and reflexes (Enoka, 2015). These mechanisms—and which combination of them applies to a given movement—remain controversial.

Let's consider one of the mechanisms, elastic energy, in some detail. During the eccentric phase, elastic energy is stored in the titin strands and connective tissue sheaths, then released during the following concentric action. The muscle's normal force production is augmented by the addition of the released elastic energy. The added benefit is that the storage and release of the elastic energy did not require any additional ATP. The concentric action needs to immediately follow the eccentric action; otherwise some of the stored energy will be dissipated in the form of heat.

In general, the more ballistic the cycle, the greater the amount of energy stored. If the joint range of motion is too small, stored energy will be minimal. If the range of motion is too great, more energy may be stored, but the muscle's leverage may be compromised. For example, consider the technique used to maximize a vertical jump. Normally the jumper takes a few quick steps; unweighs by flexing the hips, knees, and ankles; and then explodes upward into the jumping, or concentric, phase. If the unweighing phase is too deep, the leverage of the hip and knee extensors will be compromised, making it difficult to explode vertically.

Strength and power athletes routinely use the stretch–shortening cycle in their training by performing plyometrics. **Plyometrics** is a form of exercise that uses the force-enhancing characteristics of the stretch–shortening cycle. Walking and running also utilize stretch–shortening cycles. Because of greater joint range of motion and more rapid force generation in running, however, runners typically benefit more from the stretch–shortening cycle than do walkers.

Applying the Concept

Feats of Muscular Strength

Something in our nature seems drawn to the limits of human performance. We are fascinated by world records, both traditional and unusual. The limits of human strength can be seen in reports of a frantic parent who lifts an automobile off a trapped child. In more measured settings, limits of strength are determined by records set in weight-lifting competitions such as the Olympic Games.

One classic measure of human strength is the deadlift, in which a weight is lifted from the ground to a position with knees locked and body erect. The record for the deadlift, relative to body weight, is held by Lamar Gant, who, in 1985, deadlifted an astonishing 661 lb (300 kg) to become the first person ever to deadlift *five times* his own body weight. Gant weighed a mere 132 lb (60 kg).

Records continue to be broken, which raises tantalizing questions such as, Are there limits to human performance? If so, what are those limits?

Muscle Hypertrophy

Muscle hypertrophy involves an increase in the size and number of myofibrils, which increases the size of the muscle fibers, the fascicles, and, ultimately, the overall muscle belly. In addition, the connective tissue must also expand to accommodate the increasing size of the muscle.

In response to repeated overload stress, new actin and myosin filaments are added to the periphery of the sarcomeres within a myofibril, under the sarcoplasmic reticulum, until the myofibril increases to approximately twice its original diameter. Then, for reasons still not completely understood, the myofibril splits down its longitudinal axis, forming two parallel myofibrils. Studies have also shown that fast-twitch fibers tend to hypertrophy more quickly, and to a greater extent, than slow-twitch fibers. People with a predominance of fast-twitch fibers, therefore, may hypertrophy at a greater rate and achieve more mass than a person endowed with primarily slow-twitch fibers. During hypertrophy, the sarcoplasmic reticulum must increase in size to match the increase in diameter of the myofibrils, and the sarcolemma must grow to match the increase in size of the muscle fiber.

Although isometric, concentric, and eccentric actions can all produce muscle hypertrophy, eccentric actions provide the greatest stimulus. When a weight is lowered with gravity in an eccentric action, the nervous system recruits fewer muscle fibers. This produces more muscle and connective tissue damage to the fibers involved and therefore provides a greater stimulus for muscle growth.

Muscle Names

With the basics of muscle function now in place, we continue with a detailed examination of specific skeletal muscles, in particular those responsible for producing and controlling movements of the trunk and limbs. Less than one-third of the approximately 600 skeletal muscles in the human body are responsible for most body movements (figure 4.9). We will now examine these muscles in detail.

Muscles are named in various ways, and often something in the name helps describe the muscle. A muscle's name might describe its size, shape, location, action, or attachment sites; the number of muscle bellies; or the direction of its fibers, as the following examples show:

- Size: The pectoralis major is a large muscle on the anterior pectoral, or chest, region. In addition to *major*, other size-related terms include *maximus* (largest, e.g., gluteus maximus), *minimus* (smallest, e.g., gluteus minimus), *longus* (long, e.g., peroneus longus), and *brevis* (short, e.g., peroneus brevis).

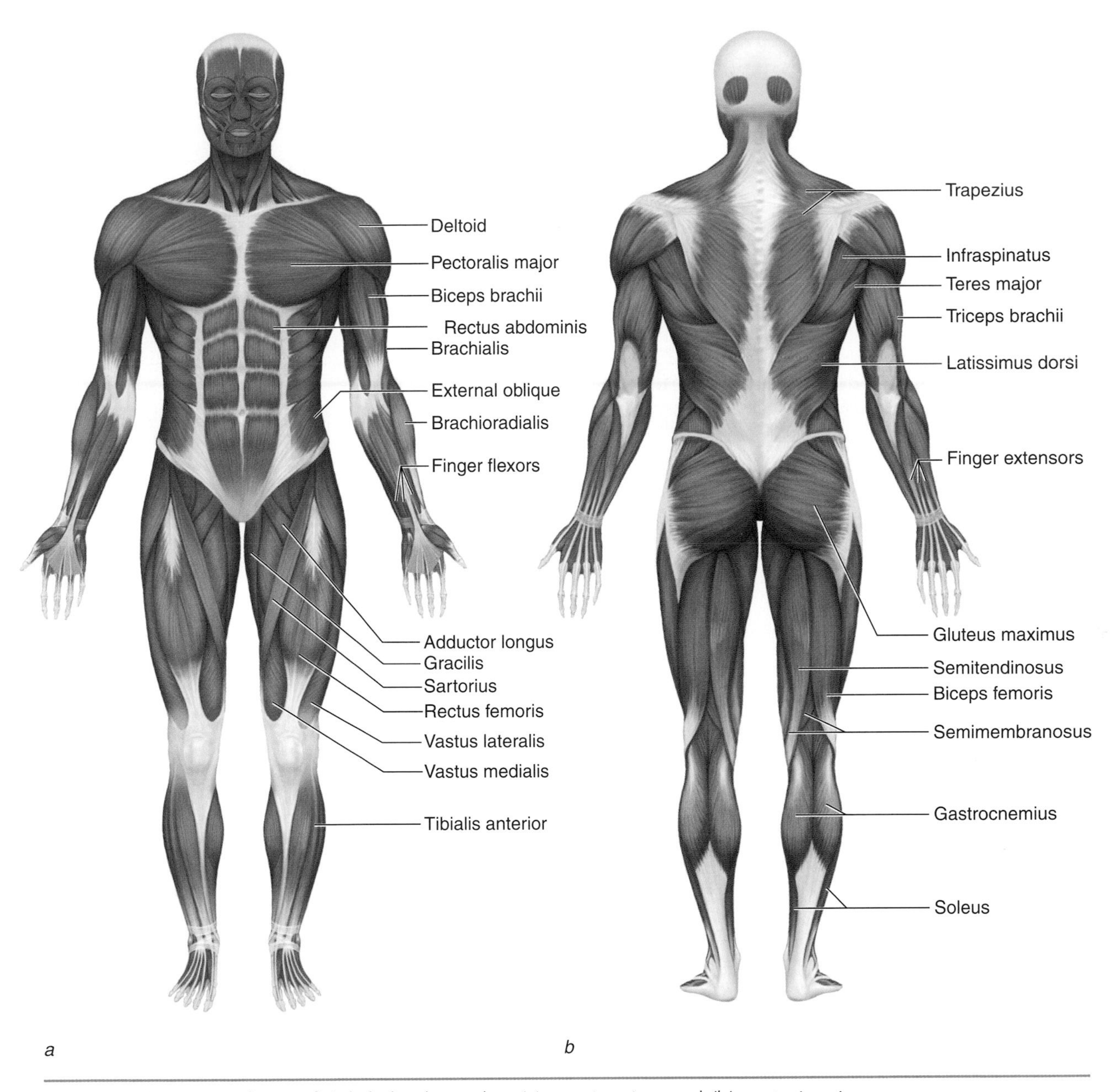

FIGURE 4.9 Principal superficial skeletal muscles: *(a)* anterior view and *(b)* posterior view.

- Shape: The deltoid is a shoulder muscle shaped like a triangle whose name is derived from the triangularly shaped Greek letter delta Δ. The trapezius is named for its trapezoidal (four-sided) shape.
- Location: The tibialis anterior lies on the anterolateral side of the lower leg, adjacent to the tibia.
- Action: The flexor digitorum muscle flexes the fingers or toes.
- Attachment sites: The sternocleidomastoid attaches from the sternum and clavicle to the mastoid process.
- Number of origins, or muscle bellies: The biceps brachii on the anterior side of the upper arm (brachium) has two heads, or bellies. The triceps brachii has three heads.

- Fiber direction: The rectus abdominis is a muscle whose fibers run parallel along the midline of the trunk's anterior surface (*rectus* = straight). Other directional terms include *transversus* (perpendicular to the body's midline, e.g., transversus abdominis) and *oblique* (diagonal to the midline, e.g., external oblique).

See table 4.2 for a more complete list of muscle terminology.

Muscles sometimes are collectively referred to by group names. When using group names, keep in mind that the group name represents a collection of muscles rather than a single muscle. For example, quadriceps, or quads, is a group name for four muscles on the anterior thigh. The **quadriceps** group includes the vastus medialis, vastus lateralis, vastus intermedius, and rectus femoris. Similarly, the **hamstring** group contains the biceps femoris, semitendi-

TABLE 4.2 Muscle Terminology

Factor	Term	Meaning
Size	Brevis	Short
	Gracilis	Slender
	Lata	Wide
	Latissimus	Widest
	Longissimus	Longest
	Longus	Long
	Magnus	Large
	Major	Larger
	Maximus	Largest
	Minimus	Smallest
	Minor	Smaller
	Vastus	Great
Shape	Deltoid	Triangular
	Orbicularis	Circular
	Pectinate	Comblike
	Piriformis	Pear shaped
	Platys-	Flat
	Pyramidal	Pyramid shaped
	Rhomboideus	Rhomboidal
	Serratus	Serrated
	Splenius	Bandage shaped
	Teres	Long and round
	Trapezius	Trapezoidal
Location or direction relative to body axes	Anterior	Front
	Externus	Superficial
	Extrinsic	Outside
	Inferioris	Inferior
	Internus	Internal or deep
	Intrinsic	Inside
	Lateralis	Lateral
	Medialis	Medial or middle
	Medius	Medial or middle
	Obliquus	Oblique
	Posterior	Back
	Profundus	Deep
	Rectus	Straight or parallel
	Superficialis	Superficial
	Superioris	Superior
	Transversus	Transverse

nosus, and semimembranosus. Other common group names include the **rotator cuff** group (subscapularis, supraspinatus, infraspinatus, and teres minor) and the **triceps surae** group (soleus and gastrocnemius). Careful consideration of the last group name, the triceps surae, may seem incongruous with the fact that the group contains only two muscles while the name (triceps) would suggest three. This is reconciled by noting that the gastrocnemius has two heads (medial and lateral). The two heads of the gastrocnemius together with the single head of the soleus form a triceps group, or triceps surae.

When learning the many muscles of the human body, paying attention to clues within a muscle's name can facilitate the learning process. In addition, try to visualize each muscle's location, attachments, and movements as you study. Being able to see the muscle and its action can be helpful in learning about the muscular system and its function. Students who only memorize the tables and lists place themselves at a disadvantage and make the task of learning anatomy more difficult.

Factor	Term	Meaning
Location relative to body region	Abdominis	Abdomen
	Anconeus	Elbow
	Brachialis	Brachium (upper arm)
	Capitis	Head
	Capri	Wrist
	Cervicis	Neck
	Cleido/clavius	Clavicle
	Costalis	Ribs
	Cutaneous	Skin
	Femoris	Femur
	Glosso/glossal	Tongue
	Hallucis/hallux	Big (great) toe
	Ilio-	Ilium
	Lumborum	Lumbar region
	Nasalis	Nose
	Oculo-	Eye
	Oris	Mouth
	Pollicis	Thumb
	Popliteus	Behind the knee
	Radialis	Radius
	Scapularis	Scapula
	Temporalis	Temples
	Thoracis	Thoracic region
	Tibialis	Tibia
	Ulnaris	Ulna
Action	Abductor	
	Adductor	
	Depressor	
	Extensor	
	Flexor	
	Levator	
	Pronator	
	Rotator	
	Supinator	
	Tensor	
Number of origins or muscle heads/bellies	Biceps	Two heads
	Triceps	Three heads
	Quadriceps	Four heads

Functional Actions of Muscles

You can determine the action of a muscle in several ways. One of the simplest methods, termed **palpation**, involves feeling the muscle action through the skin. When you flex your elbow, for example, the bulging of the biceps brachii is easily palpated. Obviously, palpation works only for superficial muscles, not for deep muscles that are obscured by different muscles or other tissues.

Another method involves assessment of a muscle's location and its attachments. The action of a muscle can be deduced by constructing a simple line drawing of the muscle and its **origin** and **insertion** (figure 4.10). Draw a straight line or, if necessary, a curved line (e.g., gluteus maximus) connecting the origin to the insertion. Recall that the origin usually remains relatively fixed (immovable) and that the insertion typically moves toward the origin. Place an arrow along the muscle line of action you have drawn, with the head of the arrow directed toward the origin. Move the bone attached at the insertion toward the bone attached to the origin, and observe the movement produced. It's just that simple.

For biarticular (two-joint) muscles (e.g., semitendinosus, rectus femoris), repeat the procedure, but this time direct the arrow toward the insertion. Now, rotate the bone with the origin attachment toward the bone of insertion attachment and observe the motion at the second joint.

When you can visualize joint movements in this way, you are far along the road to understanding human movement. Visualization of movement is much easier than rote memorization. Learning muscles and their functions will be easier if you say the names, write down their attachments and functions, and, if possible, palpate the muscles on your own body. Most important, do your best to create vivid images in your mind and see the movements with your imagination.

Another way of assessing muscle function is through the use of **electromyography** (EMG). EMG measures the electrical activity of muscles and allows researchers and clinicians to

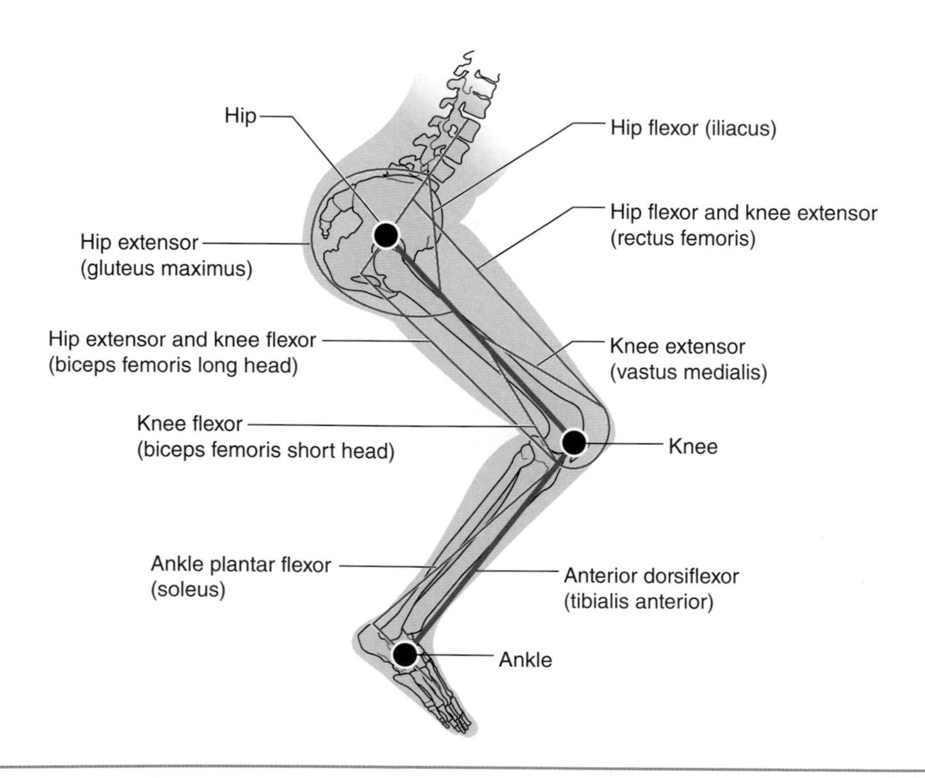

FIGURE 4.10 Model of one-joint and two-joint muscles.

Adapted from G.J. van Ingen Schenau, M.F. Bobbert, and A.J. van Soest, "The unique action of bi-articular muscles in leg extensions" in *Multiple Muscle Systems: Biomechanics and Movement Organization*, edited by J. Winters, S.L.-Y. Woo (Springer, Berlin, 1990).

Research in Mechanics

Biarticular Muscle Function

Numerous researchers, too many to mention by name here, have explored uniarticular versus biarticular muscle function over the past 40 or more years. Much of the work began in the early 1980s, and researchers often used the triceps surae group to study the uniarticular soleus in comparison with the biarticular gastrocnemius. Among the many findings are that the gastrocnemius transfers 25% of the energy from the knee to the ankle in jumping and 28% in sprinting during plantar flexion (Jacobs et al., 1996), and that two-joint muscles (rectus femoris at the hip and knee; gastrocnemius at the knee and ankle) transfer energy from proximal joint to distal joints during squat vertical jumps and jump landings (Prilutsky & Zatsiorsky, 1994).

See the references for the full citations:

Jacobs, Bobbert, & van Ingen Schenau, 1996.

Prilutsky & Zatsiorsky, 1994.

explore the action of muscles in a wide variety of movements. Much of the information on muscle action presented in this and later chapters comes from electromyography studies.

Before we continue, commit to memory the following notes on terminology and presentation format:

- Muscles are said to cross a joint, meaning they have action, or can produce movement, at the joint being crossed. Some muscles have action at a single joint and are termed **uniarticular**. Other muscles cross two joints (**biarticular**), three joints (**triarticular**), or more than three joints (**multiarticular**). Biarticular muscles are common in the human body. Triarticular muscles are rare. Multiarticular muscles are primarily found in the distal regions of the limbs (e.g., muscles that move the fingers and toes).
- Joint actions, or movements, typically are described in one of two ways: by identifying either the joint being moved (e.g., flexion of the elbow) or the segment being moved (e.g., flexion of the forearm). Both methods are correct, and each is used by authors of other texts.
- Muscle actions refer to what each muscle does when acting concentrically.
- All muscles acting at a particular joint do not participate equally in performing or controlling a movement. The relative contribution of each muscle is determined by many factors, including muscle size (cross-sectional area), level of neural stimulation, muscle fiber–type composition, fatigue, joint position, and the type of movement being performed. The muscles making the greatest contribution are termed **prime movers**. In plantar flexion at the ankle, for example, the relatively large gastrocnemius and soleus (prime movers) play a much greater role than does the diminutive plantaris.

Muscles of Major Joints

To see how muscles act together, tables 4.3 through 4.7 and figures 4.11 through 4.15 present the primary movements at each of the major joints in the body and the muscles acting to produce and control those movements. (Note that in the tables, nonitalicized muscles are considered prime movers. Italicized muscles represent assistant movers.) As a reminder, the muscle actions described are for *concentric* action. The *eccentric* action of a muscle controls the opposite motion listed for its respective concentric action. For example, the biceps brachii acts concentrically to produce elbow flexion and works eccentrically to control elbow extension.

Go to the web study guide to access muscle origin, insertion, and innervation tables.

Vertebral Column

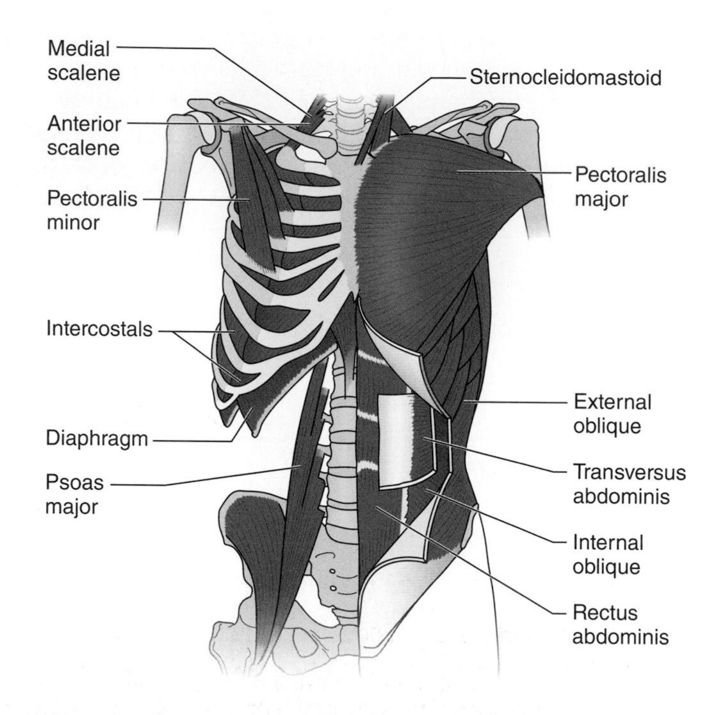

FIGURE 4.11 Muscles of the vertebral column.

TABLE 4.3 Vertebral Column Muscle Actions

Flexion	Extension
Rectus abdominis External oblique Internal oblique *Psoas major (lumbar region)*	Erector spinae group
Rotation to the same side	**Rotation to the opposite side**
Internal oblique	External oblique
Lateral flexion	
External oblique Internal oblique Quadratus lumborum *Rectus abdominis*	

Shoulder Girdle

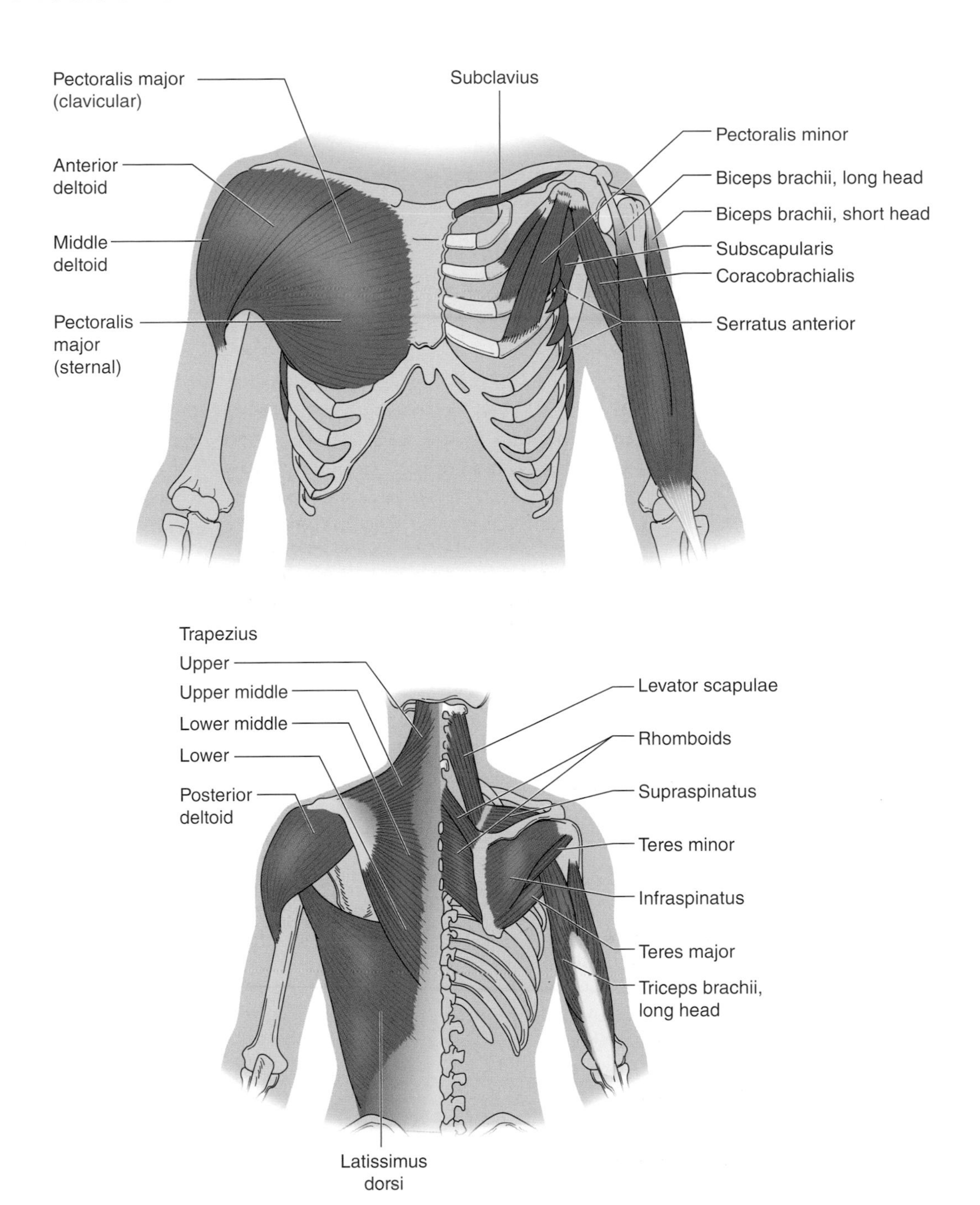

FIGURE 4.12 Muscles of the shoulder girdle and shoulder joint.

TABLE 4.4 Shoulder Girdle and Shoulder Joint Muscle Actions

SHOULDER GIRDLE	
Elevation	**Depression**
Levator scapula Upper trapezius Rhomboids	Lower trapezius Pectoralis minor
Adduction	**Abduction**
Rhomboids Middle trapezius	Pectoralis minor Serratus anterior
Upward rotation	**Downward rotation**
Trapezius Serratus anterior	Rhomboids Levator scapula Pectoralis minor
SHOULDER (GLENOHUMERAL) JOINT	
Flexion	**Extension**
Pectoralis major, clavicular portion Anterior deltoid *Biceps brachii, short head* *Coracobrachialis*	Pectoralis major, sternal portion Latissimus dorsi Teres major *Posterior deltoid* *Triceps brachii, long head*
Adduction	**Abduction**
Latissimus dorsi Teres major Pectoralis major, sternal portion *Biceps brachii, short head* *Triceps brachii, long head*	Middle deltoid Supraspinatus Anterior deltoid *Biceps brachii*
Medial (internal) rotation	**Lateral (external) rotation**
Latissimus dorsi Teres major Subscapularis *Anterior deltoid* *Pectoralis major* *Biceps brachii, short head*	Teres minor Infraspinatus *Posterior deltoid*
Horizontal flexion (adduction)	**Horizontal extension (abduction)**
Pectoralis major Anterior deltoid *Coracobrachialis* *Biceps brachii, short head*	Middle deltoid Posterior deltoid Teres minor Infraspinatus *Teres major* *Latissimus dorsi*

Elbow, Radioulnar, and Wrist Joint

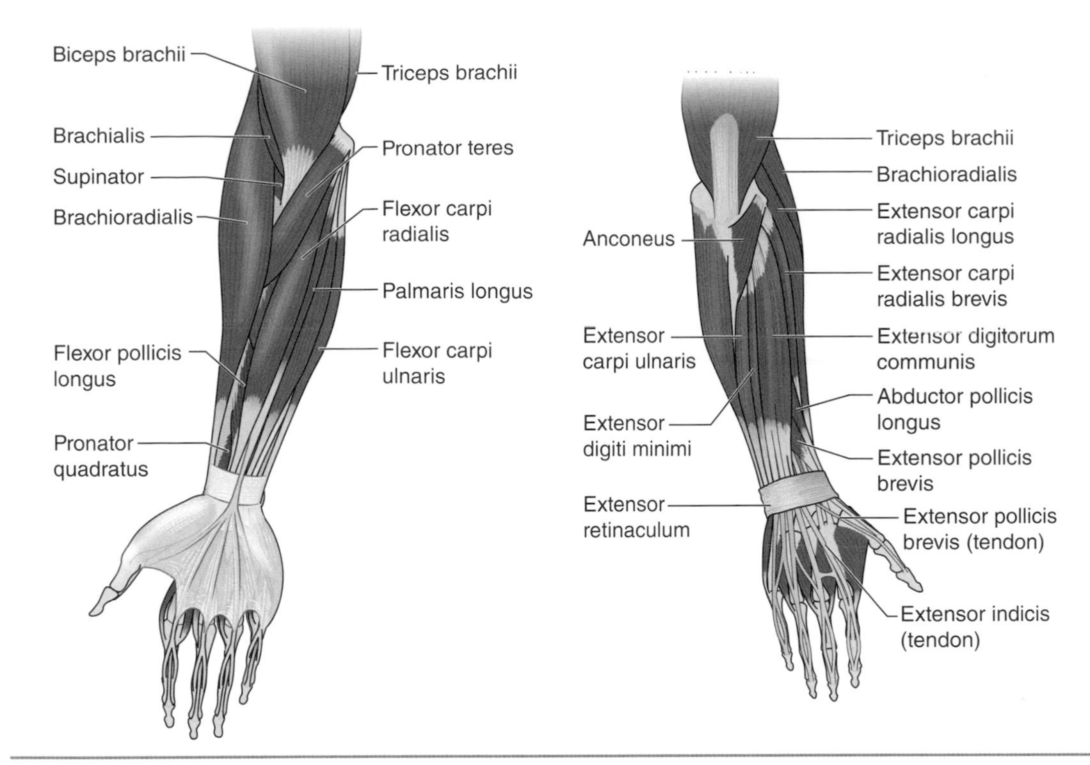

FIGURE 4.13 Muscles of the elbow, radioulnar, and wrist joints.

TABLE 4.5 Elbow, Radioulnar, and Wrist Joint Muscle Actions

ELBOW JOINT	
Flexion	**Extension**
Biceps brachii Brachialis Brachioradialis	Triceps brachii *Anconeus*
RADIOULNAR JOINT	
Supination	**Pronation**
Biceps brachii Supinator *Brachioradialis** *functions to move the forearm to the mid- or neutral position	Pronator teres Pronator quadratus *Brachioradialis** *functions to move the forearm to the mid- or neutral position
WRIST JOINT	
Flexion	**Extension**
Flexor carpi radialis Flexor carpi ulnaris *Flexor digitorum superficialis* *Flexor digitorum profundus* *Palmaris longus*	Extensor carpi radialis longus Extensor carpi radialis brevis Extensor carpi ulnaris *Extensor indicis* *Extensor digiti minimi* *Extensor digitorum*
Radial deviation (abduction)	**Ulnar deviation (adduction)**
Flexor carpi radialis Extensor carpi radialis longus Extensor carpi radialis brevis	Flexor carpi ulnaris Extensor carpi ulnaris

Hip and Knee

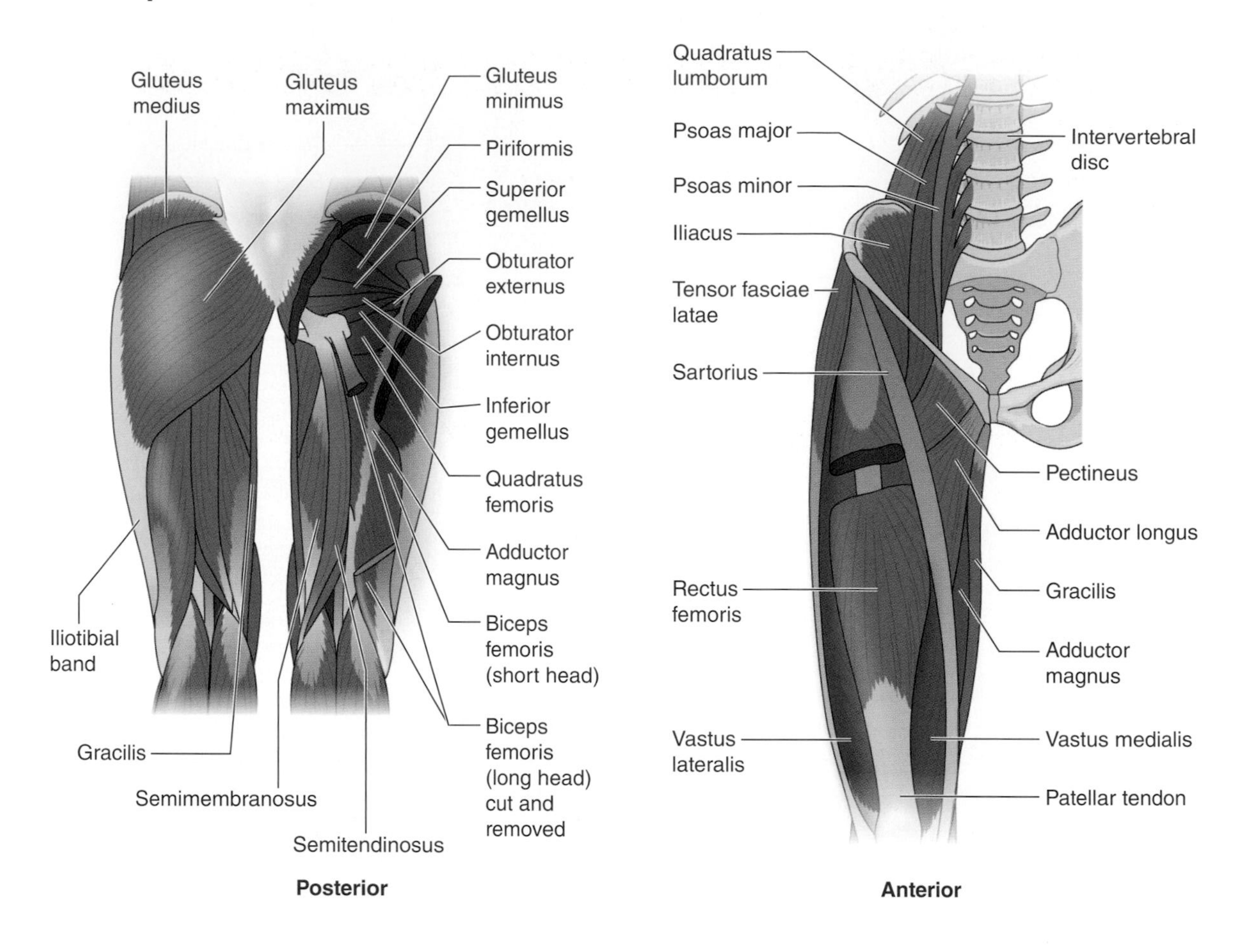

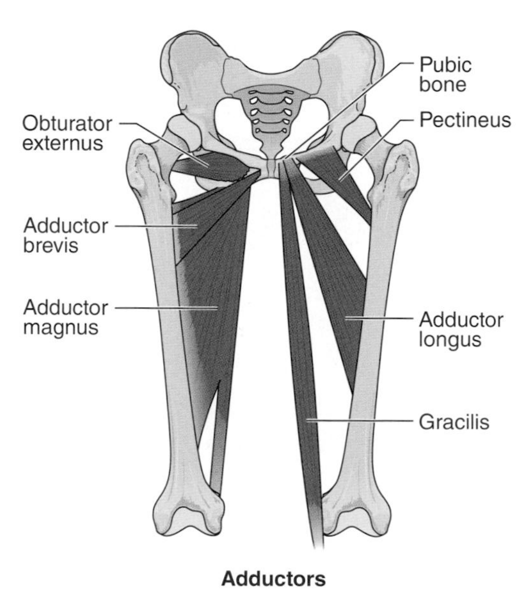

FIGURE 4.14 Muscles of the hip and knee joints.

TABLE 4.6 Hip and Knee Joint Muscle Actions

HIP JOINT	
Extension	**Flexion**
Gluteus maximus Semitendinosus Semimembranosus Biceps femoris, long head Adductor magnus, posterior fibers	Psoas major Iliacus Pectineus Rectus femoris Adductor brevis Adductor longus Adductor magnus, anterior upper fibers Tensor fasciae latae Sartorius *Gracilis*
Abduction	**Adduction**
Gluteus medius *Gluteus minimus* *Tensor fasciae latae* *Gluteus maximus, superior fibers* *Psoas major* *Iliacus* *Sartorius*	Pectineus Adductor brevis Adductor longus Adductor magnus Gracilis *Gluteus maximus, inferior fibers*
HIP JOINT	
Medial rotation	**Lateral rotation**
Gluteus minimus *Tensor fasciae latae* *Pectineus* *Adductor brevis* *Adductor longus* *Adductor magnus, anterior upper fibers* *Semitendinosus* *Semimembranosus*	Gluteus maximus Piriformis Gemellus superior Obturator internus Gemellus inferior Obturator externus Quadratus femoris *Psoas major* *Iliacus* *Sartorius* *Biceps femoris, long head*
KNEE JOINT	
Extension	**Flexion**
Vastus medialis Vastus intermedius Vastus lateralis Rectus femoris	Semimembranosus Semitendinosus Biceps femoris *Sartorius* *Gracilis* *Popliteus* *Gastrocnemius* *Plantaris*
Medial rotation*	**Lateral rotation***
Popliteus Semimembranosus Semitendinosus *Sartorius* *Gracilis*	Biceps femoris

*Rotation can occur only when the knee is flexed.

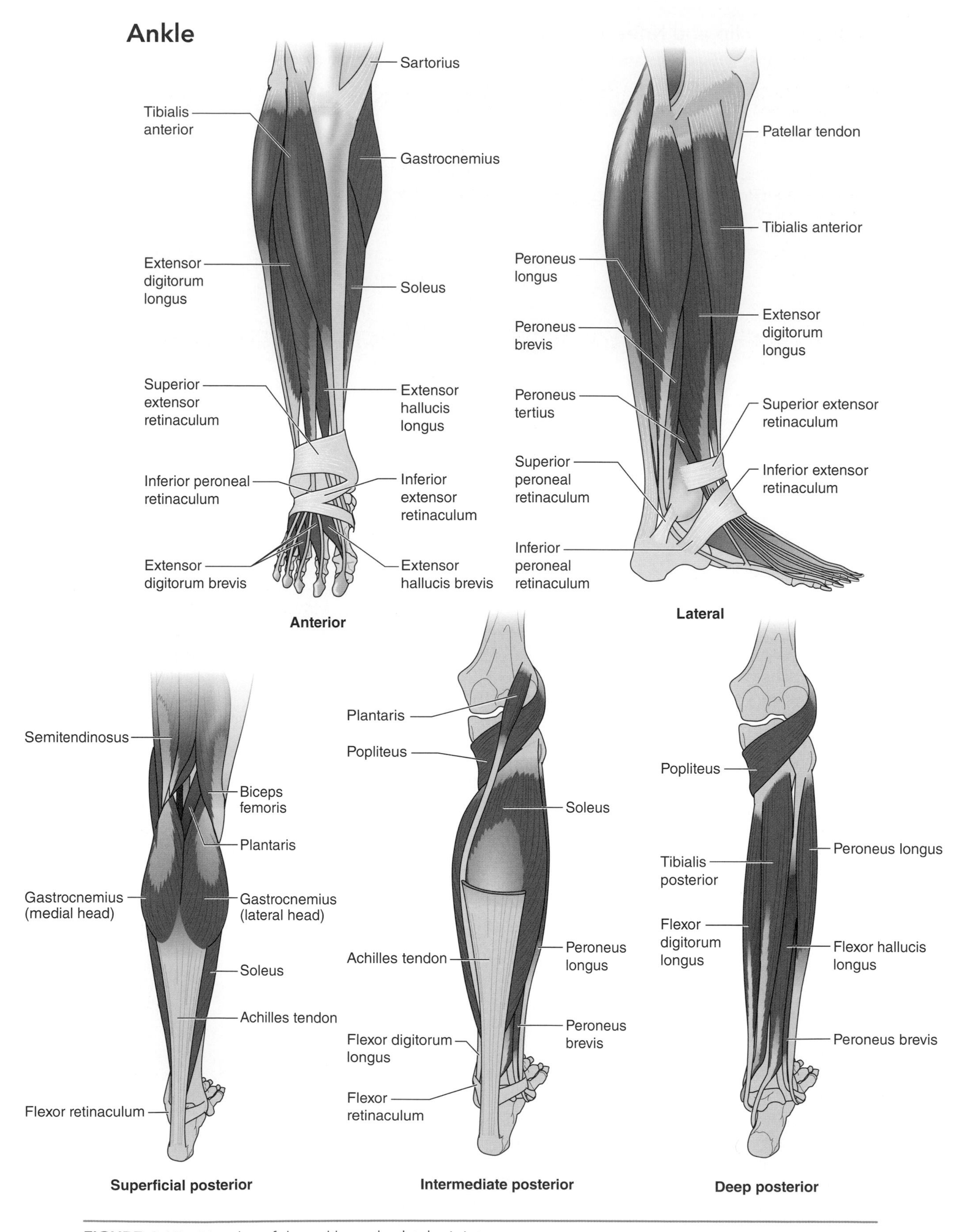

FIGURE 4.15 Muscles of the ankle and subtalar joints.

TABLE 4.7 Ankle and Subtalar Joint Muscle Actions

ANKLE JOINT	
Plantar flexion	**Dorsiflexion**
Gastrocnemius Soleus *Plantaris* *Tibialis posterior* *Flexor hallucis longus* *Flexor digitorum longus* *Peroneus longus* *Peroneus brevis*	Tibialis anterior Extensor digitorum longus Peroneus tertius *Extensor hallucis longus*
SUBTALAR JOINT	
Inversion	**Eversion**
Tibialis anterior Tibialis posterior *Flexor hallucis longus* *Flexor digitorum longus* *Gastrocnemius* *Soleus* *Plantaris*	Peroneus longus Peroneus brevis Peroneus tertius Extensor digitorum longus

Muscle Injury, Pain, and Soreness

Although skeletal muscles are capable of generating high forces without injury, too much force transmitted through the **musculotendinous unit** can, and often does, produce injury. Musculotendinous injuries are called strain injuries, or **strains**. (Note: The term *strain* also is used to describe the physical deformation of a material, whether or not injury or damage occurs. This mechanical strain is explained in chapter 5.)

Strain injuries can occur as tearing where the tendon inserts into the bone (osteotendinous junction), within the body of the tendon, where the tendon joins with the muscle (myotendinous or musculotendinous junction), or within the belly (substance) of the muscle itself. Recent research has shown that the myotendinous junction and the area immediately adjacent to it are common sites of strain injury.

Injury typically occurs during forced lengthening, or eccentric, muscle action employed to control or decelerate high-velocity movements (e.g., sprinting, throwing, jumping) and is hastened by many factors, including fatigue, muscle imbalances, inflexibility, and insufficient warm-up.

Certain muscles seem more prone to strain injuries. The muscles of the hamstring group (on the posterior side of the thigh) are especially susceptible to injury. This may, in part, be due to the fact that these muscles are biarticular (i.e., have action at two joints). This structural arrangement dictates that muscle length is determined by the combined action of the hip and knee joints. Hip flexion and knee extension both cause lengthening of the hamstring muscles. Simultaneous hip flexion and knee extension place the hamstrings in a lengthened state that contributes to the muscles' risk of injury. If these movements are made quickly, as when a sprinter swings her leg forward through the air and plants her foot on the ground, the chance of injury increases considerably. Hamstring injuries of this type often happen at the myotendinous junction. So a muscle's gross and microscopic anatomical structure, biarticular arrangement, and involvement in controlling high-velocity movement all contribute to its susceptibility to strain injuries.

Muscle pain and soreness felt during and immediately after exercise are normally attributed to tissue swelling (**edema**) and the accumulation of lactate and other metabolites. These

Applying the Concept

Delayed Onset Muscle Soreness

After strenuous exercise, it is not uncommon for an individual to experience muscle soreness. Although muscle soreness may occur more frequently among those just starting a training program, it can also occur in well-trained athletes if they significantly increase their training volume. We normally do not perceive this **delayed onset muscle soreness** (DOMS) until 48 to 72 hours after the exercise session. Electron micrograph images of skeletal muscles taken after intense exercise show tears in the connective tissue, sarcolemma, sarcoplasmic reticulum, actin and myosin filaments, and Z lines of muscle fibers. DOMS is typically accompanied by edema, tenderness, and stiffness.

Although we do not understand all the factors that produce DOMS, muscle damage appears to be the major cause. Because eccentric actions produce more muscle damage than either concentric or isometric actions, they are the principal cause of DOMS. Adequate recovery after intense training, therefore, allows the muscle to adapt to the training stress without producing continual soreness. If athletes dramatically increase training volume, however, they may again experience DOMS. Because this training-induced soreness is not indicative of any serious injury or problem, continued training at a reduced intensity is generally recommended. Lighter training speeds the repair process by increasing metabolism, supplying required nutrients, and removing waste products from the sore muscles.

substances diffuse out of the muscle and either directly stimulate free nerve endings or cause edema, resulting in pain. Edema in this case is typically the result of fluid from blood plasma moving into the muscle tissue. The good news is that this pain normally disappears within hours after exercise.

Concluding Comments

Muscles are the engines that power all of our body's movements. Among muscle's remarkable characteristics are its abilities to generate force and adapt to training. This chapter provides the basics for understanding how human skeletal muscle generates force and what factors affect this force production. We review the fundamentals of muscle structure, from the macroscopic whole-muscle level down to the microscopic events involved in muscle contraction at the actin–myosin level, and discuss how muscle produces force through the excitation–contraction coupling process. We discuss many factors that help modulate force output, including contraction type, fiber type, recruitment, fiber arrangement, muscle length, and velocity.

No other tissue in the human body responds to intense training more noticeably than muscle. An elite bodybuilder, for example, conjures an entirely different image than a world-class marathoner. The size of an athlete's muscles readily lets us know, or at least enables us to guess, whether they have trained for endurance, strength, power, or hypertrophy.

Our gross movements are controlled by a relatively small subset of the body's more than 600 skeletal muscles. Sets of muscles at each major joint in the body work together in coordinated fashion to produce and control our movements. In most cases, they do so with little conscious attention on our part. Our neuromuscular system's ability to control movements is remarkable. Often only when something affects our ability to move, such as injury or disability, and disrupts our regular movements do we become aware of how effectively our muscles work on a day-to-day basis.

In subsequent chapters, we use this information to examine how muscles work in specific situations to produce everyday movements and how misuse and overuse can result in muscle injury.

Go to the web study guide to access critical thinking questions for this chapter.

Suggested Readings

Behnke, R.S. (2012). *Kinetic anatomy* (3rd ed.). Champaign, IL: Human Kinetics.

Enoka, R.M. (2015). *Neuromechanics of human movement* (5th ed.). Champaign, IL: Human Kinetics.

Floyd, R.T. (2018). *Manual of structural kinesiology* (20th ed.). New York: McGraw-Hill.

Jenkins, D.B. (2008). *Hollinshead's functional anatomy of the limbs and back* (9th ed.). Philadelphia: Saunders.

Levangie, P.K., & Norkin, C.C. (2011). *Joint structure and function: A comprehensive analysis* (5th ed.). Philadelphia: Davis.

Lieber, R.L. (2009). *Skeletal muscle structure, function and plasticity: The physiological basis of rehabilitation* (3rd ed.). Philadelphia: Lippincott Williams & Wilkins.

Martini, F.H., Tallitsch, R.B., & Nath, J.L. (2018). *Human anatomy* (9th ed.). London: Pearson.

MacKinnon, P., & Morris, J. (2005). *Oxford textbook of functional anatomy* (Vols. 1-3). Oxford: Oxford University Press.

McArdle, W.D., & Katch, F.I. (2014). *Exercise physiology: Energy, nutrition, and human performance* (8th ed.). Baltimore: Lippincott Williams & Wilkins.

Neumann, D.A. (2016). *Kinesiology of the musculoskeletal system: Foundations for physical rehabilitation* (3rd ed.). St. Louis: Mosby.

Tortora, G.J., & Derrickson, B.H. (2011). *Principles of anatomy and physiology* (13th ed.). New York: Wiley.

PART II

Biomechanics and Movement Control

Human movement must obey the laws of mechanics as set forth by Sir Isaac Newton and others. We therefore begin part II with chapter 5 (Biomechanics) on the mechanics of movement. This chapter presents a qualitative summary of essential mechanical terms and concepts. With that foundation in place, chapter 6 (Muscular Control of Movement and Movement Assessment) presents the muscle control formula, a set of simple steps that identifies the muscles responsible for producing and controlling any human movement. In addition, we consider important movement concepts and assessment techniques important to a full understanding of human movement.

5

Biomechanics

Objectives

After studying this chapter, you will be able to do the following:

- Identify the major areas of biomechanics relevant to human movement: movement mechanics, fluid mechanics, joint mechanics, and material mechanics
- Explain general biomechanical concepts and measures, including linear and angular motion, center of gravity, stability, mobility, and movement equilibrium
- Explain concepts of movement mechanics: kinematics, kinetics, force, pressure, lever systems, torque (moment of force), Newton's laws of motion, work, power, energy, momentum, and friction
- Explain concepts of fluid mechanics: fluid flow and resistance
- Explain concepts of joint mechanics: range of motion, joint stability, and joint mobility
- Explain concepts of material mechanics: stress, strain, stiffness, bending, torsion, and viscoelasticity

Movement fascinates us today in much the same way as it has our ancestors over many millennia. Artists have used human movement as inspiration for artistic expression. Scientists have made human movement the subject of extensive scientific inquiry. From a scientific point of view, movement can be considered from many perspectives, including those of anatomy, physiology, psychology, and physics. One branch of physics, mechanics, is particularly applicable to the assessment and appreciation of movement. With respect to human movement, the mechanical perspective falls within the domain of **biomechanics**, broadly defined as the application of mechanical principles to the study of biological organisms and systems.

Why is the study of biomechanics important in understanding human movement? Perhaps the most important reason is that our movement potential and limitations often are dictated by mechanical properties and events. How fast we can run, how high we can jump, and how much we can lift are determined by the forces acting both inside and outside our bodies. Many of the details of how biomechanics affects human movement are presented in this chapter, with a continuing theme of mechanical influence evident throughout the following chapters as well.

This chapter considers several areas of biomechanics that are relevant to understanding the dynamics of human movement. We start with some basic biomechanical concepts and then move on to a brief discussion of movement mechanics, fluid mechanics, joint mechanics, and material mechanics.

Biomechanical Concepts

Many excellent texts are devoted solely to the discussion of human biomechanics. In a single chapter, we are limited to presenting only a brief summary of mechanical concepts most relevant to understanding human movement. Although biomechanics is inherently a quantitative discipline, the approach taken here is to limit mathematical considerations and focus instead on providing a conceptual framework for understanding the mechanics of movement.

In discussing biomechanics, we often refer to objects as *bodies*. The word **body** in this context is taken to mean any collection of matter. It may refer to the entire human body, a body segment (e.g., thigh or upper arm), or any other collected mass (e.g., a block of wood).

The amount of matter, or substance, constituting a body is known as its **mass**. A body's mass is *not* the same as its **weight**. Mass certainly is related to weight, but the two terms should not be used as synonyms. For a body with a given mass, its weight is determined by the effect of gravity. In the absence of gravity (e.g., in outer space), a body is weightless, but still retains its mass. The gravitational pull on Earth gives the body a certain weight. The same body's weight on our moon would be less than on Earth due to the moon's smaller gravitational pull. Similarly, the same body on a larger planet (e.g., Jupiter) would have a much greater weight than on Earth, but the same mass. This relationship between mass and weight is shown in figure 5.1.

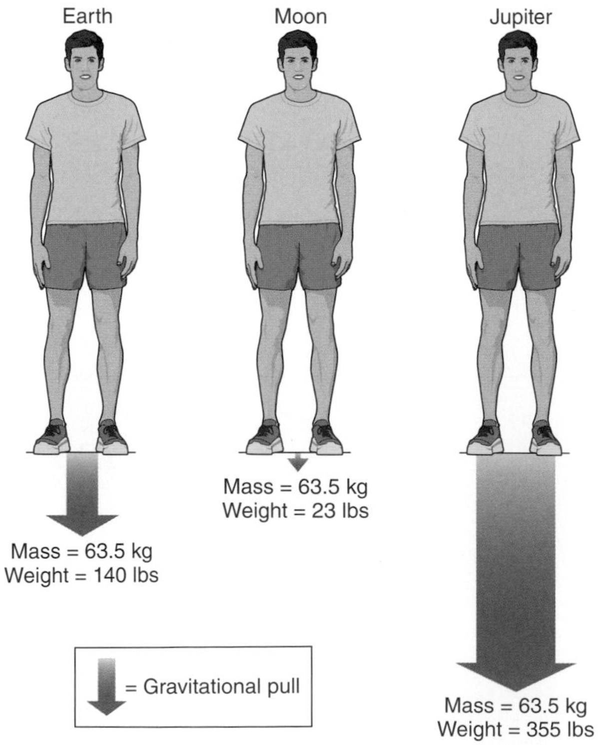

FIGURE 5.1 Relationship between mass and weight.

Linear and Angular Motion

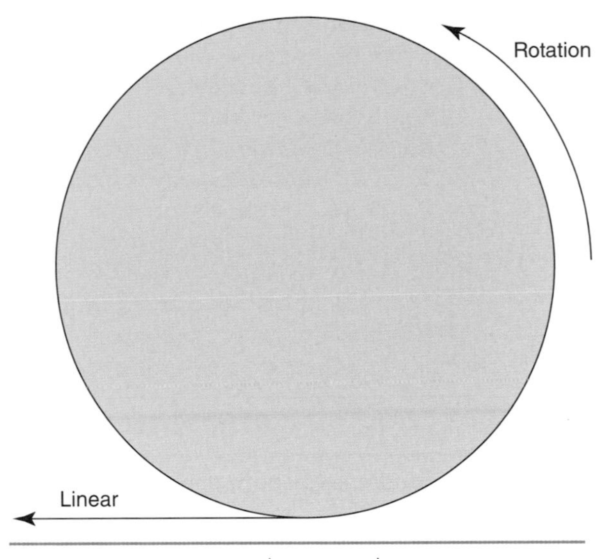

FIGURE 5.2 Linear and rotational motions.

In mechanical terms, there are two basic forms of movement: (1) **linear** (also **translational**) **motion**, in which a body moves along a straight line (**rectilinear motion**) or a curved line (**curvilinear motion**), and (2) **angular** (or **rotational**) **motion**, in which a body rotates about a fixed line called the axis of rotation (figure 5.2). In walking, for example, the thigh rotates in the sagittal plane about an axis of rotation through the hip, the lower leg (shank) rotates about an axis through the knee, and the foot rotates about an axis through the ankle.

Many human movements combine linear and angular motion in what is termed **general motion**. The movement of a person's thigh during walking, for example, involves linear motion of the entire thigh segment in a forward direction and angular motion at the hip in alternating periods of flexion and extension.

The movement of inanimate objects can also exhibit general motion. The flight of a thrown softball, for example, consists of linear motion (the curved path, or arc, of the ball) and angular motion (the ball's spin).

As we explore the mechanics of movement, the concepts of linear and angular motion recur often. Combination of these two simple movement forms results in the vast array of movement patterns we perform and observe on a daily basis.

Center of Gravity

Every body contains a point, known as the **center of mass** or **center of gravity**, about which that body's mass is equally distributed. Although there is a technical difference in the definitions of *center of mass* and *center of gravity*, in practical terms the two points are coincident, and we therefore use the terms interchangeably.

If the body's mass were concentrated into a **point mass** at the center of gravity, this point would move in exactly the same way as the body would in its original distributed state. This concept is important in discussing human movement because all points within a given body often are not moving in the same direction or at the same speed. Figure 5.3, for example,

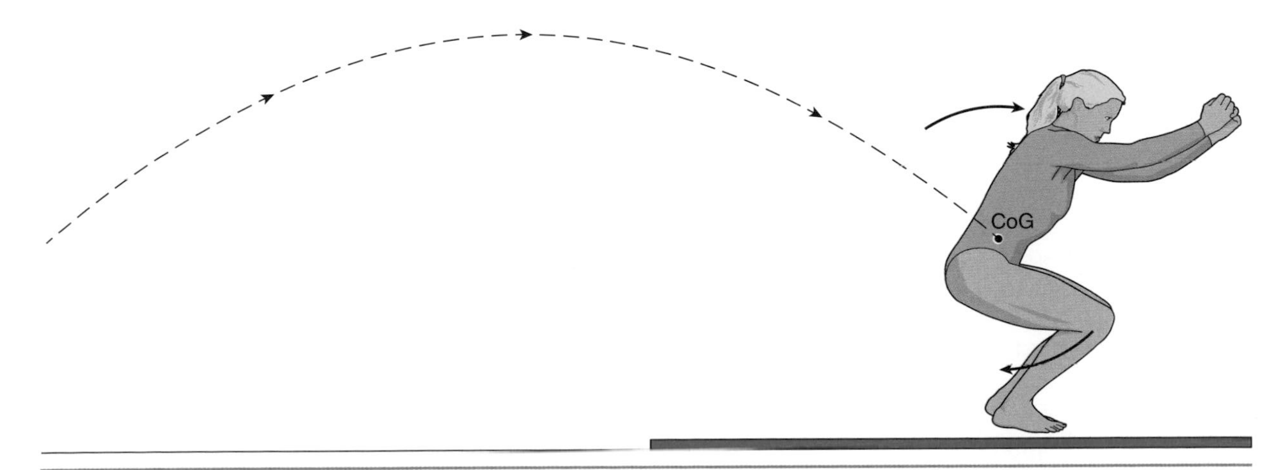

FIGURE 5.3 Linear and angular motion as seen in a gymnast during her dismount.

shows a gymnast performing a vault. The twists and turns she executes happen around her center of gravity while her center of gravity moves along a smooth curve, or arc.

In normal standing (or anatomical position), the body's center of gravity is located just anterior to the second sacral vertebra. From a frontal view, the center of gravity normally is located along the body's midline, typically 55% to 57% of a person's height above the ground (figure 5.4).

Movement of body segments (e.g., lifting the arms, flexing the knee) causes the center of gravity to move within the body. For example, raising (abducting) the arms from anatomical position to an overhead position results in superior movement of the center of gravity along the body's midline. If only one arm is raised, the center of gravity similarly moves superiorly but also deviates from the midline slightly toward the raised-arm side.

Most of the time, the center of gravity lies within the body. However, in the same way as a doughnut's center of gravity lies outside the body of the doughnut (i.e., in the hole), the center of gravity of a human body can lie outside the body when a person assumes a bent-over, or pike, position, as in gymnastics or diving (figure 5.5).

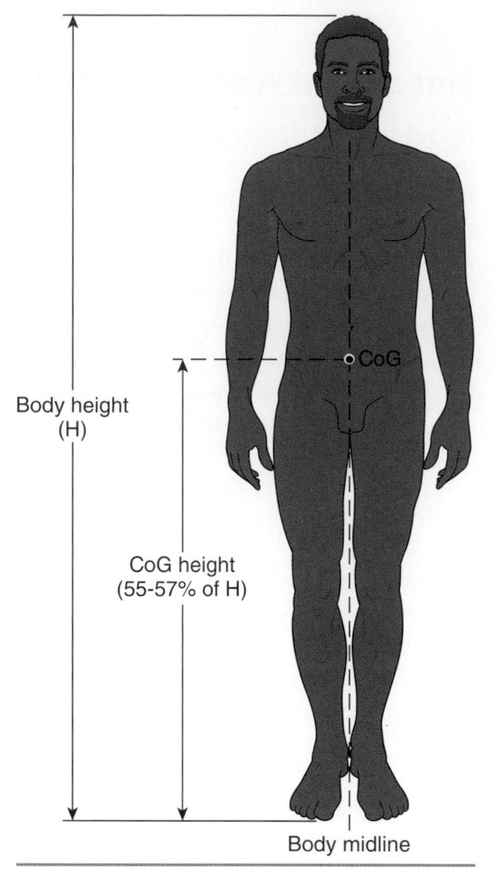

FIGURE 5.4 Center of gravity location in the human body.

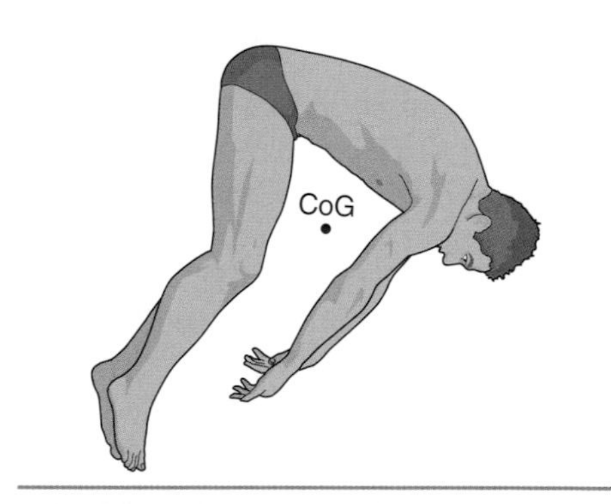

FIGURE 5.5 Center of gravity located outside the body when the body is in a bent-over, or pike, position.

Stability, Mobility, and Movement Equilibrium

When standing, we typically have two feet in contact with the ground. If our feet are close together, we feel less stable than when the feet are spread apart. Increasing the distance between the feet increases what is termed our **base of support**, or the area within an outline of all ground contact points. Several bases of support are illustrated in figure 5.6. The importance of the base of support in determining our stability and ability to move effectively is discussed in this section and again, in more detail, in chapters 8 and 9, where we explore movements such as walking, running, jumping, and throwing.

As we prepare to move in any situation, with little or no conscious thought, we place ourselves along what might be termed a *stability–mobility continuum*. In situations of imminent contact, we try to enhance our **stability**; when we want to move quickly, we try to increase our **mobility**. In preparation for impending contact by an opponent, for example, an American football player will try to brace himself by widening his base of support and bending his knees. If, on the other hand, the player decides to run away from the collision, he would adopt a different body posture that would enhance his mobility.

From a mechanical perspective, five factors determine our levels of stability and mobility.

1. *Size of the base of support in the direction of force or impending force.* In general, increasing the size of the base of support increases stability. In preparation for an impact, we tend to spread our feet apart. We do so, however, in the direction of the force. If you were

about to be struck from the front, would you widen your base of support by abducting at the hips to spread your feet apart to the side? Probably not. Most likely you would increase your base of support by staggering your feet front-to-back, or in an anterior–posterior orientation. Merely increasing the size of your base of support will not necessarily make you more stable. The increase must be made in the direction of force or impending force. Increases in base of support can be made by placing the feet in a certain position, as in the previous example, or by adding ground contact points. Additional contact points can be added by using other body parts, as when a baby creeps along the ground on hands and knees or when an athlete assumes a three-point or four-point stance (figure 5.7). Older or injured people also can enhance their stability by using a cane or crutch to add contact points to the system, thereby increasing their bases of support (see figure 5.6*c*).

2. *Height of the center of gravity above the base of support.* When you squat down to improve your stability, you lower your center of gravity, or decrease the height of the center of gravity above the base of support. Conversely, standing up straight raises the center of gravity above the base of support and decreases stability.

3. *Location of the center of gravity projection within the base of support.* Imagine that you drop a plumb line (i.e., a string with a weight on the end) straight down from your center of gravity. That line is referred to as the *vertical projection*, or **projection**, of your center of gravity within the base of support. If the projection moves outside the base of support, you become unstable and will fall without corrective muscle action. In normal standing, when the center of gravity projection lies at or near the center of the base of support, you are more stable than when the projection lies near the edge of the base of support. When another

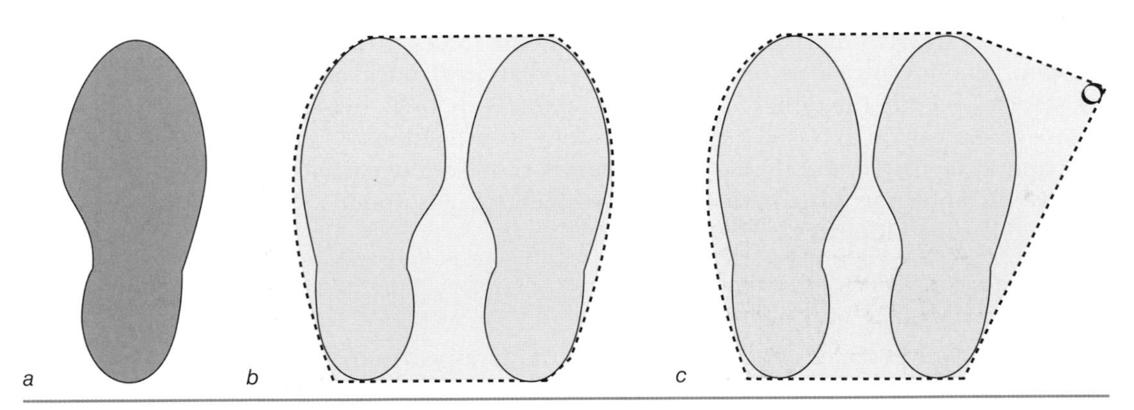

FIGURE 5.6 Examples of bases of support as seen from above: *(a)* one-leg standing, *(b)* two-leg standing, and *(c)* standing with a cane.

FIGURE 5.7 Examples of *(a)* three-point and *(b)* four-point stances.

body is about to collide with yours, you tend to lean toward the colliding body. This lean moves the projection near the edge of the base of support so that at impact, the center of gravity projection has a greater distance to travel before leaving the base of support on the opposite side and causing you to fall.

4. *Body mass or body weight.* A body's mass (or weight) contributes to stability. Simply stated, heavier bodies are harder to move and hence are more stable. Lighter bodies are moved more easily and are less stable.

5. *Friction.* The amount of frictional resistance at the interface between the ground and any contact points (e.g., foot or shoe) contributes to stability and mobility. A young basketball player trying out her new shoes on a freshly polished gymnasium floor would encounter relatively high friction that would improve her stability. A teenager running on an icy sidewalk in the middle of winter, in contrast, would have much lower stability because of the low friction and would be more likely to slip and fall.

In summary, high stability (low mobility) is characterized by a large base of support, a low center of gravity, a centralized center of gravity projection within the base of support, a large body mass, and high friction at the ground interface. Low stability (high mobility), in contrast, occurs with a small base of support, a high center of gravity, a center of gravity projection near the edge of the base of support, a small body mass, and low friction.

Movement Mechanics

Human movement results from mechanical factors that produce and control movement from inside the body (*internal mechanics*) or affect the body from without (*external mechanics*). Examples of internal mechanical factors include the forces produced by muscle action and the joint stability provided by ligaments. External mechanical factors include gravity, air resistance, and other external forces acting on the body (e.g., direct impact).

Movement can be assessed descriptively or by investigation of the underlying forces. The description of spatial and timing characteristics of movement without regard to the forces involved is known as **kinematics**. The assessment of movement with respect to the forces involved is called **kinetics**.

Kinematics

Kinematics involves the description of movement with respect to space and time without regard to the forces that produce or control the movement. Kinematics involves five primary variables:

1. Timing, or temporal, characteristics of movement
2. Position or location
3. Displacement (measuring the movement from starting point to ending point)
4. Velocity (a measure of how fast something moves)
5. Acceleration (an indicator of how quickly the velocity changes)

Kinematics can be considered for movements viewed two dimensionally or three dimensionally. Some movement patterns, such as walking, are essentially planar, so two-dimensional assessment may be sufficient. Other patterns, such as throwing, are multiplanar movements requiring three-dimensional consideration.

Time

The kinematic variable of time provides a measure of the duration of a particular event. Noting that during a single step a person's foot is in contact with the ground for 0.5 second would be an example of such a timing (temporal) measure. The duration, or time, may be long (e.g., a marathon that may take several hours) or very short (e.g., the duration of force application when a soccer player kicks the ball).

Position

The position of a person's whole body, or a segment of the body, at any instant in time plays an important role in determining the mechanical characteristics of the body system. The position of a body segment can be described qualitatively (e.g., leg is abducted) or quantitatively (e.g., forearm is positioned with the elbow flexed 90°).

Displacement

When a body moves from one place to another, we measure this displacement in a straight line from the starting point to the ending position. This is termed the **linear displacement**. A body rotating about an axis experiences **angular displacement**, which is measured as the number of degrees of rotation (e.g., the knee flexed through an angular displacement of 35°). In kinematics, a distinction is drawn between displacement and distance. As already defined, displacement is measured along a straight line from one point to another. **Distance**, in contrast, represents the overall measure of how far the body moves in getting from a starting point to a finishing point. Figure 5.8 shows the difference between displacement and distance.

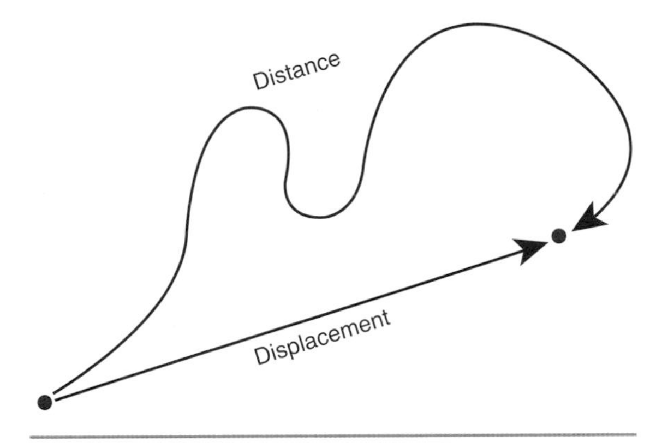

FIGURE 5.8 Displacement versus distance.

Velocity

Velocity is a measure of how fast the body is moving and in what direction. For example, an Olympic sprinter might run with a linear velocity of 10 m/s in a straight line from the starting blocks to the finish line. Velocity can also be used to measure angular motion, as when a softball pitcher swings her arm with an angular velocity of 1,000 degrees per second in a counterclockwise direction. In common usage, the terms *velocity* and *speed* often are used interchangeably. In mechanical terms, however, they have distinct—though related—meanings. Velocity is a vector quantity (magnitude and direction), while speed is a scalar (magnitude only) measure. The speed of a runner might be 5 m/s. To transform the movement measurement to velocity, we must indicate the running direction (e.g., 5 m/s due north).

Acceleration

Acceleration measures the change in a body's velocity. Linear acceleration is measured as the change in linear velocity divided by the change in time. Similarly, angular acceleration is the change in angular velocity divided by the change in time. One of the most common accelerations affecting human movement is the acceleration due to gravity. Gravitational acceleration acts downward, as seen in the simple example of dropping a ball. When the ball is being held, its velocity is zero. Once dropped, the ball accelerates by increasing its downward velocity. As long as gravity is acting, the ball falls faster and faster until it reaches terminal velocity. Linear acceleration often is expressed in units of *g*s, where 1 *g* is the acceleration created by the Earth's gravitational pull (~9.81 m/s^2). Thus, if a boxer's head, struck by his opponent's punch, experiences 5 *g*s, the boxer has been hit by a force strong enough to accelerate the head at five times the acceleration caused by the force of gravity.

Kinetics

Kinematic description is an important first step in analyzing any movement. Kinematic analyses, however, are limited to describing the spatial and timing characteristics of movement without investigating the underlying forces involved. Because forces cause movement, kinetics (the study of forces and their effects) is an area worthy of consideration. We now present important force-related concepts.

Inertia and Moment of Inertia

Bodies at rest have a tendency to resist being moved. Because there are two forms of movement (linear and angular), there logically are two forms of resistance to motion. The resistance to linear motion is termed *inertia*, while the resistance to angular motion is termed *moment of inertia*.

As noted earlier, mass is the quantity of matter. In SI units (système international d'unités), mass is measured in kilograms (kg). Common sense suggests that the greater a body's mass, the more difficult it is to move. This resistance to being moved linearly, termed **inertia**, is the property of matter by which it remains at rest or in motion. To linearly move an object at rest, we must overcome its inertia, or its tendency to stay where it is. To slide a box from a resting position across the floor, you must push or pull it with enough force to overcome its inertia.

In the same way that bodies resist linear motion, they also tend to resist all forms of angular movement, including rotation, bending, and twisting (torsion). The general term used to describe this resistance to a change in angular state of position or motion is **moment of inertia**. A word of caution is warranted here because confusion sometimes arises from using the word *inertia* (which we just defined as resistance to change in a body's state of linear position or motion) in the term *moment of inertia*, used to describe angular resistance.

Three types of moment of inertia correspond with three forms of angular movement. The first, **mass moment of inertia**, describes the resistance to a body's rotation about an axis. A body at rest with a fixed axis (e.g., a pendulum) will resist being moved rotationally, just as a body that is already rotating at a constant angular velocity will tend to maintain that angular velocity and will resist change in its velocity. This concept holds, of course, for human movements. In normal standing, for example, the arms hang straight down. To abduct or flex the limbs from this position, their resistance to being moved angularly (i.e., their mass moment of inertia) must be overcome by muscle action at the shoulder (glenohumeral) joints.

Recall that the magnitude of resistance in the case of linear movement is determined by the mass of the object. In the case of angular movement, the magnitude of the resistance is determined by the mass and the mass' location, or distribution, with respect to the axis of rotation. As the mass is moved farther from the axis, the resistance, or mass moment of inertia, increases. For example, a bat that is held at its end (figure 5.9*a*) will have a greater mass moment of inertia (i.e., greater resistance to rotation) compared to a bat that is held with the hands further up the bat (i.e., choking up on the bat), as shown in figure 5.9*b*.

Moving a limb segment's mass closer to or farther away from a joint axis has a profound effect on overall movement. A sprinter, for example, when swinging his leg through the air in preparation for the foot hitting the ground, will flex his knee as much as possible to bring the combined mass of the lower leg and foot closer to the axis of rotation at the hip joint, thus reducing the resistance to rotation (mass moment of inertia) and allowing the entire leg to swing through its arc as quickly as possible.

The second form of angular resistance, **area moment of inertia**, is the resistance to bending. Because our limb segments are relatively rigid (i.e., they bend very little) during most normal movements, the area moment of inertia is of minor concern. In case of injury, however, when a segment and its enclosed bones are severely bent, the area moment of inertia plays a prominent role. If the bones have insufficient strength to resist the bending, they will fracture.

The third form of resistance, **polar moment of inertia**, describes the resistance to twisting, or torsion. As with bending, torsional loading of segments is usually quite low, and the effects of the polar moment of inertia are minimal. But consider the case of a skier whose ski is violently twisted during a fall. The twisting ski transfers a torsional force (torque) to the lower leg and may result in fracture of the tibia. In this case, the polar moment of inertia of the tibia determines whether injury occurs and how severe the fracture will be.

Subsequent sections on material mechanics discuss the resistance to bending and torsion more fully.

FIGURE 5.9 Effect of grip location by a softball batter on the bat's mass moment of inertia.

Force

Force, a fundamental element in human movement mechanics, is a mechanical action or effect applied to a body that tends to produce acceleration. In simpler terms, force can be described as a push or pull. The standard (SI) unit of force is the newton (N), which is the force required to accelerate a 1 kg mass at 1 m/s^2 in the direction of the force (1 N = 1 kg · m · s^{-2}). One lb of force equals 4.45 N, so someone who weighs 180 lb as measured in British units would weigh 801 N in SI units.

As a prelude to a more general discussion of force, we introduce the concept of an **idealized force vector**. If we consider, for example, the forces acting on the head of the femur while a person assumes a standing posture, an infinite number of individual force vectors could be distributed over the joint surface. Each of these vectors would have its own magnitude and direction. To analyze the effect of all these vectors would be a complex task, requiring sophisticated instrumentation and computer modeling. We can create a single force vector (idealized force vector) that represents the net effect of all the other vectors, essentially idealizing the situation through simplification. What is lost in information describing the distribution of forces is gained by creating a model with a single vector from which calculations and evaluations can be made.

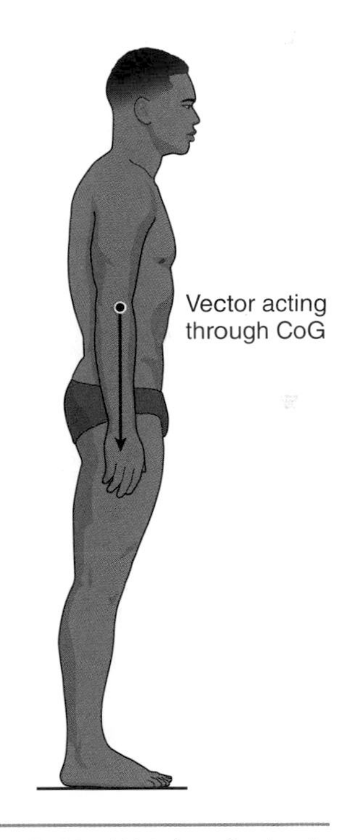

FIGURE 5.10 Weight of a body represented by a single idealized force vector acting at the body's center of gravity.

Reprinted by permission from W.C. Whiting and R.F. Zernicke, *Biomechanics of musculoskeletal injury*, 2nd ed. (Champaign, IL: Human Kinetics, 2008), 48.

This notion of an idealized force vector is useful in many situations. For example, consider the concept of a body's center of gravity. Using an idealized force vector, we can represent the weight of the body by a single vector projecting down from the body's center of gravity (figure 5.10).

The forces inherent to movement analysis are those that act in or on the human body. Among these are the force of gravity (a downward force tending to accelerate objects at ~9.81 m/s^2), the impact of the feet on the ground, objects colliding with the body (e.g., a thrown ball or another body), musculotendinous forces, frictional forces with the ground, ligament forces acting to stabilize joints, and compressive forces exerted on long bones of the lower extremities.

The result of forces on movement depends on the combined effect of seven force-related factors (Whiting & Zernicke, 2008):

1. Magnitude (How much force is applied?)
2. Location (Where on the body or structure is the force applied?)
3. Direction (Where is the force directed?)
4. Duration (Over what time interval is the force applied?)
5. Frequency (How often is the force applied?)
6. Variability (Is the magnitude of the force constant or variable over the application interval?)
7. Rate (How quickly is the force applied?)

Applying the Concept

Forces and Joint Prostheses

A **prosthesis** is an artificial device that replaces or supplements a missing or impaired part of the body. Prosthetic devices are precisely designed and constructed to accommodate, as closely as possible, the forces seen in the movements of a healthy limb. A person with an amputated lower leg, for example, might wear a specially designed artificial limb to replace the missing shank and foot. The prosthesis must be able to absorb and transmit forces exerted by the ground during the stance phase of gait in a way that allows the person to walk or run effectively.

In severe cases of degenerative osteoarthritis (OA), the joint may need to be replaced with a prosthetic, or artificial, joint. The process of joint replacement is termed **arthroplasty**. OA most commonly affects load-bearing joints of the lower extremity. Total hip replacement (THR), also called total hip arthroplasty (THA), and total knee replacement (TKR), also termed total knee arthroplasty (TKA), are the most common joint replacement procedures.

In all joint replacements, the prosthetic device mimics a healthy joint. At the hip, for example, the prosthetic device's acetabular component and femoral head form an artificial joint (see figure 5.11) that closely approximates the function of a healthy hip and accommodates the joint forces during load-bearing activities such as walking, running, and jumping.

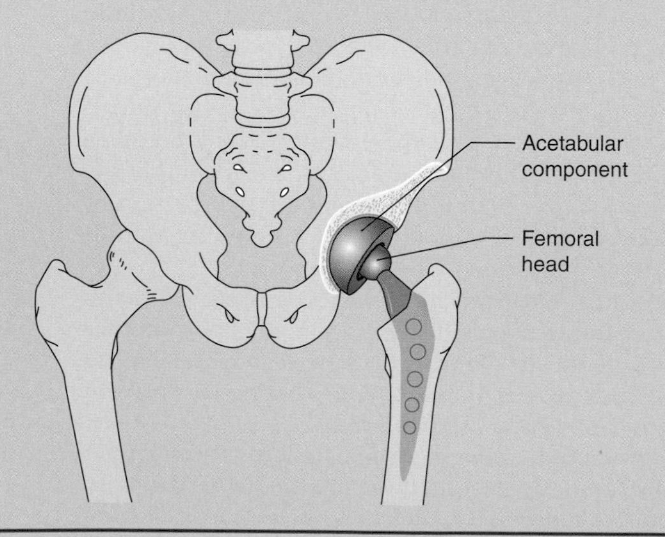

FIGURE 5.11 Total hip replacement.

Pressure

In many situations, it is important to know how the force is distributed over the surface being contacted. In walking, for example, the area of foot contact (e.g., heel versus mid-foot) affects the force distribution and the potential for injury. A general principle of injury mechanics suggests that as the area of force application increases, the likelihood of injury decreases. A sharp object contacting the skin with a certain amount of force, for instance, will have a different effect than a blunt object contacting the skin with the same force. The former condition might result in a puncture wound, while the latter might leave a bruise.

The measure of force and its distribution is pressure (*p*). Pressure is the total applied force (*F*) divided by the area (*A*) over which the force is applied ($p = F/A$). The standard unit of pressure, the pascal (Pa), is equal to a 1 N force applied to an area 1 m square ($1 \text{ Pa} = 1 \text{ N/m}^2$). In the British system, pressure is measured in pounds per square inch (psi).

Lever Systems

Most motion at the major joints results from the body's structures acting as a system of mechanical levers. A **lever** is a rigid structure, fixed at a single point, to which two forces are applied at two different points. One of the forces is commonly referred to as the *resistance force* (*R*), with the other termed the *applied*, or effort, *force* (*F*). The fixed point, known as the axis (also fulcrum or pivot), is the point (or line) about which the lever rotates. In the human body, these three components typically involve an externally applied resistance force (*R*) such as gravity, a muscle effort force (*F*), and a joint axis of rotation (*A*).

These three lever-system components may be spatially related to one another in three different configurations that give rise to three lever classes. Distinctions among the classes are determined by the location of each component relative to the other two (figure 5.12). In a **first-class lever**, the axis (*A*) is located between the resistance (*R*) and the effort force (*F*). A **second-class lever** has *R* located between *F* and *A*, while a **third-class lever** has *F* between *R* and *A*. Joints in the human body are predominantly third-class levers, with some first-class levers and few second-class levers.

Lever systems in the human body serve two important functions. First, they increase the effect of an applied force because the applied force and the resistance force usually act at different distances from the axis, as seen in the leverage advantage gained by using a long bar to pry a large rock loose from the ground. In such a first-class lever, increasing the distance from the axis to the effort force increases the effective force seen on the other side of the pivot point (i.e., it is easier to move the rock).

The second function of levers is to increase the effective speed (or velocity) of movement. During knee extension (figure 5.13), for example, a given angular displacement produces different linear displacements for points along the lower leg. Points farther away from the knee joint axis move a greater distance along the curved arc than do points closer to the axis. Because all points along the lower leg move with the same *angular* velocity, the more distal points have higher *linear* velocities. This is easily seen in baseball and volleyball. Baseball pitchers with longer arms have the potential to throw the ball faster than those with shorter arms. In volleyball, the hand speed while spiking the ball may be higher for players with longer arms.

The human body makes effective use of both the force and speed advantages provided by lever systems in accomplishing the many tasks it performs on a daily basis.

Torque (Moment of Force)

In the case of linear motion, force is the mechanical agent creating and controlling movement. For angular motion, the agent is known as **torque** (*T*), or moment of force, or **moment** (*M*). Torque is the effect of a force that tends to cause rotation or twisting about an axis (figure 5.14*a*). The mathematical definitions of *torque* and *moment* are the same; however, there is a technical difference between the two. Torque typically refers to the twisting movement created by a force (figure 5.14*b*), whereas moment is related to the bending or rotational action of a force (figure 5.14*c*). Despite this difference, the two terms often are used interchangeably.

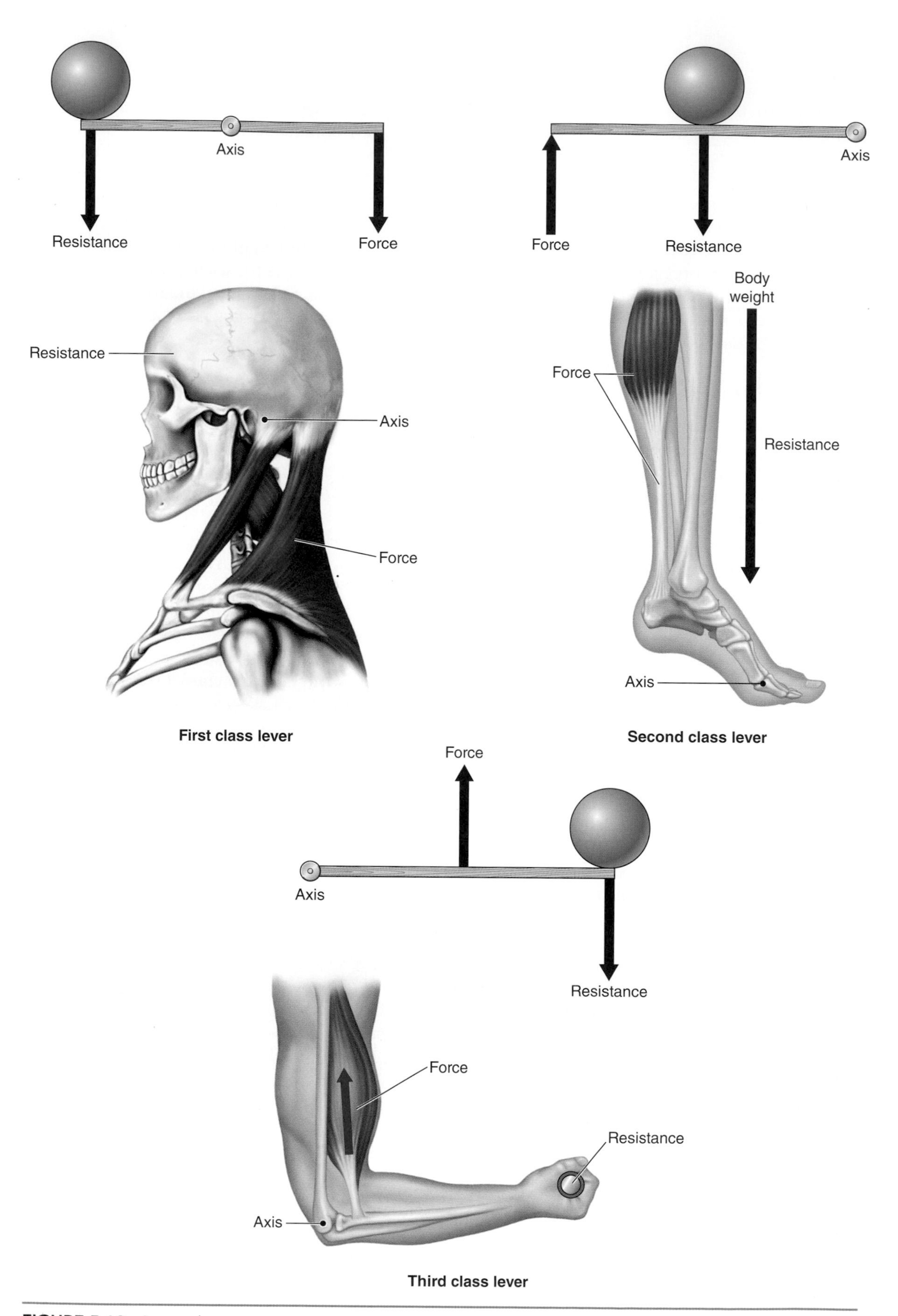

FIGURE 5.12 Lever classes.

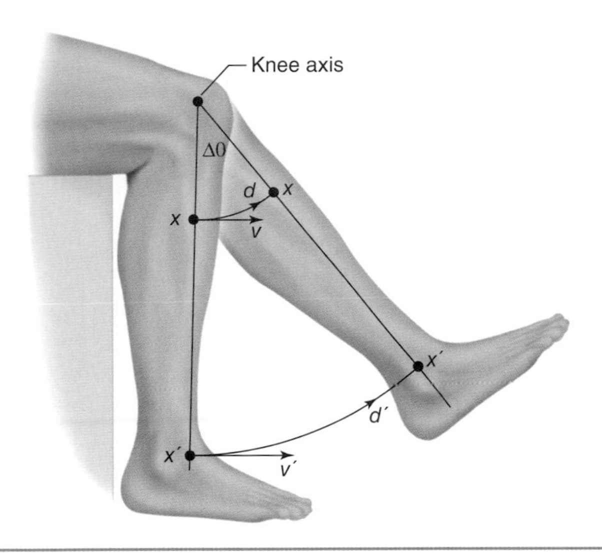

FIGURE 5.13 Effect of levers in increasing the speed of distal points.

Reprinted by permission from W.C. Whiting and R.F. Zernicke, *Biomechanics of musculoskeletal injury,* 2nd ed. (Champaign, IL: Human Kinetics, 2008), 78.

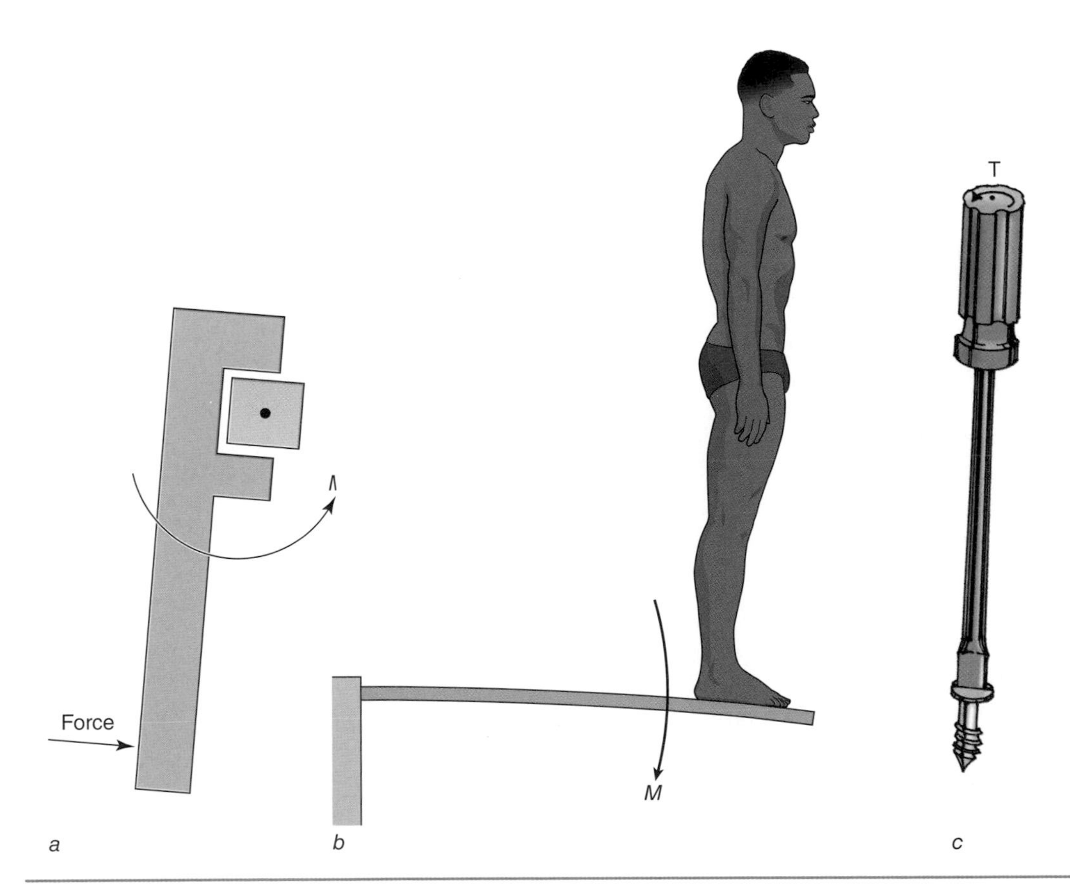

FIGURE 5.14 Applied examples of moment of force (*M*), or torque (*T*). *(a)* Force applied to a wrench creates a moment to turn a nut on a bolt. *(b)* The body weight of a diver creates a moment that bends the diving board. *(c)* Torque shown by the twisting action of a screwdriver.

Reprinted by permission from W.C. Whiting and R.F. Zernicke, *Biomechanics of musculoskeletal injury,* 2nd ed. (Champaign, IL: Human Kinetics, 2008), 61.

The magnitude of a torque is equal to the applied force (*F*) multiplied by the perpendicular distance (*d*) from the axis of rotation to the **line of force action**. This perpendicular distance is known as the **moment arm**, **torque arm**, or **lever arm** (figure 5.15).

Biomechanical examples of torque include the biceps creating a flexion moment about the elbow, the moment created by a weight on the lower leg during a knee extension exercise, and the torque applied to the tibia in a skiing fall (figure 5.16). The standard unit of torque or moment comes from the product of the two terms: force in newtons (N) multiplied by moment arm in meters (m). The resulting unit is a newton-meter (Nm), or foot-pound (ft-lb) in the British system.

Closer examination of the moment equation (moment = force × moment arm) reveals several important general principles when applying moment concepts to movement biomechanics. First, there is an obvious interaction between the force and the moment arm that directly affects the magnitude of the applied moment. To increase the moment, we have the option of increasing the force, increasing the moment arm, or both. To decrease the moment, we can decrease either the force or moment arm, or both.

A second moment-related concept, while simple in statement, is powerful in its application. That is, when a force is applied *through* the axis of rotation, or parallel to the axis, no moment (torque) is produced. This concept follows directly from the moment equation. If the force passes through the axis, the moment arm is zero. Hence, no moment is produced.

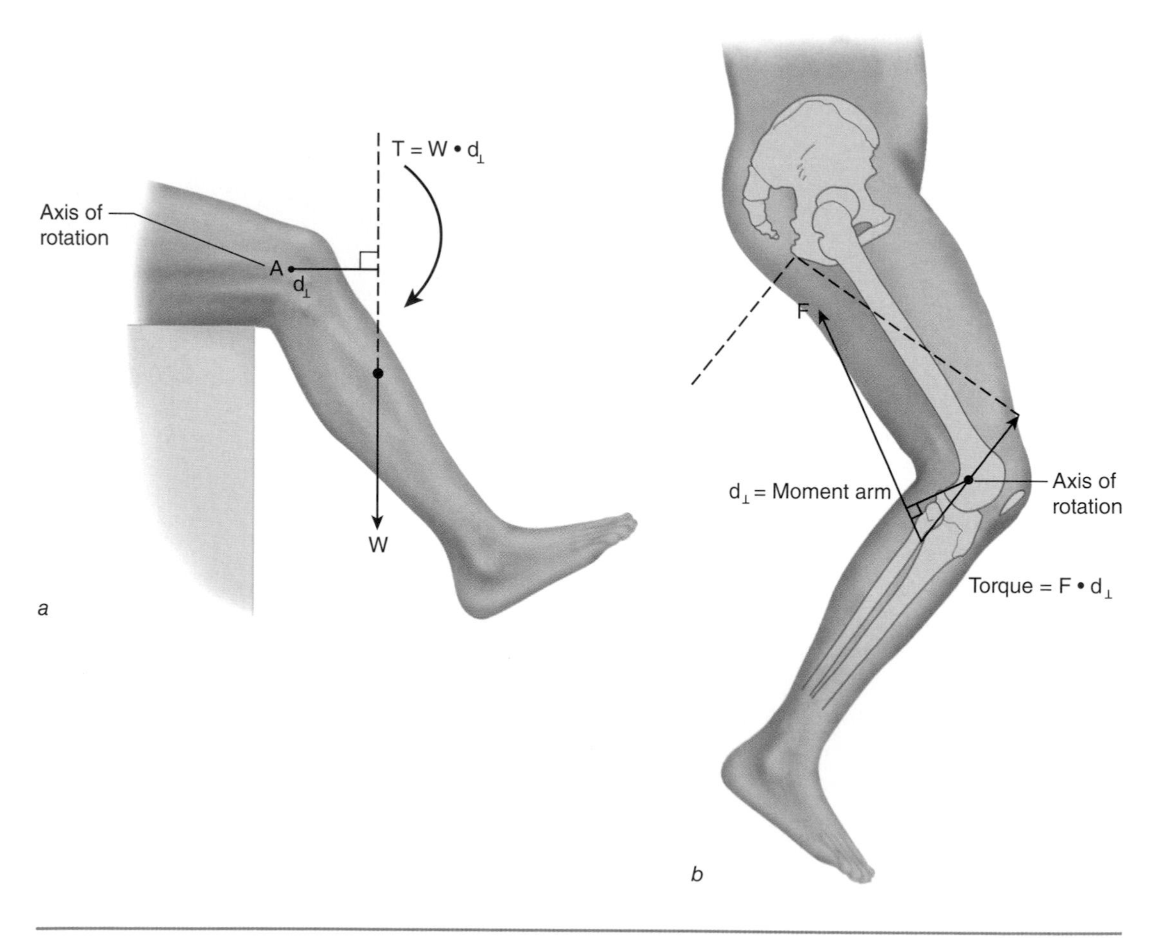

FIGURE 5.15 *(a)* Torque calculation at the knee when the line of force action is perpendicular. *(b)* Torque calculation when the force is not perpendicular involves using trigonometric functions.

Reprinted by permission from W.C. Whiting and R.F. Zernicke, *Biomechanics of musculoskeletal injury,* 2nd ed. (Champaign, IL: Human Kinetics, 2008), 62.

This can be seen by pushing on the hinges of a door. The hinges serve as the door's axis, and forces applied to the hinges will not create any moment; the door will not move.

At joints in the human body, this principle creates a situation where tissues can be exposed to extremely high forces, but no moment is created. Compressive forces acting through the center of a vertebral body, for example, will cause no vertebral rotation but may increase the risk of a compressive vertebral fracture.

A third moment concept emerges from the fact that in most situations, more than one moment is being applied. The system's response is based on the **net moment** (also **net torque**) or the result of adding together all the moments acting about the given axis. An example of this concept is seen in a simple glenohumeral abduction exercise (figure 5.17). Gravity, acting on both the arm and dumbbell, creates a moment about the glenohumeral axis of rotation that tends to adduct the arm.

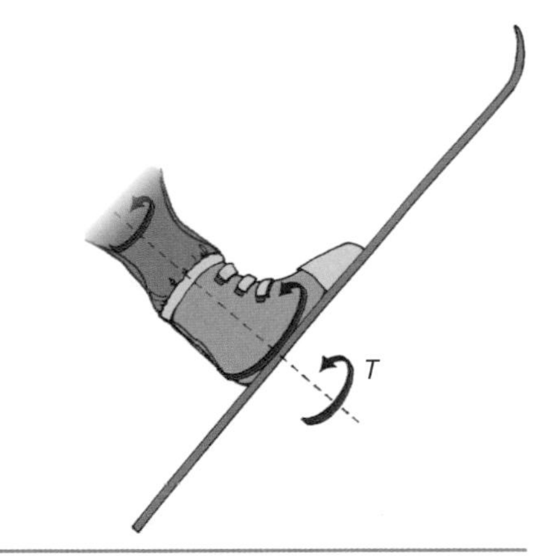

FIGURE 5.16 Torsional load being transferred from the ski and applied to the skier's lower leg.

Reprinted by permission from W.C. Whiting and R.F. Zernicke, *Biomechanics of musculoskeletal injury* (Champaign, IL: Human Kinetics, 1998), 50.

If this were the only moment, the arm would immediately adduct under the effect of gravity. To hold the arm in an abducted position, the abductor muscles acting about the glenohumeral joint must create a moment that acts in the opposite direction to counterbalance the moment created by gravity. This counterbalancing moment is termed a *countermoment*, or **countertorque**. The countermoment in this case will tend to abduct the arm.

The movement that results depends on the relative magnitudes of these two moments. Adding together all of the moments acting about a joint creates a net moment. If the two moments are equal in magnitude (but opposite in direction), the net moment is zero and no movement occurs. If the moment created by gravity exceeds that created by the abductors, the net moment favors gravity and the arm will adduct. On the other hand, if the moment created by the abductors is greater than that created by gravity, the net moment favors the muscle action and the arm will abduct.

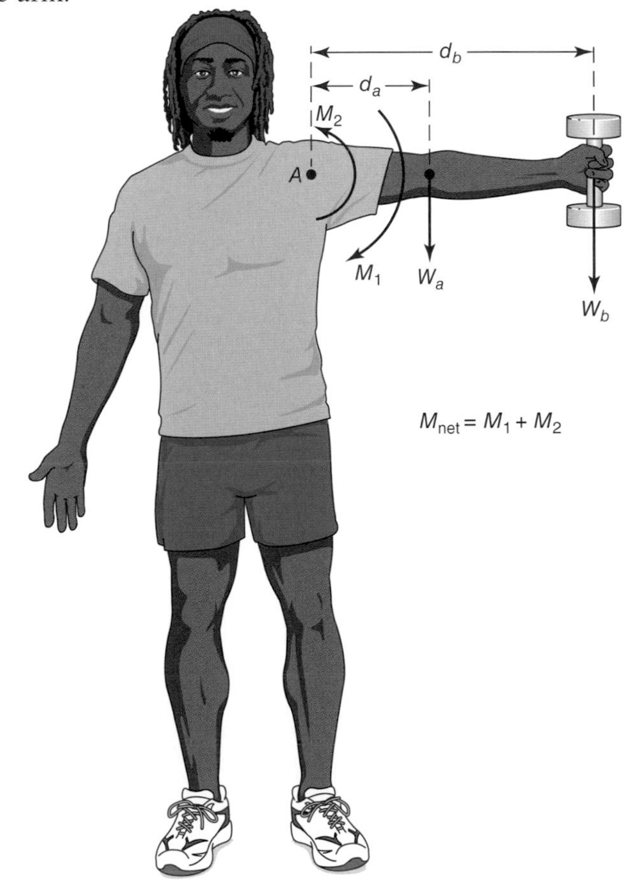

FIGURE 5.17 Net moment, calculated as the sum of all component moments acting about a joint axis.

The importance of these moment- or torque-related concepts will become evident in forthcoming chapters as we explore the details of the production and control of human movements.

Research in Mechanics

Kinematics and Kinetics in Vertical Jumping

Numerous research studies have examined mechanical aspects of vertical jumping, a fundamental movement skill common to many sports, including basketball and volleyball. As an example, Chiu and colleagues (2014) analyzed 16 female volleyball players performing vertical jumps with and without arm swing. The researchers used force platforms (to measure ground reaction forces) and a motion analysis system (to measure body segment kinematics). Among their findings were that successful vertical jump performance requires certain motor skills, specifically the use of a proximal-to-distal sequencing strategy. This strategy enhanced arm swing during the jump. In addition, they found that use of a proximal-to-distal strategy allows for the generation of larger hip extensor, knee extensor, and ankle plantar flexor net joint moments, which results in large angular accelerations of lower-extremity joints and maximized acceleration of the pelvis.

See the references for the full citation:

Chiu, Bryanton, & Moolyk, 2014.

Newton's Laws of Motion

Of Sir Isaac Newton's (1642-1727) many notable scientific contributions, the laws of motion are among his most important. Newton's three laws of motion form the basis for classical mechanics and provide the rules that govern the physics of how we move.

- First law of motion (law of inertia): A body at rest, or in motion, will tend to remain at rest, or in motion, unless acted upon by an external force.
- Second law of motion (law of acceleration): A force acting on a body will produce an acceleration proportional to the force, or mathematically, $F = m \times a$ (i.e., force = mass times acceleration).
- Third law of motion (law of action and reaction): For every action, there is an equal and opposite reaction.

Newton's three laws of motion apply to all human movements. The essence of Newton's first law is that forces are required to start, stop, or modify the movement of a body. In the absence of forces (e.g., gravity, friction), a body will persist in its state of rest or motion. For example, a skater slides across the ice until the friction between the skates and the ice eventually brings her to a stop. A dancer leaping through the air begins his flight by exerting force against the ground. Once in the air, his upward speed is slowed by the effect of gravitational forces until he reaches the peak of his jump; he then comes back to the ground under the continuing influence of gravity.

Newton's first law also is very much in evidence in the movement of astronauts in space. In the absence of gravity, the astronauts appear to float around the space vehicle's cabin. They actually are just continuing the motion created by the forces used to push or pull themselves relative to the cabin walls.

Common lifting tasks provide a good example of Newton's second law of motion. In lifting a box from the floor, the lifter must exert enough force to overcome the force of gravity and accelerate the box upward. The second law of motion determines the magnitude of the acceleration in response to the applied force ($F = m \times a$). For a box of a given mass (m), a greater force (F) will create a proportionally greater acceleration (a).

In a similar fashion, application of a moment, or torque, changes a body's angular state of motion. To accelerate the arm when throwing a ball, the muscles of the shoulder and arm must generate moments about the respective joint axes to make the arm move faster.

Newton's third law of motion states that every force action creates an equal and opposite reaction. This can be seen in a long-distance runner whose feet contact the ground many thousands of times. At each foot contact, the force that the foot exerts on the ground is equally and oppositely resisted by the ground, giving rise to the term **ground reaction force** to describe the force of the ground acting on the foot. Increasing the magnitude and frequency of ground reaction forces increases the chance of injury.

Equilibrium

The term **equilibrium** suggests a balanced situation. From a mechanical perspective, equilibrium exists when forces and moments (torques) are balanced. In general, equilibrium exists for a body at rest or for one moving with constant linear and angular velocities. The net forces and net moments acting on the body equal zero. In this case, the body is said to be in **static equilibrium**. Bodies in motion experiencing external forces and moments that cause acceleration are in **dynamic equilibrium**. The concepts of static and dynamic equilibrium are important in our discussion of posture and balance in chapter 7.

Work and Power

The term *work* is used in many ways, in various contexts referring to physical labor ("I'm working hard."), physiological energy expenditure ("I worked off 200 calories."), or a place of employment ("I'm going to work."). In mechanical terms, however, work has a very specific meaning. Mechanical **work** is performed by a force acting through a displacement in the direction of the force. By definition, linear work (W) is equal to the product of force (F) and the displacement (d) through which the body is moved (figure 5.18*a*). The standard (SI) unit of work is the joule (1 J = 1 Nm). If the entire force does not act in the direction of motion (figure 5.18*b*), only the component of force acting in the direction of movement is used in calculating the work done.

In the example depicted in figure 5.19, the work performed in lifting the barbell from point A to point B is equal to the product of the barbell's weight (W_b) and the displacement from A to B (d_{AB}). If, for example, the barbell weighs 800 N (~180 lb) and is lifted 0.5 m (~20 in.), the work done would be 400 J (800 N × 0.5 m).

The calculation of work often does not completely describe the mechanics of a body's movement or task performance. Take the case, for example, of two material handlers whose job requires them to lift boxes from the ground onto a moving conveyor belt. If the boxes each weigh 200 N (~45 lb) and the conveyor belt is 1.5 m above the ground, then 300 J of work is performed in lifting each box. However, the first lifter can lift each box and place it on the conveyor belt in 1.2 seconds, while the second lifter takes 1.4 seconds to complete the task. Because the boxes weigh the same and are moved through the same distance, each lifter performs the same amount of mechanical work. One lifter can complete the task more quickly, however, so there is some mechanical difference in how they lift. The difference is in the rate at which the work is performed. The rate of work, termed **power**, is calculated

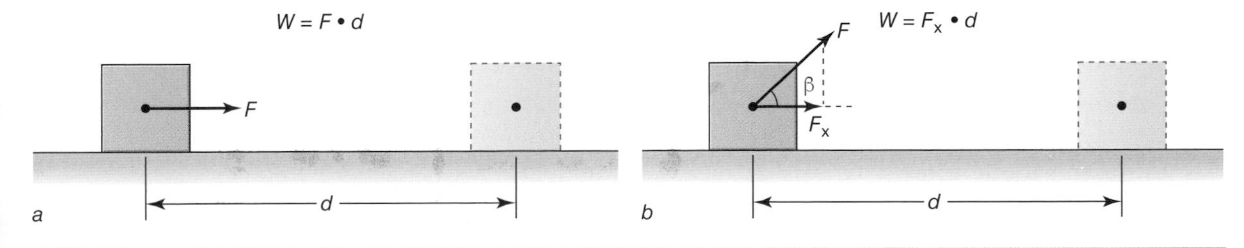

FIGURE 5.18 Mechanical work: *(a)* linear work when the force is in the direction of displacement; *(b)* linear work when only a portion of the force acts in the direction of displacement.

Reprinted by permission from W.C. Whiting and R.F. Zernicke, *Biomechanics of musculoskeletal injury*, 2nd ed. (Champaign, IL: Human Kinetics, 2008), 67.

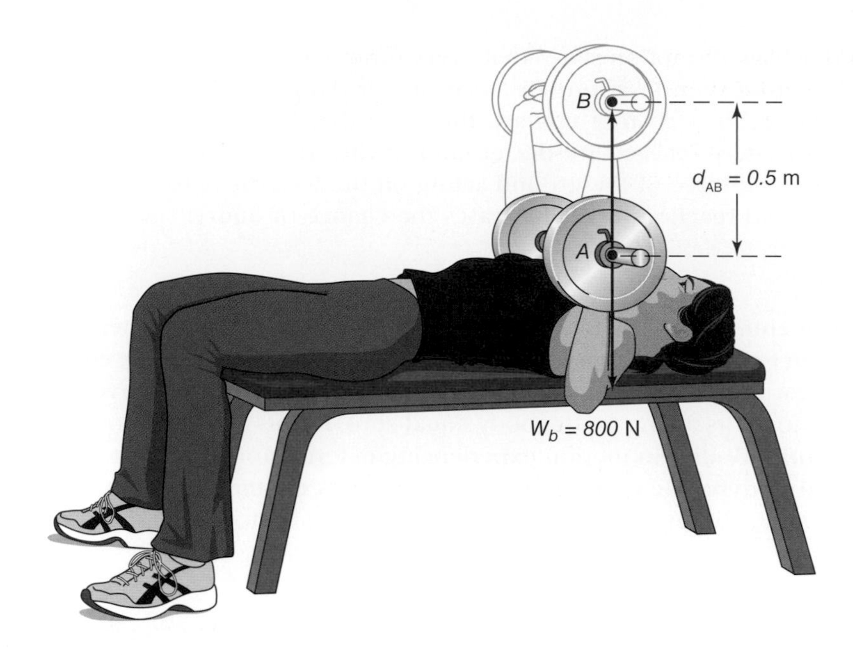

FIGURE 5.19 Bench press illustrating mechanical work performed.

Reprinted by permission from W.C. Whiting and R.F. Zernicke, *Biomechanics of musculoskeletal injury,* 2nd ed. (Champaign, IL: Human Kinetics, 2008), 67.

as the amount of work done divided by the time it takes to do the work (power = work ÷ time). Power is expressed in units of watts (1 W = 1 J/s).

In the previous example, both workers perform 300 J of work. The worker who lifts a box onto the conveyor belt in 1.2 seconds lifts with a power of 250 W (300 J ÷ 1.2 s), while the worker taking 1.4 seconds lifts with a power of about 214 W (300 J ÷ 1.4 s). In general, a given amount of work performed in a shorter time yields a greater power output.

Power may also be expressed as the product of force and velocity. Successful performance of many movement tasks (e.g., jumping, throwing) requires high power output. To be powerful, the performer must generate high forces and do so quickly (i.e., high velocity). A shot-putter, for example, must be both strong (high force) and fast (high velocity) to be successful. Other examples of tasks requiring high power output are discussed in subsequent chapters.

Energy

Energy is another term with a variety of meanings. One could have a high-energy personality or, if tired, run out of energy. As with work, though, energy has a more specific meaning. Energy can assume many forms, including thermal, chemical, nuclear, electromagnetic, and mechanical. Mechanical energy is the type most commonly used to describe or assess human movement. Mechanical **energy** is the capacity, or ability, to perform mechanical work.

- The mechanical energy of a body can be classified according to its **kinetic energy** (energy of motion) or its **potential energy** (energy of position or deformation).
- **Linear kinetic energy** (E_k) for a given body is defined as $E_k = 1/2 \times m \times v^2$, where m equals mass and v equals linear velocity of the center of mass.
- **Angular kinetic energy** ($E_{/k}$) is defined as $E_{/k} = 1/2 \times I \times \omega^2$, where I equals mass moment of inertia and ω equals angular velocity. Energy is measured in joules (J), the same units used to measure work.

In terms of human movement, one of the most important concepts to emerge from the kinetic energy equations is that the velocity term (v or ω) is squared. As movement speed increases, energy multiplies as a function of the velocity squared. For example, consider a downhill skier with a body mass of 60 kg who increases her downhill velocity from 20 m/s to 25 m/s. Her velocity has increased by 25%, but her linear kinetic energy has increased by more than 56%. What might at first appear to be a relatively small increase in velocity considerably enhances her kinetic energy. And if the skier unexpectedly (and unfortunately) collides with another skier or a tree, the higher energy can result in serious injury.

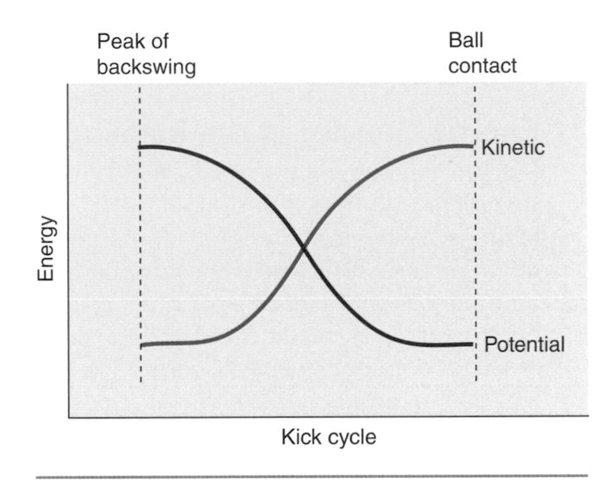

FIGURE 5.20 Components of energy as seen during a kicking motion.

Potential energy can take two forms. The gravitational form (potential energy of position) measures the potential to perform work as a function of the height a body is elevated above some reference level, most typically the ground. The equation describing **gravitational potential energy** is $E_p = m \times g \times h$, where m equals mass, g equals gravitational acceleration (~9.81 m/s^2), and h equals height (in meters) above the reference level.

The second form of potential energy is **deformational** (also **strain**) **energy** that is stored in a body when it is deformed. The deformation can take the form of a body being stretched, compressed, bent, or twisted. Common examples of strain energy include a stretched rubber band or tendon, a pole-vaulter's bent pole, a drawn bow before arrow release, and a compressed intervertebral disc. When the force causing the deformation is removed, some of the stored strain energy is returned to use, while the rest is lost as heat energy.

Scientists studying whole-body or limb segment movement dynamics often assume that each body segment is rigid (i.e., nondeformable). When this simplifying assumption is made, there is no strain energy component in the system. In these cases, the **total mechanical energy** is simply the sum of the linear kinetic, angular kinetic, and positional potential energies. (Note that total mechanical energy also includes a thermal energy term, but because it usually is negligible compared with the other terms, we omit it here.)

Thus, total mechanical energy equals linear kinetic energy plus angular kinetic energy plus positional potential energy. Consider, for example, a soccer player swinging her leg to kick the ball. Each of the lower-limb segments (thigh, shank, foot) has a continuously changing total mechanical energy. Let's focus on the shank (lower leg) to illustrate how to measure the total mechanical energy. As the values in figure 5.20 show, at the peak of the leg's backswing, the kinetic energy is essentially zero because the shank is almost motionless and the positional potential energy is at its highest. As the leg swings forward toward the ball, the potential energy decreases (i.e., the shank is lowered) and the kinetic energy increases as the leg swings faster. At contact, much of the leg's energy is transferred to the ball, and the ball accelerates toward its target. More details of kicking dynamics are presented in chapter 9.

Momentum

In an athletic contest, when one team is playing well and seems to be getting all of the breaks, we say that the players have momentum on their side. From a mechanical perspective, however, **momentum** is defined differently and can be characterized as a measure of a body's quantity of motion. In general, the larger the body and the faster it is going, the higher its momentum. In mechanical terms, **linear momentum** is the product of mass (m) and velocity (v). Increasing either a body's size (mass) or speed (velocity) will increase its linear momentum.

Similarly, **angular momentum**, or the quantity of angular motion, is the product of mass moment of inertia (I) and angular velocity (ω).

Research in Mechanics

Energy Transfer in the Running Gait of People with Amputations

Studies have shown that the walking and running mechanics of people with amputation differ from those of people without amputation. A study by Czerniecki and colleagues (1996), for example, compared energy transfer mechanisms in runners with amputations with those of nonamputee controls. Five participants with below-knee amputations and wearing a SACH (solid-ankle, cushioned-heel) prosthesis and five control subjects ran at a controlled speed of 2.8 m/s while kinematic and ground reaction force data were collected. The researchers measured energy flow across the hip joint in both groups, and concluded that

> Energy transfer out of the intact swing phase limb combined with the temporal characteristics of this energy flow suggests that energy transfer may be an adaptive mechanism that allows energy redistribution to the trunk which may partially compensate for the reduced power output of the stance phase prosthetic limb. (p. 717)

In essence, the healthy limb helps mechanically compensate for deficits in the limb with amputation.

See the references for the full citation:

Czerniecki, Gitter, & Beck, 1996.

Principles of Conservation and Transfer

Two principles govern the effect of both energy and momentum on human movement: conservation and transfer. First, let's look at how these principles apply to momentum. **Conservation of momentum** indicates how much of a system's combined linear and angular momentum, or quantity of motion, is conserved and how much is gained or lost during a given time period. This principle is a consequence of Newton's first law of motion, which implies that a body's momentum is conserved (i.e., remains the same) unless the body changes its mass or is acted on by an external force.

A person running along the beach, for example, has some amount of momentum based on his body size (mass) and how fast he is running (velocity). If he stops, he loses all of his momentum because his velocity drops to zero. In contrast, when a sprinter is in the starting blocks awaiting the starter's gun, her momentum is zero because she is motionless. When the gun sounds, she pushes against the starting blocks and accelerates. As her velocity increases, so does her momentum. In neither of these cases is momentum conserved. In the first case, momentum is lost; in the second, momentum is gained.

The companion principle, **transfer of momentum**, is the mechanism by which momentum from one body is transferred to another. This can take many forms in human movement. Transfer during a throwing motion can occur as momentum moves from a proximal segment (e.g., upper arm) to a more distal segment (e.g., forearm, hand) as the throw progresses.

Momentum can also be transferred between different bodies, as in the case of an automobile crash or an American football player blocking or tackling an opponent. Transfer of momentum in these cases often results in injury when the quantity of motion transferred exceeds the tolerance of the tissues in one or both of the bodies.

The principles of conservation and transfer also can be applied to energy. Conservation of energy dictates that a body's total energy is conserved (i.e., remains constant) unless the body changes its mass or is acted on by an outside force. Transfer of energy, in a manner similar to transfer of momentum, explains how energy is passed from one body to another.

Friction

Newton's first law of motion tells us that bodies in motion tend to remain in motion unless acted on by an outside force. That force may be an abrupt one, such as a collision, or of lower magnitude and greater duration, such as the force of friction. **Friction** is the resistance created at the interface between two bodies in contact with one another and acting in a direction opposite to impending or actual movement. Frictional resistance results from microscopic irregularities, known as asperities, on the opposing surfaces. Asperities tend to adhere to each other, and efforts to move the bodies result in very small resistive forces that oppose the motion.

In the simple case of a body at rest on a surface, **static friction** resists movement until a force sufficient for overcoming the frictional resistance is applied. As the force applied to a body at rest increases, a level is reached at which the static resistance is overcome, and the body begins to slide along the surface. Once the body begins moving, the friction decreases slightly and then is known as **kinetic friction**.

The magnitude of the frictional force (F_f) depends on two factors. The first is the coefficient of friction (μ), a unitless number, typically between 0 and 1, that serves as an index of the material properties of the two surfaces in contact with each other. Values of μ near 0 indicate very little friction (e.g., an ice skater sliding across a frozen pond), while values near 1 are characteristic of high friction (e.g., a basketball player wearing new shoes making a jump stop on a clean gym floor). The second factor is termed the *normal force (N)*, or *reaction force (R)*. In general, the normal force is the component of the total force acting

Applying the Concept

Artificial Versus Natural Turf

The type and quality of playing surface in many sports determine, at least in part, performance level and risk of injury. Since the invention of artificial turf in 1965, there has been controversy, both anecdotally and in the research literature, about whether artificial turf (AT) or natural turf (NT) is better in terms of player performance and injury risk. The first generation of AT, first named ChemTurf but quickly renamed Astroturf, proved useful but also had its problems. Second-generation AT in the 1980s was much improved, and current third-generation AT design is even better. Nonetheless, controversy has continued.

In terms of injury, one of the most important variables is surface friction. Higher friction on AT surfaces, as compared to NT fields, may improve performance (e.g., due to greater traction), but potentially may increase the risk of injury due to more rapid deceleration. The research is mixed at best. One study (Dragoo et al., 2012) concluded that NCAA football players between 2004 and 2009 "experienced a greater number of ACL injuries . . . when playing on artificial surfaces" (p. 990) but highlighted the need for confirmation of this finding by additional studies. A recent systematic review by Balazs and others (2015) on the risk of anterior cruciate ligament (ACL) injury in athletes playing on synthetic surfaces concluded that "high quality studies support an increased rate of ACL injury on synthetic playing surfaces in football, but there is no apparent increased risk in soccer. Further study is needed to clarify the reason for this apparent discrepancy" (p. 1798). And finally, a meta-analysis by Williams and colleagues (2013) concluded that "it appears that the risk of sustaining an injury on AT under some conditions might be lowered compared to NG [natural grass]" (p. 1). This study added a qualifying statement: "However, until more is known about how issues . . . [may] affect injury incidence, it is difficult to make firm conclusions regarding the influence of AT on player safety" (p. 1). In summary, stay tuned.

Applying the Concept

Friction in Synovial Joints

Synovial joints (e.g., hip, knee, ankle, elbow) are composed of an outer fibrous joint capsule whose inner lining, the synovial membrane, produces synovial fluid. This fluid serves as a lubricant to minimize friction between the two joint surfaces and facilitate joint movement. How well does synovial fluid work in minimizing friction? Remarkably well! Studies have estimated the μ_s in synovial joints at about 0.01 and μ_k as low as 0.003. By comparison, ice sliding across ice has a μ of between 0.02 and 0.09.

perpendicular to the surface. Thus, on level ground, the normal force is simply the weight of the object. On an inclined surface, however, the perpendicular component is less than the weight of the object. The steeper the incline, the lower the normal force.

In equation form, the frictional force, $F_f = \mu \times N$, where μ is the coefficient of friction and *N* is the normal force.

As noted previously, kinetic friction is less than maximal static friction. As a consequence, the coefficient of static friction (μ_s) is higher than the coefficient of kinetic friction (μ_k). In simple terms, $\mu_s > \mu_k$.

If the tires of a car are prevented from rotating (e.g., when one slams on the brakes), the vehicle slides along the road. In this case, kinetic friction (as just described) is in effect. Most of the time, however, the wheels are free to rotate and the vehicle rolls forward. Even in rolling movement, friction is present. Rolling resistance is not as obvious as sliding resistance because rolling friction is much lower, often by a factor of 100 to 1,000 times. The actual value of resistance in both sliding and rolling friction depends on the material properties of the body and the surface and on forces acting between them.

In some cases, friction works to our advantage. In fact, we would be unable to walk or run without friction acting between our shoes and the ground. Too little friction, such as when someone walks on an icy surface, may lead to a slip and fall. Too much friction, on the other hand, may contribute to injuries of another type. High levels of friction lead to abrupt deceleration, which causes high forces and extreme loading of body tissues.

Our examples thus far have focused on friction acting externally on the body. Friction also plays an important role within the human body. During normal limb movements, for example, the friction in synovial joints is extremely low, allowing for freedom of movement with minimal resistance (see sidebar).

Fluid Mechanics

Fluid mechanics, the branch of mechanics dealing with the properties and behavior of gases and liquids, plays an important role in our consideration of human movement. Areas as diverse as performance biomechanics (study of movement mechanics), tissue biomechanics (study of the mechanical response of tissues), and hemodynamics (study of blood circulation) all rely on the basic principles of fluid mechanics.

We move in various fluid environments, where air is the principal gas and water is the predominant liquid (e.g., swimming). We consider two important mechanical properties: flow and resistance. Fluid flow refers to the characteristics of a fluid, whether liquid or gas, that allow it to move and govern the nature of this movement. Blood flow through an artery exemplifies fluid flow. Fluids also provide resistance, such as the resistance we might experience while running against the wind or swimming in a pool. Understanding fluid flow and resistance is essential to our understanding of human movement.

Fluid Flow

Fluid flow can exhibit many movement patterns. **Laminar flow** is characterized by a smooth, essentially parallel pattern of movement, as seen in the waters of a slow-moving river. **Turbulent flow**, in contrast, exhibits a more chaotic flow pattern, characterized by areas of turbulence and multidirectional fluid movement. Arterial blood flow provides a good example of these different types of flow. In the middle of the artery, blood flow may be laminar. Blood in contact with the arterial wall or at a branching in the artery may exhibit turbulence. Factors contributing to turbulent flow include the roughness of the surface over which the fluid passes, the diameter of the vessel, obstructions, and speed of flow.

Fluid Resistance

Fluid resistance takes many forms, some of which are advantageous to human movement and others that may prove detrimental. Examples of positive effects of fluid resistance include **buoyant forces**, which allow a person to float in water; **aerodynamic forces**, which keep an object in flight; and **magnus forces**, which affect the trajectories of objects spinning through the air. Negative effects of fluid resistance are evident in the extra physiological effort required by a cyclist moving into the wind or by the severe and unpredictable forces acting on a hang glider during a storm.

Resistance to flow is termed **viscosity**. Sometimes described as "fluid friction," viscosity enables a fluid to develop and maintain a resistance to flow, dependent on the flow's velocity. The effect of viscosity and its dependence on velocity can be seen in a familiar example. When you move your hand slowly through water, the resistance is minimal. Increasing the speed of movement markedly increases the resistance. The increase in resistance is due to the liquid's viscosity-related, or viscous, properties.

Fluid mechanical effects play an integral role in human movement tasks such as swimming. Specific application of fluid mechanics principles to swimming is presented in chapter 11.

Joint Mechanics

Human movement relies on the action of hundreds of articulations (joints) in the human body. As discussed in earlier chapters, the amount of movement depends on each joint's structure. No two joints are the same in structural terms; each has its own distinct combination of tissues, tissue configuration, and movement potential. This variety of joint structure and function allows for complex movement patterns.

Each joint in the body has an operational range of motion (ROM) that determines the joint's mobility. The allowable ROM is both joint specific and person specific. Joints with an ability to move in more than one plane have ROMs specific to each particular plane of movement. Note that ROMs vary considerably from one person to another, and thus individual measurement is the surest method of determining accurate joint ROMs. Refer to table 3.4 for average ranges of joint motion for specific joints. Integrally related to ROM is the notion of joint stability, or "the ability of a joint to maintain an appropriate functional position throughout its range of motion" (Burstein & Wright, 1994, p. 63).

One way of viewing joint stability is to consider the joint's ability to resist dislocation. Stable joints have a high resistance to dislocation. Unstable joints tend to dislocate more easily. By way of review from chapter 3, recall that in general, joints can be classified along a stability–mobility continuum, which specifies that joints with a tight bony fit and numerous ligamentous and other supporting structures, or surrounded by large muscle groups, will be very stable and relatively immobile. Joints with a loose bony fit, limited extrinsic support, or minimal surrounding musculature tend to be very mobile and unstable. The classic exception to this categorization is the hip joint, which is both very mobile—with large ROM potential in all three primary planes—and very stable, as evidenced by the rarity of its dislocation.

Material Mechanics

So far, our discussion has focused primarily on the movement of bodies and the external forces affecting those movements. In this section, we shift our attention briefly to the human body's internal mechanics, focusing on the internal response of tissues to externally applied loads. Although not directly related to movement mechanics, fundamental principles of **material mechanics** do influence the integrity of musculoskeletal structures and their susceptibility to injury. Injury-related tissue damage can compromise movements in many ways. Thus, an understanding of the basics of material mechanics is essential to our study of human movement.

Anthropometric and mechanical properties of biological materials (e.g., tissues) influence the material's response to forces. Among these properties are size, shape, area, volume, and mass. The tissue's structural constituents and form, discussed in earlier chapters, also play an important role in its mechanical response.

Externally applied forces, termed **loads**, come in three types: tension, compression, and shear (figure 5.21). Tensile (**tension**) loads tend to pull ends apart. Compressive (**compression**) loads tend to push ends together. **Shear** loads tend to produce horizontal sliding of one layer over another.

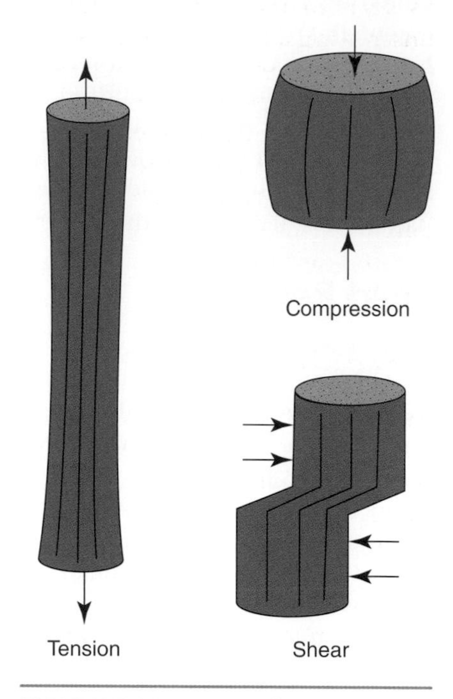

FIGURE 5.21 Types of loads.

Stress and Strain

Tissues mechanically loaded by an external force develop an internal resistance to the load. In the case of a thin rubber band, this resistance is minimal. In contrast, the resistance developed by a steel rod is considerable. This internal resistance to loading, called mechanical **stress**, is common to all materials. Materials change their shape, though sometimes imperceptibly, in response to external loads. This change in shape is termed deformation, or mechanical **strain**.

A direct relationship exists between stress and strain, and the consequences of this relationship in a tissue determine its response to loading. Bone loaded by compression, for example, develops high resistance while deforming very little. Skin, in contrast, deforms considerably more at substantially lower forces. The stress–strain responses of tendons, ligaments, and cartilage fall somewhere between those of bone and skin.

The stress–strain relationship can be summarized in a single measure as the ratio of the two values. This stress–strain ratio defines a material's **stiffness**. The opposite, or inverse, of stiffness is known as **compliance**. Stiff materials, such as bone, have high stress–strain ratios. More compliant materials, such as skin, have lower ratios.

Bending and Torsion

In our earlier discussion of moment of inertia, we briefly mentioned the resistance to both bending and twisting (or torsion). Any relatively long and slender structure (e.g., a long bone) may be considered in mechanical terms as a beam. Any force or moment tends to bend the beam. In bending, the material on the concave (inner) surface of the structure experiences compressive stress, while that on the convex (outer) surface is subject to tensile stress (figure 5.22). These tensile and compressive stresses are maximal at the outer surfaces of the beam and diminish toward the middle of the beam.

Any twisting action applied to a structure results in torsional loading, as seen in the simple example of unscrewing a lid from its jar. Angular resistance is involved when torsional loads are applied to a body. For example, when a skier's leg is twisted, his tibia experiences torsional loading (or torque). The internal stresses developed in response to the torsional loading produce resistance to the applied torque.

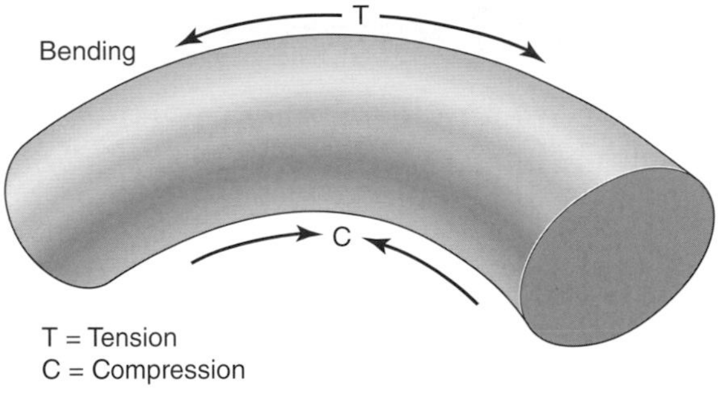

FIGURE 5.22 Bending with tension created on the convex surface and compression developed on the concave surface.

Viscoelasticity

As noted at the beginning of our discussion of material mechanics, the mechanical response of a material depends on its constituent matter, which in the case of biological tissues includes a fluid component (e.g., water). A tissue's viscosity provides resistance to flow and makes the mechanical response of the tissue partially dependent on the rate at which it is loaded. As a result, we say that the tissue's mechanical response is **strain-rate dependent**, which means that the response depends on the rate at which the tissue is deformed. A tendon stretched quickly, for example, will be stronger and stiffer than the same tendon stretched more slowly. This strain-rate dependency is important because tissue strain rate is related to the speed of movement. Rapid movements elicit different tissue responses than do slower movements.

Another important tissue property is elasticity, which is a tissue's ability to return to its original shape, or configuration, when a load is removed. Biological tissues, largely by virtue of their fluid content, have combined viscous and elastic responses. We therefore often describe tissues as having **viscoelasticity**. The viscoelastic properties of tissues play an important role in their mechanical response and indirectly can have profound effects on movement. A rapidly stretched tendon, for example, can store strain energy, some of which is released as the tendon subsequently shortens. This return of stored energy can improve movement performance. A person performing a vertical jump can jump higher, for instance, if she rapidly stretches the musculotendinous structures in the leg by performing a quick squat immediately before jumping upward. More examples of how tissue properties affect human movements are discussed in later chapters.

Concluding Comments

Many of the mechanical concepts presented in this chapter are evident as we explore human movement in the following chapters. Keep in mind that although the information here is divided into separate sections for presentation, the concepts interrelate in complex and fascinating ways. One of the essential challenges in studying the biomechanics of human movement is to identify the salient aspects of movement, fluid, joint, and material mechanics and blend them into an understandable and useful whole.

Go to the web study guide to access critical thinking questions for this chapter.

Suggested Readings

Enoka, R.M. (2015). *Neuromechanics of human movement* (5th ed.). Champaign, IL: Human Kinetics.

Flanagan, S. (2013). *Biomechanics: A case-based approach*. Burlington, MA: Jones & Bartlett Learning.

Hall, S.J. (2014). *Basic biomechanics* (7th ed.). New York: McGraw-Hill.

Hamill, J., Knutsen, K., & Derrick, T. (2014). *Biomechanical basis of human movement* (4th ed.). Philadelphia: Lippincott Williams & Wilkins.

Knudson, D. (2007). *Fundamentals of biomechanics* (2nd ed.). New York: Springer.

McGinnis, P.M. (2013). *Biomechanics of sport and exercise* (3rd ed.). Champaign, IL: Human Kinetics.

Neumann, D.A. (2016). *Kinesiology of the musculoskeletal system: Foundations for rehabilitation*. St. Louis: Mosby.

Nordin, M., & Frankel, V.H. (2012). *Basic biomechanics of the musculoskeletal system* (4th ed.). Philadelphia: Lippincott Williams & Wilkins.

Whiting, W.C., & Zernicke, R.F. (2008). *Biomechanics of musculoskeletal injury* (2nd ed.). Champaign, IL: Human Kinetics.

6

Muscular Control of Movement and Movement Assessment

Objectives

After studying this chapter, you will be able to do the following:

- Describe concepts of muscle function: agonist action, neutralization, stabilization, antagonist action, and coactivation
- Use the muscle control formula to determine muscle action for any movement
- Explain movement concepts of coordination, efficiency, and economy
- Describe sources of movement inefficiency
- Describe methods of kinematic, kinetic, and electromyographic movement assessment

As we learned in chapter 4, the amount of force generated by a given muscle is influenced by many factors, including the nature of the task, speed of movement, amount of external resistance, level of activation, and muscle length and velocity. In addition, the force needed by the muscle must be determined in the context of other muscles' actions and the constraints imposed by the principles of mechanics and muscle physiology.

In most movements, mechanical and physiological variables change continuously throughout the movement and thus make the nervous system's job complex. Although the hows and whys of muscle control remain somewhat a mystery, the fact that the neuromuscular system *does* control our actions is indisputable. Does it do so perfectly? Hardly. But in most cases, our bodies do a commendable job of movement control.

The story of muscular control of movement is analogous to the children's story of the three bears. The moral of that story is that things shouldn't be too hot or too cold but should be "just right." The same rule applies to muscle action. Too much or too little force compromises movement control. In the words of futurist Alvin Toffler, "Overcontrol is just as dangerous as undercontrol" (1990, p. 463).

At times, generating too much force can be just as detrimental to performance as not producing enough force, as in the case of someone just learning to perform a task. Novices, in their unfamiliarity with a particular movement, may excessively recruit muscles and create hesitant, jerky, and inefficient movements. Skilled performers, in contrast, make good use of their muscles and execute confident, fluid, and coordinated movements. In terms of movement control, sometimes less is more.

Too much conscious thought can compromise the effectiveness of movement, a principle sometimes jokingly referred to as "paralysis by analysis." In matters of the mind and its effect on movement control, noted Russian actor and director Konstantin Stanislavsky (1863-1938) observed, "At times of great stress, it is especially necessary to achieve a complete freeing of the muscles" (1948, p. 94).

Muscle Function

The control of even the simplest joint movement typically requires the cooperative action of several muscles working together as a single unit. This cooperative action is called **muscle synergy**. Synergistic muscles work together, but other muscles with opposite functions may work against a particular movement. The overall, or net, effect of all muscles acting at a joint determines the ultimate mechanical effect, or movement.

Several concepts of muscle function are important for understanding how muscles cooperate and compete to control movement: agonists, neutralization, stabilization, antagonists, and coactivation.

- *Agonists.* Muscles that actively produce or control a single joint movement or maintain a single joint position are called **agonists**. In most movements, several muscles act together as agonists, with some playing a greater role than others.
- *Neutralization.* Muscles often perform more than one movement function at a given joint. At the ankle complex, for example, a muscle might act as both a plantar flexor and invertor (e.g., tibialis posterior). To produce pure plantar flexion, another muscle whose action produces plantar flexion and eversion (e.g., peroneus longus) would also need to be involved. The eversion action of the second muscle would cancel out, or neutralize, the inversion action of the first muscle. This process of canceling out an unwanted secondary movement is called **neutralization**.
- *Stabilization.* During concentric action, a muscle attempts to shorten by pulling its two bony attachment sites together. In most cases, the bone with the least resistance to movement (inertia) will move. When the inertia of both bones is similar in magnitude, both ends tend to move. If movement of only one end is desired, the other end must be prevented from moving, or stabilized. This **stabilization** is provided by other muscles or an external force.

Applying the Concept

Superhuman Strength

Occasionally we hear reports of someone exhibiting tremendous strength, such as in lifting a car off a trapped person. These prodigious feats involve what is termed "superhuman strength" or "hysterical strength." Are such feats of extreme muscle strength possible? The answer is not entirely clear. In certain situations such as life-or-death moments or when a person is under the influence of certain drugs, humans can develop far more muscle force than is used in a typical voluntary maximal exertion. Why do we have this reserve strength? Primarily for reasons of safety. Extreme forces can easily injure muscle, tendon, ligaments, and bone.

In everyday situations, the mind (i.e., psychological factors) often is a limiting factor in muscle recruitment. When the mind is overridden by panic or drugs, for example, the body can produce supernormal forces, largely due to supranormal levels of adrenaline. That said, can a person actually lift an entire car weighing 3,000 lb (1,364 kg)? Very likely not. In instances of such claims, the lifter probably is lifting only one corner of the car, or only a fraction of the car's total weight. Many of the questions related to superhuman strength will remain unanswered, since controlled scientific studies of such instances (e.g., simulating a life-or-death situation) are not feasible.

As an example, consider hip flexion created by the anterior thigh musculature. In attempting to move the femur in flexion, the hip flexors also tilt the pelvis anteriorly. If pelvic tilt is unwanted, then the abdominal musculature must act isometrically to stabilize the pelvis and prevent its movement.

- *Antagonists.* Muscles acting against a movement or position are called **antagonists**. To perform a movement most effectively, when the agonists actively shorten in concentric action, the corresponding antagonists *passively lengthen*. When agonists actively lengthen in eccentric action, the associated antagonists *passively shorten*. In many movements, then, the agonists and stabilizers are active while the antagonists are passive.

- *Coactivation.* Simultaneous action of both agonists *and* antagonists is called **coactivation** (also **co-contraction**). Coactivation might occur, for example, when an unskilled performer is unsure of the necessary muscle recruitment strategy. Skilled performers, however, do not exhibit an absence of coactivation. At least four possible explanations exist for coactivation in skilled performers: (1) less overall effort may be required in agonist–antagonist pairings for movements that involve changes of direction when the muscles maintain some level of activity, as opposed to working in an on–off manner; (2) coactivation increases joint stiffness and consequently joint stability, which may be desired for movements involving heavy loads; (3) coactivation of a single-joint muscle (e.g., gluteus maximus) and a two-joint muscle (e.g., rectus femoris) can increase the torque at a joint (e.g., knee) acted on by the two-joint muscle; and (4) given the neural complexities of the forearm and hand, fine movements of the fingers require complex coactivation strategies (Enoka, 2002).

Note that the term *coactivation* is limited to the concurrent action of agonists and antagonists and should *not* be used to describe the simultaneous action of multiple agonists. For instance, concurrent activity of the biceps brachii and triceps brachii during an elbow curl exercise would be considered coactivation. On the other hand, if the triceps was passive (i.e., inactive), simultaneous action of the three elbow flexors (biceps brachii, brachialis, brachioradialis) would *not* be considered coactivation.

With these concepts in mind, we now consider a simple yet fundamental question: How do we determine which muscles are active in producing or controlling a given movement?

Research in Mechanics

Muscle Coactivation

Coactivation is a common neuromuscular control strategy used in a variety of movements. One of the primary purposes of coactivation is to provide or enhance joint stabilization. Numerous research studies over the past several decades have examined coactivation across many populations, tasks, and conditions, including walking (Falconer & Winter, 1985) and running (Osternig et al., 1986), gait in children with cerebral palsy (Dimiano et al., 2000; Unnithan et al., 1996), age-related balance changes (Benjuya et al., 2004), post-stroke gait (Lamontagne et al., 2000), and drop jump landing (Kellis et al., 2003).

See the references for the full citations:

Benjuya, Melzer, & Kaplanski, 2004.

Dimiano, Martellotta, Sullivan, Granata, & Abel, 2000.

Falconer & Winter, 1985.

Kellis, Arabatzi, & Papadopoulos, 2003.

Lamontagne, Richards, & Malouin, 2000.

Osternig, Hamill, Lander, & Robertson, 1986.

Unnithan, Dowling, Frost, Volpe Ayub, & Bar-Or, 1996.

Muscle Action

One of the most fundamental and important goals of movement analysis is identifying which specific muscles are active in producing and controlling movement at a particular joint. In chapter 4, we presented specific muscles and their concentric actions. However, we know that muscles can act in three modes: isometric, concentric, and eccentric. The task at hand, therefore, is to determine for a given joint movement (1) the specific muscles involved in controlling the movement and (2) the type of muscle action.

The following **muscle control formula** provides a step-by-step procedure for determining the involved muscles and their action for any joint movement. This formula may appear a bit cumbersome and complex at first glance. However, with practice, you should be able to get through it quickly. Eventually (with enough practice), the process will become automatic and instinctive, and you will be able to analyze movements without consciously going through each step in the formula. It helps, though, to use the formula until you develop these movement analysis instincts.

Muscle Control Formula

We begin the muscle control formula with a statement of the problem: Given a specific joint movement (or position), identify the name of the movement (or position), the plane of movement, the effect of the external force acting on the system, the type of muscle action (i.e., shortening or concentric, lengthening or eccentric, or isometric), and the muscles involved (i.e., which muscle or muscles are actively involved in producing or controlling the movement or in maintaining a position).

Now, we move on to the formula itself, which involves six steps:

Step 1: Identify the *joint movement* (e.g., flexion, abduction) or position.

Step 2: Identify the *effect of the external force* (e.g., gravity) on the joint movement or position by asking the following question: What movement would the external force produce in the absence of muscle action (i.e., if there were no active muscles)?

Step 3: Identify the *type of muscle action* (concentric, eccentric, isometric) based on the answers to step 1 (#1) and step 2 (#2) as follows:

a. If #1 and #2 are in *opposite* directions, then the muscles are actively shortening in a *concentric action*. Speed of movement is *not* a factor.

b. If #1 and #2 are in the *same* direction, then ask yourself, "What is the speed of movement?"

c. If the movement is *faster* than what the external force would produce by itself, then the muscles are actively shortening in a *concentric action*.

d. If the movement is *slower* than what the external force would produce by itself, then the muscles are actively lengthening in an *eccentric action*.

e. If no movement is occurring, yet the external force would produce movement if acting by itself, then the muscles are performing an *isometric action*.

f. Movements *across gravity* (i.e., parallel to the ground) are produced by a *concentric action*. When gravity cannot influence the joint movement in question, shortening (concentric) action is needed to pull the bone against its own inertia. The speed of movement is *not* a factor.

By this point, we have identified the type of muscle action. The next steps identify which muscles control the movement.

Step 4: Identify the *plane of movement* (frontal, sagittal, transverse) and the *axis of rotation* (i.e., line about which the joint is rotating). The purpose of this step is to identify which side of the joint the muscles controlling the movement cross (e.g., flexors cross one side of a joint, while extensors cross the opposite side).

Step 5: Ask yourself, "On which side of the joint axis are muscles lengthening and on which side are they shortening during the movement?"

Step 6: Combine the information from steps 3 and 5 to *determine which muscles must be producing or controlling the movement* (or position). For example, if a concentric (shortening) action is required (from step 3) and the muscles on the anterior side of the joint are shortening (from step 5), then the anterior muscles must be actively producing the movement. The information in chapter 4 allows us to name the specific muscles.

Application of the Muscle Control Formula

Let's see how to apply the muscle control formula by going through the step-by-step process for some simple single-joint movements. For all these examples, we assume *no* coactivation of antagonists, which allows us to simplify the analysis by eliminating the need to consider the effect of antagonist muscles. Coactivation of agonist–antagonist groups stiffens the joint and makes joint motion more difficult.

WWW **Go to the web study guide to access interactive practice problems that let you apply the muscle control formula.**

Example 1: **Biceps Curl**

Consider the simple movement (figure 6.1) of elbow flexion as the person moves the joint from position A (elbow fully extended) to position B (elbow flexed).

Step 1: The movement is flexion.

Step 2: The external force (gravity) tends to extend the elbow.

Step 3: The movement (flexion) is opposite of that created by the external force, so the muscles are actively shortening in *concentric action*.

Step 4: Movement occurs in the sagittal plane about an axis through the elbow joint.

Step 5: The muscles on the joint's anterior surface are shortening during the movement, while the muscles on the posterior side are lengthening.

FIGURE 6.1 Biceps curl: elbow flexion from position A to B; elbow extension from position B to A.

Step 6: The muscle action is concentric (from step 3), and muscles on the anterior side of the joint are shortening (from step 5). Thus, the anterior muscles actively produce the movement. Using the information from chapter 4, we know that the biceps brachii, brachialis, and brachioradialis are the muscles responsible for the movement.

Now consider the movement (figure 6.1) of elbow extension as the person moves the joint from position B (elbow flexed) to position A (elbow fully extended). The movement speed is slow, meaning the movement occurs *more slowly* than what the external force would produce acting by itself (i.e., in the absence of muscle action).

Step 1: The movement is extension.

Step 2: The external force (gravity) tends to extend the elbow.

Step 3: The movement (extension) is the same as that created by the external force, so ask yourself, "What is the speed of movement?" The speed is slow, which dictates that the controlling muscles are actively lengthening in an *eccentric action*.

Step 4: Movement occurs in the sagittal plane about an axis through the elbow joint.

Step 5: The muscles on the joint's anterior surface are lengthening, while the muscles on the posterior side are shortening.

Step 6: The muscle action is eccentric (from step 3), and muscles on the anterior surface are lengthening (from step 5). Thus, the anterior muscles actively control the movement. Again, using the information from chapter 4, we identify the biceps brachii, brachialis, and brachioradialis as the muscles responsible for the movement.

In this example, we see that the muscles normally identified as elbow flexors (i.e., biceps brachii, brachialis, brachioradialis) act concentrically to produce elbow flexion and also act eccentrically to control elbow extension. This principle that "flexors" can control extension may, at first, seem counterintuitive. Why wouldn't the so-called elbow extensors (e.g., triceps brachii) control elbow extension? The answer lies in the speed of the movement. Movements that are slower than what the external force would create (in the absence of muscle action) are controlled by eccentric action of muscles on the side opposite of those that would produce the movement concentrically. This principle of eccentric control cannot be overemphasized.

Now consider the same movement (figure 6.1) of elbow extension as the person moves from position B (elbow flexed) to position A (elbow fully extended), but this time by moving fast, meaning that the movement occurs *faster* than what the external force would produce acting by itself (i.e., in the absence of muscle action). In this action, the arm is rapidly snapped into extension.

Step 1: The movement is extension.

Step 2: The external force (gravity) tends to extend the elbow.

Step 3: The movement (extension) is the same as that created by the external force, so ask yourself, "What is the speed of movement?" The speed is fast, which dictates that the controlling muscles are actively shortening in a *concentric action.*

Step 4: Movement occurs in the sagittal plane about an axis through the elbow joint.

Step 5: The muscles on the joint's anterior surface are lengthening during the movement, while the muscles on the posterior side are shortening.

Step 6: The muscle action is concentric (step 3), and muscles on the posterior side of the joint are shortening (step 5). Thus, the posterior muscles actively produce the movement. Therefore, the elbow extensors, primarily the triceps brachii, are responsible for the movement.

It is important to note that in this case, the speed of movement (fast) determines that the movement (elbow extension) is produced by concentric action of the elbow extensors. In the previous case, the speed of movement (slow) dictates that eccentric action of the elbow flexors is required to control joint extension.

Example 2: Leg Extension

Consider the leg extension movement shown in figure 6.2. The knee joint is extended from position A (knee flexed) to position B (knee fully extended).

Step 1: The movement is extension.

Step 2: The external force (gravity) tends to flex the knee.

Step 3: The movement (extension) is opposite of that created by the external force, so the muscles are actively shortening in *concentric action.*

Step 4: Movement occurs in the sagittal plane about an axis through the knee joint.

Step 5: The muscles on the joint's anterior surface are shortening during the movement, while the muscles on the posterior side are lengthening.

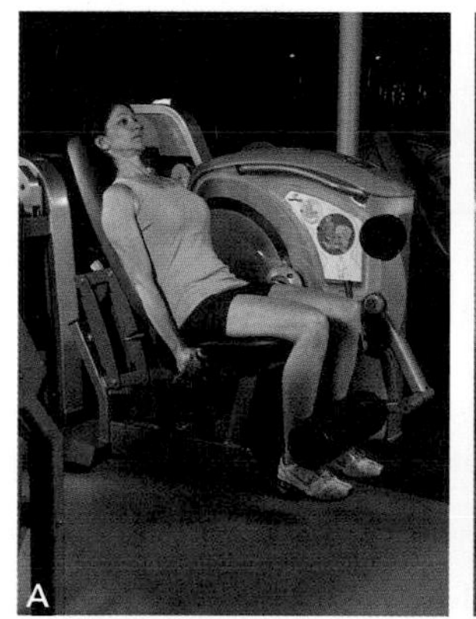

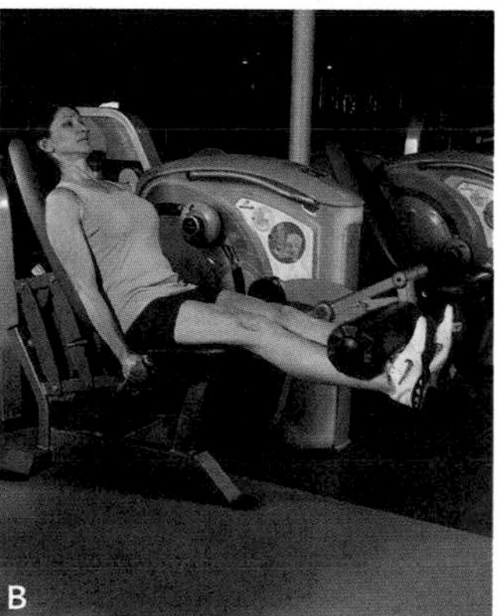

FIGURE 6.2 Leg extension: knee extension from position A to B; knee flexion from position B to A.

Step 6: The muscle action is concentric (step 3), and muscles on the anterior side of the joint are shortening (step 5). Thus, the anterior muscles actively produce the movement. Therefore, the muscles of the quadriceps group (vastus medialis, vastus lateralis, vastus intermedius, rectus femoris) produce the movement.

Now consider the movement (figure 6.2) of knee flexion as the joint moves from position B (knee extended) to position A (knee flexed). The movement speed is slow, meaning the movement occurs *more slowly* than what the external force would produce acting by itself (i.e., in the absence of muscle action).

Step 1: The movement is flexion.

Step 2: The external force (gravity) tends to flex the knee.

Step 3: The movement (flexion) is the same as that created by the external force, so ask yourself, "What is the speed of movement?" The speed is slow, which dictates that the controlling muscles are actively lengthening in an *eccentric action.*

Step 4: Movement occurs in the sagittal plane about an axis through the knee joint.

Step 5: The muscles on the joint's anterior surface are lengthening, while the muscles on the posterior side are shortening.

Step 6: The muscle action is eccentric (step 3), and muscles on the anterior surface are lengthening (step 5). Thus, the anterior muscles actively control the movement. Therefore, the vastus medialis, vastus lateralis, vastus intermedius, and rectus femoris control the movement.

Consider the same movement (figure 6.2) of knee flexion as the joint moves from position B (knee extended) to position A (knee flexed), but this time moving fast, meaning the movement occurs *faster* than what the external force would produce acting by itself (i.e., in the absence of muscle action). In this action, the leg is rapidly snapped into flexion.

Step 1: The movement is flexion.

Step 2: The external force (gravity) tends to flex the knee.

Step 3: The movement (flexion) is the same as that created by the external force, so ask yourself, "What is the speed of movement?" The speed is fast, which dictates that the controlling muscles are actively shortening in a *concentric action.*

Step 4: Movement occurs in the sagittal plane about an axis through the knee joint.

Step 5: The muscles on the joint's anterior surface are lengthening during the movement, while the muscles on the posterior side are shortening.

Step 6: The muscle action is concentric (step 3), and muscles on the posterior side of the joint are shortening (step 5). Thus, the posterior muscles actively produce the movement. Therefore, the knee flexors (e.g., biceps femoris, semitendinosus, semimembranosus) produce the movement.

Example 3: Dumbbell Lateral Raise

This example involves raising the arms from the side of the body (position A) to parallel with the ground (position B) as shown in figure 6.3.

Step 1: The movement is abduction.

Step 2: The external force (gravity) tends to adduct the arm.

Step 3: The movement (abduction) is opposite of that created by the external force, so the muscles are actively shortening in *concentric action.*

Step 4: Movement occurs in the frontal plane about an axis through the glenohumeral joint.

Step 5: The muscles on the joint's superior surface are shortening during the movement, while the muscles on the inferior side are lengthening.

Step 6: The muscle action is concentric (step 3), and muscles on the superior side of the joint are shortening (step 5).

Thus, the superior muscles (the abductors) actively produce the movement. Therefore, the anterior and middle deltoid and supraspinatus produce the movement.

Now consider the movement (figure 6.3) of glenohumeral adduction as the person moves the joint from position B (abducted) to position A (adducted), this time slowly.

FIGURE 6.3 Dumbbell lateral raise: arm abduction from position A to B; arm adduction from position B to A.

Step 1: The movement is adduction.

Step 2: The external force (gravity) tends to adduct the arm.

Step 3: The movement (adduction) is the same as that created by the external force, so ask yourself, "What is the speed of movement?" The speed is slow, which dictates that the controlling muscles are actively lengthening in an *eccentric action.*

Step 4: Movement occurs in the frontal plane about an axis through the glenohumeral joint.

Step 5: The muscles on the joint's superior surface are lengthening, while the muscles on the inferior side are shortening.

Step 6: The muscle action is eccentric (step 3), and muscles on the superior surface are lengthening (step 5). Thus, the superior muscles actively control the movement. Therefore, the deltoid and supraspinatus control the movement.

Consider the same movement (figure 6.3) of glenohumeral adduction, this time moving fast. In this action, the arm is rapidly snapped to the side of the body from its abducted position.

Step 1: The movement is adduction.

Step 2: The external force (gravity) tends to adduct the arm.

Step 3: The movement (adduction) is the same as that created by the external force, so ask yourself, "What is the speed of movement?" The speed is fast, which dictates that the controlling muscles are actively shortening in a *concentric action.*

Step 4: Movement occurs in the frontal plane about an axis through the glenohumeral joint.

Step 5: The muscles on the joint's superior surface are lengthening during the movement, while the muscles on the inferior side are shortening.

Step 6: The muscle action is concentric (step 3), and muscles on the inferior side of the joint are shortening (step 5). Thus, the inferior muscles actively produce the movement. Therefore, the glenohumeral adductors (e.g., pectoralis major, latissimus dorsi, teres major) produce the movement.

Example 4: Abdominal Curl-Up

When performing a curl-up exercise, the exerciser flexes the trunk from position A (trunk fully extended) to position B (trunk flexed), as shown in figure 6.4.

Step 1: The movement is flexion.

Step 2: The external force (gravity) tends to extend the trunk.

Step 3: The movement (flexion) is opposite of that created by the external force, so the muscles are actively shortening in *concentric action.*

Step 4: Movement occurs in the sagittal plane about multiple axes through the different vertebrae of the spine.

Step 5: The muscles on the joint's anterior surface are shortening during the movement, while the muscles on the posterior side are lengthening.

Step 6: The muscle action is concentric (step 3), and muscles on the anterior side of the joint are shortening (step 5). Thus, the anterior muscles actively produce the movement. Therefore, the rectus abdominis and the internal and external obliques produce the movement.

FIGURE 6.4 Abdominal curl-up: trunk flexion from position A to B; trunk extension from position B to A.

To return to the starting position, the exerciser extends the trunk from position B (trunk flexed) to position A (trunk fully extended) in a slow and controlled manner.

Step 1: The movement is extension.

Step 2: The external force (gravity) tends to extend the trunk.

Step 3: The movement (extension) is the same as that created by the external force, so ask yourself, "What is the speed of movement?" The speed is slow, which dictates that the controlling muscles are actively lengthening in an *eccentric action.*

Step 4: Movement occurs in the sagittal plane about multiple axes through the different vertebrae of the spine.

Step 5: The muscles on the joint's anterior surface are lengthening, while the muscles on the posterior side are shortening.

Step 6: The muscle action is eccentric (step 3), and muscles on the anterior surface are lengthening (step 5). Thus, the anterior muscles actively control the movement. Therefore, the rectus abdominis, assisted by the internal and external obliques, again controls the movement.

This analysis of muscle action indicates that curl-up (and related) exercises require almost continuous abdominal muscle activity, concentrically during the up phase and eccentrically during the controlled down phase. Snapping the trunk back to the ground at a speed faster than gravity would be both difficult and inadvisable because of the risk of injury. We therefore do not analyze that condition.

Example 5: Standing Calf Raise

A common exercise for strengthening the calf muscles on the posterior aspect of the lower leg (shank) involves isolated plantar flexion of the ankle, often with added resistance in the form of carried weights or a weight machine designed for this purpose. This exercise (commonly called heel raises or calf raises) is typically performed with the ball of the foot on a raised surface to allow the rear of the foot to drop lower on the down phase. Consider first the movement from position A (ankle in neutral anatomical position) to position B (ankle plantar flexed) as shown in figure 6.5.

Step 1: The movement is plantar flexion.

Step 2: The external force (gravity) tends to dorsiflex the ankle.

Step 3: The movement (plantar flexion) is opposite of that created by the external force, so the muscles are actively shortening in *concentric action*.

Step 4: Movement occurs in the sagittal plane about an axis through the ankle joint.

Step 5: The muscles on the joint's posterior surface are shortening during the movement, while the muscles on the anterior side are lengthening.

Step 6: The muscle action is concentric (step 3), and muscles on the posterior side of the joint are shortening (step 5). Thus, the posterior muscles actively produce the movement. Therefore, the soleus and gastrocnemius, with slight assistance by the other plantar flexors (peroneus longus, peroneus brevis, tibialis posterior, flexor hallucis longus, flexor digitorum longus, plantaris), produce the movement.

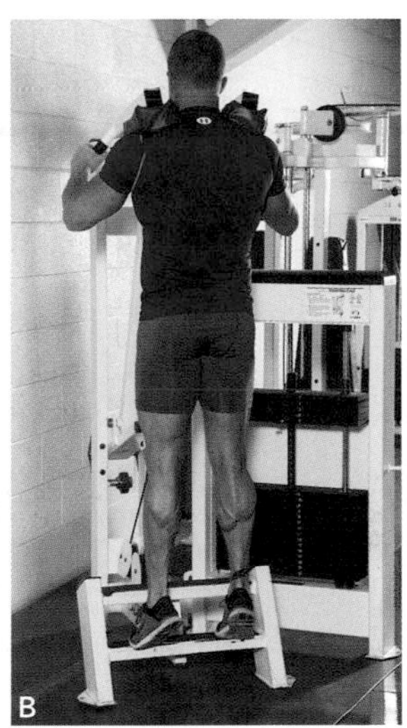

FIGURE 6.5 Standing calf raise: ankle plantar flexion from position A to B; ankle dorsiflexion from position B to A.

During the lowering phase of the movement, the ankle *slowly* moves from position B (ankle plantar flexed) to position A (ankle neutral).

Step 1: The movement is dorsiflexion.

Step 2: The external force (gravity) tends to dorsiflex the ankle.

Step 3: The movement (dorsiflexion) is the same as that created by the external force, so ask yourself, "What is the speed of movement?" The speed is slow, which dictates that the controlling muscles are actively lengthening in an *eccentric action*.

Step 4: Movement occurs in the sagittal plane about an axis through the ankle joint.

Step 5: The muscles on the joint's posterior surface are lengthening, while the muscles on the anterior side are shortening.

Step 6: The muscle action is eccentric (step 3), and muscles on the posterior surface are lengthening (step 5). Thus, the posterior muscles actively control the movement. Therefore, the soleus and gastrocnemius (again with assistance from the other plantar flexors) control the movement.

Example 6: **Shoulder Rotation**

In the examples considered so far, gravity has been the external force. Consider now an example where the external force is other than gravity, as might be the case of an astronaut using elastic (bungee) cords in the microgravity of outer space or a patient undergoing physical therapy who uses elastic bands for resistance. In this example, the subject is trying to strengthen rotator muscles acting at the glenohumeral joint. Working against the resistance provided by an elastic band (figure 6.6), the subject moves her arm from position A (glenohumeral joint externally rotated) to position B (internally rotated).

Step 1: The movement is internal rotation.

Step 2: The external force (elastic band) tends to externally rotate the glenohumeral joint.

Step 3: The movement (internal rotation) is opposite of that created by the external force, so the muscles are actively shortening in *concentric action.*

Step 4: Movement occurs in the transverse plane about a vertical axis through the glenohumeral joint.

Step 5: The muscles on the joint's anterior surface are shortening during the movement, while the muscles on the posterior side are lengthening.

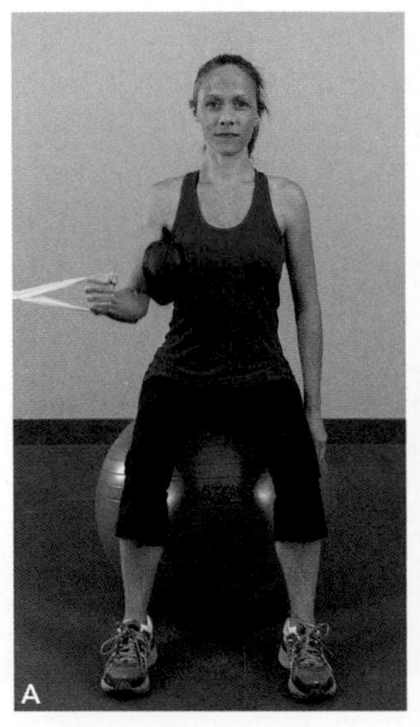

FIGURE 6.6 Shoulder rotation: glenohumeral internal rotation from position A to B; glenohumeral external rotation from position B to A.

Step 6: The muscle action is concentric (step 3), and muscles on the anterior side of the joint are shortening (step 5). Thus, the anterior muscles actively produce the movement. Therefore, the subscapularis, pectoralis major, anterior deltoid, latissimus dorsi, and teres major produce the movement.

If the subject *slowly* returns from position B to the starting position A, the muscle action is determined as follows:

Step 1: The movement is external rotation.

Step 2: The external force (elastic band) tends to externally rotate the glenohumeral joint.

Step 3: The movement (external rotation) is the same as that created by the external force, so ask yourself, "What is the speed of movement?" The speed is slow, which dictates that the controlling muscles are actively lengthening in an *eccentric action.*

Step 4: Movement occurs in the transverse plane about a vertical axis through the glenohumeral joint.

Step 5: The muscles on the joint's anterior surface are lengthening, while the muscles on the posterior side are shortening.

Step 6: The muscle action is eccentric (step 3), and muscles on the anterior surface are lengthening (step 5). Thus, the anterior muscles actively control the movement. Therefore, the same internal rotators (subscapularis, pectoralis major, latissimus dorsi, anterior deltoid, teres major) control the movement.

If the external rotation was performed quickly, the muscles would cause external rotation faster than the elastic band could develop tension. The muscles basically would defeat the purpose of the elastic band in providing resistance, and the band would become slack. The muscle action in this case would be determined as follows:

Step 1: The movement is external rotation.

Step 2: The external force (elastic band) tends to externally rotate the glenohumeral joint.

Step 3: The movement (external rotation) is the same as that created by the external force, so ask yourself, "What is the speed of movement?" The speed is fast, which dictates that the controlling muscles are actively shortening in a *concentric action*.

Step 4: Movement occurs in the transverse plane about a vertical axis through the glenohumeral joint.

Step 5: The muscles on the joint's anterior surface are lengthening during the movement, while the muscles on the posterior side are shortening.

Step 6: The muscle action is concentric (step 3), and muscles on the posterior side of the joint are shortening (step 5). Thus, the posterior muscles actively produce the movement. Therefore, the external rotators (infraspinatus, teres minor, posterior deltoid) produce the movement.

Placing the elastic band on the other side (i.e., so that it pulls from a medial direction) reverses the muscle actions. The external rotators (infraspinatus, teres minor, posterior deltoid) would act concentrically to externally rotate the arm against the elastic band's resistance, which tends to internally rotate the joint. In returning to the starting (internally rotated) position slowly, the same external rotators would act eccentrically to control the internal rotation.

Example 7: Standing Flys

Some movements occur parallel to the ground and, therefore, aren't directly affected by gravity. Consider an exercise (figure 6.7) in which the exerciser uses elastic bands with arms held in an abducted position (A). Note that although muscle action of the glenohumeral abductors (deltoid and supraspinatus) is required to keep the arms in an elevated position, movements in the transverse plane are not directly affected by gravity. The muscle control formula is applied as follows.

Step 1: The movement is horizontal adduction (also called horizontal flexion).

Step 2: Gravity does not affect this motion because the movement is parallel to the ground (no effect of gravity for movement in the transverse plane). The elastic bands provide a force that tends to horizontally abduct the arms.

Step 3: The movement (horizontal adduction) occurs *across gravity*, so the muscles are actively shortening in *concentric action* to overcome the inertia of the limb segments and the resistance of the elastic bands.

FIGURE 6.7 Standing flys: glenohumeral horizontal adduction from position A to B; glenohumeral horizontal abduction from position B to A.

Step 4: Movement occurs in the transverse plane about a vertical axis through the glenohumeral joint.

Step 5: The muscles on the shoulder joint's anterior surface are shortening during the movement, while the muscles on the posterior side are lengthening.

Step 6: The muscle action is concentric (step 3), and muscles on the anterior side of the joint are shortening (step 5). Thus, the anterior muscles actively produce the movement. Therefore, the pectoralis major and anterior deltoid primarily produce the movement.

In returning from position B to the starting position A (figure 6.7), the muscle action is determined as follows:

Step 1: The movement is horizontal abduction (also called horizontal extension).

Step 2: The external force (elastic band) tends to horizontally abduct the glenohumeral joint.

Step 3: The movement (horizontal abduction) is the same as that created by the external force, so ask yourself, "What is the speed of movement?" The speed is slow, which dictates that the controlling muscles are actively lengthening in an *eccentric* action.

Step 4: Movement occurs in the transverse plane about a vertical axis through the glenohumeral joint.

Step 5: The muscles on the joint's anterior surface are lengthening during the movement, while the muscles on the posterior side are shortening.

Step 6: The muscle action is eccentric (step 3), and muscles on the anterior side of the joint are lengthening (step 5). Thus, the anterior muscles actively control the movement. Therefore, the same glenohumeral adductors (pectoralis major and anterior deltoid) actively control the movement.

Example 8: Back Extension

Let's consider one final example in this section to cover the case when no motion occurs, but muscle force is needed to maintain joint position. What muscle action is required for the person in figure 6.8 to hold the trunk in the position shown? (Note: This exercise can place high loads on the spine and so is not recommended for general use. It is presented here only to show how the muscle control formula can be used to determine muscle function.)

Step 1: The trunk's position is fully extended (i.e., anatomical position).

Step 2: The external force (gravity) tends to flex the trunk.

Step 3: No movement is occurring, yet the external force would produce movement if allowed to act by itself, so the muscles are performing an *isometric action*.

Step 4: If movement was allowed, gravity would flex the trunk in the sagittal plane about multiple axes through the different vertebrae of the spine.

FIGURE 6.8 Back extension: trunk held in neutral position while suspended above the ground.

Step 5: The muscles on the joint's anterior surface would shorten during the movement (if allowed to occur), while the muscles on the posterior side would lengthen.

Step 6: The muscle action is isometric (step 3). To prevent muscle length changes on both sides of the trunk (step 5), the posterior muscles (erector spinae) must be active to prevent trunk flexion and counteract the action of gravity.

Single-Joint Versus Multijoint Movements

Our examples thus far have involved movement at a single joint. We now consider two multijoint movements to show how multiple applications of the muscle control formula can fully describe muscle actions in more complex multijoint movements. More examples of multijoint movements are discussed in subsequent chapters.

Example 9: Squat

Many activities of daily living, as well as many athletic movements, require us to move from an upright standing position into some form of a squat. A typical squatting movement is shown in figure 6.9 as the person moves from position A (upright standing) to position B (squat). This movement invariably is done *slowly*. The squat is a multijoint movement involving primarily the hip, knee, and ankle joints.

As the exerciser moves from position A to position B, the muscle action is determined as follows:

Step 1: The movements are hip and knee flexion and ankle dorsiflexion.

Step 2: The external force (gravity) tends to flex the hip and knee and dorsiflex the ankle.

Step 3: The movements are the same as those created by the external force, so ask yourself, "What is the speed of movement?" The speed is slow, which dictates that the controlling muscles are actively lengthening in an *eccentric action* at all three joints.

Step 4: Movement occurs in the sagittal plane about axes through the hip, knee, and ankle joints.

Step 5: The muscles on the hip joint's anterior surface are shortening, while the muscles on the posterior side are lengthening. At the knee, the anterior muscles are lengthening and the posterior muscles are shortening. At the ankle, the anterior muscles are shortening and the posterior muscles are lengthening.

FIGURE 6.9 Squat: downward phase from position A to B; upward phase from position B to A.

Step 6: The muscle action is eccentric (step 3). Muscles on the posterior surface of the hip and ankle are lengthening (step 5). At the knee, muscles on the joint's anterior surface are lengthening (step 5). Therefore, the downward phase of the squat is controlled by eccentric action of the hip extensors (gluteus maximus, semitendinosus, semimembranosus, long head of the biceps femoris, posterior fibers of the adductor magnus), knee extensors (vastus medialis, vastus lateralis, vastus intermedius, rectus femoris), and ankle plantar flexors (soleus and gastrocnemius, with slight assistance of the peroneus longus and brevis, tibialis posterior, flexor hallucis longus, flexor digitorum longus, and plantaris).

In returning from the squatting position (B) to upright standing (position A), the muscle action is determined as follows:

Step 1: The movements are hip and knee extension and ankle plantar flexion.

Step 2: The external force (gravity) tends to flex the hip and knee and dorsiflex the ankle.

Step 3: The movements are opposite of those created by the external force, so the muscles are actively shortening in *concentric action* at all three joints.

Step 4: Movement occurs in the sagittal plane about axes through the hip, knee, and ankle joints.

Step 5: The muscles on the hip joint's anterior surface are lengthening, while the muscles on the posterior side are shortening. At the knee, the anterior muscles are shortening and the posterior muscles are lengthening. At the ankle, the anterior muscles are lengthening and the posterior muscles are shortening.

Step 6: The muscle action is concentric (step 3). Muscles on the posterior surface of the hip and ankle are shortening (step 5). At the knee, muscles on the joint's anterior surface (step 5) are shortening. Therefore, the upward phase of the squat is produced by concentric action of the hip extensors (gluteus maximus, semitendinosus, semimembranosus, long head of the biceps femoris, posterior fibers of the adductor magnus), knee extensors (vastus medialis, vastus lateralis, vastus intermedius, rectus femoris), and ankle plantar flexors (soleus and gastrocnemius, with slight assistance of the peroneus longus and brevis, tibialis posterior, flexor hallucis longus, flexor digitorum longus, and plantaris).

You may have noticed in several of the examples that eccentric action of muscles is immediately followed by concentric action of the same muscles. This pattern of eccentric–concentric coupling is the mechanism underlying the stretch–shortening cycle (explained in chapter 4). Additional functional applications of the stretch–shortening cycle (e.g., vertical jumping) are presented and explained in subsequent chapters.

Example 10: Overhead Press

In some situations it is necessary to lift, or press, an object from shoulder level to an overhead position. In a warehouse, for example, workers often lift boxes to place them on a high shelf. In the gym, a common exercise is an overhead press using a barbell, as shown in figure 6.10. The pressing movement involves simultaneous actions of the shoulder girdle and glenohumeral and elbow joints. The exerciser begins with the barbell at chest level (position A) and then presses the barbell overhead to position B. For this example, we restrict movement to the frontal plane and analyze only actions at the glenohumeral and elbow joints.

Step 1: The movements are glenohumeral abduction and elbow extension.

Step 2: The external force (gravity) tends to adduct the glenohumeral joint and flex the elbow.

Step 3: The movements are opposite of those created by the external force, so the muscles are actively shortening in *concentric action*.

Step 4: Movement occurs in the frontal plane about axes through the glenohumeral and elbow joints.

Step 5: The muscles on the glenohumeral joint's superior surface are shortening during the movement, while the muscles on the inferior side are lengthening. At the elbow, the posterior muscles are shortening and the anterior muscles are lengthening.

Step 6: The muscle action is concentric (step 3). Muscles on the superior surface of the glenohumeral joint and posterior surface of the elbow are shortening (step 5). Therefore, the upward phase of the overhead press is produced by concentric action of the glenohumeral abductors (supraspinatus, anterior deltoid, middle deltoid) and elbow extensors (primarily triceps brachii).

FIGURE 6.10 Overhead press: upward phase from position A to B; downward phase from position B to A.

In lowering the bar from position B back to the starting position A in a slow and controlled manner, the muscle action is determined as follows:

Step 1: The movements are glenohumeral adduction and elbow flexion.

Step 2: The external force (gravity) tends to adduct the glenohumeral joint and flex the elbow.

Step 3: The movements are the same as those created by the external force, so ask yourself, "What is the speed of movement?" The speed is slow, which dictates that the controlling muscles are actively lengthening in an *eccentric action* at both joints.

Step 4: Movement occurs in the frontal plane about axes through the glenohumeral and elbow joints.

Step 5: The muscles on the glenohumeral joint's inferior surface are shortening during the movement, while the muscles on the superior side are lengthening. At the elbow, the anterior muscles are shortening and the posterior muscles are lengthening.

Step 6: The muscle action is eccentric (step 3). Muscles on the superior surface of the glenohumeral joint and posterior surface of the elbow are lengthening (step 5). Therefore, the downward phase of the overhead press is controlled by eccentric action of the glenohumeral abductors (supraspinatus, anterior deltoid, middle deltoid) and elbow extensors (primarily triceps brachii).

For the purpose of illustration only, consider the same movement, this time moving fast. In this action, the arms and barbell are rapidly snapped back to the chest. This would *not* be a recommended movement because of the risk of injury.

Step 1: The movements are glenohumeral adduction and elbow flexion.

Step 2: The external force (gravity) tends to adduct the glenohumeral joint and flex the elbow.

Step 3: The movements are the same as those created by the external force, so ask yourself, "What is the speed of movement?" The speed is fast, which dictates that the controlling muscles are actively shortening in a *concentric action* at both joints.

Step 4: Movement occurs in the frontal plane about axes through the glenohumeral and elbow joints.

Step 5: The muscles on the glenohumeral joint's inferior surface are shortening during the movement, while the muscles on the superior side are lengthening. At the elbow, the anterior muscles are shortening and the posterior muscles are lengthening.

Step 6: The muscle action is concentric (step 3). Muscles on the inferior surface of the glenohumeral joint and anterior surface of the elbow are shortening (step 5). Therefore, rapid movement during the downward phase of the overhead press would be produced by concentric action of the glenohumeral adductors (pectoralis major, latissimus dorsi, teres major) and elbow flexors (biceps brachii, brachialis, brachioradialis).

Coordination of Movement

A skilled mover often is described as being coordinated. We, of course, have an intuitive idea of what this means, but what exactly is coordination? In general terms, **coordination** is the harmonious functioning of parts for effective results. This general definition fits movement coordination quite well. Only through the harmonious functioning of our body's parts (anatomical structures and systems) can we achieve effective results (smooth and efficient movement). More specifically, coordination requires that various muscles work together with correct timing and intensity to produce or control a movement. Muscular coordination is required for all movements, from simple to complex. Skilled performers of all kinds have a special ability to recruit the appropriate muscles, at the right time and with the right level of activation, to produce and control movement of individual segments and the body as a whole. Subsequent chapters provide many examples of coordinated actions controlled by the body's neuromuscular system.

Movement Efficiency and Economy

In biomechanical terms, **efficiency** refers to how much mechanical output (work) is produced for a given amount of metabolic input (energy). The ratio of mechanical output to metabolic input describes the efficiency of a process. Muscle, for example, has an overall efficiency of about 25%, meaning that only one-quarter of the metabolic energy goes toward performing mechanical work, while the other three-quarters is converted to heat or used in energy recovery processes. Studies have shown that muscle is more efficient during eccentric action when compared to concentric action (e.g., Ryschon et al., 1997). Efficient movements are characterized by relatively high work output with low metabolic energy expenditure.

A related measure, movement **economy**, is not synonymous with efficiency. Economy refers to how much metabolic energy is required to perform a given amount of work. In brief, efficiency applies to constant-energy conditions (i.e., how much work can be done using a given amount of energy), while economy corresponds with constant-work situations (i.e., how much energy is required to perform a constant amount of work). The relationship between efficiency and economy is shown in figure 6.11.

In most instances, movement efficiency is desirable. Common sense dictates that we want to produce as much work as possible with a minimum metabolic energy expenditure. In some cases, however, efficiency might not be the highest priority. When confronted with a dangerous situation, for example, the objective might be to move as quickly as possible or to generate as much force as possible, with little or no consideration for efficiency. In general, however, movement efficiency can be considered an important goal, and inefficiencies detract from the effectiveness of movement.

Applying the Concept

Sit-to-Stand

One of the activities of daily living is functional mobility, or *transferring*. Arising from a sitting to a standing posture (i.e., sit-to-stand, or STS), an essential transfer task, requires coordinated joint movement of the trunk, hips, arms, knees, and ankles. Typically, forward trunk lean (i.e., hip flexion) initiates arising from a seated position, followed by simultaneous hip and knee extension and ankle plantar flexion.

This coordinated pattern may be altered by a variety of conditions. For example, obese people exhibit less forward trunk lean and move their feet backward from the initial position. In doing so, obese people show greater knee joint torques than hip joint torques in contrast to nonobese people, who have higher hip joint torques than knee joint torques when performing STS (Sibella et al., 2003). Compared to nonobese controls, obese people also show 50% greater hip abduction angles during the STS movement (Huffman et al., 2015).

Medical conditions also may alter sit-to-stand strategies. Studies have shown that people with Parkinson's disease (PD) have different movement patterns than healthy controls when performing STS. For example, those with PD generate lower and more prolonged hip joint torques in STS (Mak et al., 2003) and exhibit bilateral differences in knee angle at seat-off, peak vertical ground reaction force, and peak torque (Ramsey et al., 2004). Ramsey and colleagues concluded that the inability to produce equal forces on both lower limbs may be an indicator for the increased propensity for falls in people with Parkinson's disease.

Several actions or conditions can contribute to movement inefficiency:

• *Muscular coactivation.* The net torque required at a joint to perform a given movement is determined by the sum of the torques created by all the muscles acting at that joint. The most efficient way to generate a needed torque is to activate muscles on only one side of the joint (i.e., no coactivation of muscles on the opposite side). Antagonistic coactivation works against the action of agonist muscles attempting to move the joint and, strictly speaking, contributes to inefficiency. Keep in mind, though, that coactivation can provide certain performance advantages (as explained earlier).

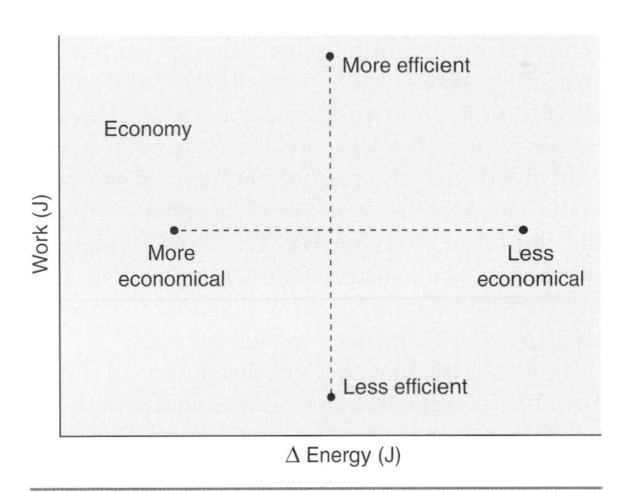

FIGURE 6.11 The relationship between efficiency and economy.

• *Jerky movements.* Movements characterized by rapid starts and stops or by alternating changes of direction are inefficient because metabolic energy is required to decelerate and accelerate the segments. If mechanical work is determined by the overall movement from one point to another, then jerky movements performed in getting there contribute to inefficiency. For example, the gait of a child with cerebral palsy may include spastic limb movements that require considerable metabolic energy expenditure, thereby increasing the effort required to walk from one place to another.

- *Extraneous movements.* Nonessential movements that do not directly contribute to completing a movement task nonetheless use metabolic energy and thus are inefficient. A runner, for instance, who adopts a peculiar running style that involves swinging his arms in a windmill fashion would run inefficiently because circumduction of the arms is extraneous to running. Some arm movement certainly is required to assist in balance and to run effectively, but excessive arm movement reduces efficiency.
- *Isometric actions.* As described in chapter 5, mechanical work is calculated as the product of force and displacement. Because isometric muscle actions involve no movement, they produce no mechanical work. Thus, the metabolic energy expended in producing isometric force is wasted in terms of producing mechanical work, but may be necessary to ensure stability.
- *Large center of gravity excursions.* Metabolic energy is required to raise and lower the body's center of gravity. In walking, for example, some vertical oscillation of the center of gravity is necessary when moving horizontally from one point to another. Excessive up and down motion of the center of gravity, however, is superfluous to the intended task and thus makes the movement less efficient.

Muscle Redundancy and Abundancy

Detailed exploration of the many mechanical and physiological factors involved in determining how much force a muscle needs to generate is beyond the scope of this text (see the suggested readings), but presentation of one such factor will give you an idea of the complexity of the nervous system's task.

Consider a simple uniplanar hinge joint such as the elbow. To flex this joint, how many muscles are *minimally* required? One, of course. Most joints in the body, however, have more than one muscle crossing each side of the joint. At the elbow, for example, three muscles (biceps brachii, brachialis, brachioradialis) perform flexion. Having more than the minimally required number of muscles creates what historically has been termed **muscle redundancy**, or more generally, *motor redundancy*, as described by Nicholas Bernstein in his classic work, *The Co-ordination and Regulation of Movements* (1967). Over the past two decades, an alternative approach to the problem has been proposed. Noted neurophysiologist Mark Latash concluded that "the famous problem of motor redundancy is wrongly formulated; it has to be replaced by the principle of abundance, which does not pose computational problems for the central nervous system" (2016, p. 7). This condition of **muscle abundancy** presents the nervous system with a challenging problem. In a task that requires less than maximal effort of all involved muscles, how does the nervous system decide how much force each muscle should provide for each movement cycle?

Consider the case of a simple elbow curl (flexion) exercise with a 10 lb (4.55 kg) dumbbell. For most people, this is a submaximal task, meaning that none of the three elbow flexors is maximally activated in flexing the joint. Curling the dumbbell requires a certain net moment, or torque. Each of the three elbow flexors supplies a portion of this moment. Each muscle's moment is determined by the product of its respective force and moment arm, as described in the previous chapter. So how does the nervous system determine how much force will be provided by the biceps brachii, brachialis, and brachioradialis? Is the responsibility divided neatly into thirds, with each muscle making an equal contribution? Does each muscle make the same contribution during each repetition of the exercise? Is there some optimizing strategy employed by the nervous system? As the muscles fatigue, how does the division of labor change to meet the continuing demands of the task? These questions, and many others, are not readily answerable. And the questions become all the more complex when considering the dynamics of multijoint movements in three-dimensional space. The area of nervous system decision making remains one of the continuing challenges in human movement studies.

Fortunately, from a practical standpoint, we don't need to worry much about *how* the nervous system accomplishes these complex tasks. In most cases, our neural circuitry handles the challenge without our conscious involvement. The body takes into account many variables, such as muscle fiber length, cross-sectional area, attachment sites and angles of pull, fiber type, and joint angle, to name but a few, and decides how much force each muscle contributes. Remarkably, the body's neuromuscular system processes all these details with little conscious effort.

Movement Assessment

We all assess movement. Some do it professionally as clinicians, therapists, teachers, coaches, or exercise scientists. Others informally assess movement just by observing humans moving, whether watching a baby's first steps, a child at play, or the slow and crouched gait of an 80-year-old. Most movement assessment is made qualitatively, as when a coach evaluates her athletes and corrects their sport technique. In some instances, however, precise quantitative measures are needed to make fine distinctions between movements. Exercise scientists and clinicians are among those who need the information provided by such quantitative analyses to diagnose a problem, assess the causes of movement dysfunction, improve performance, or increase the safety of a particular movement.

Quantitative analysis usually falls under the domain of biomechanics, the interdiscipline that applies mechanical principles to the study of biological organisms and systems. Professionals trained in biomechanics have the tools to perform detailed quantitative assessment of many aspects of human movement. Some studies measure kinematics (i.e., movement description without regard to the forces and torques involved) of a particular movement pattern. Typically, more complex studies measure the kinetic (i.e., force-related) factors involved in performing a task or movement pattern. Some of the tools used to measure the kinematics and kinetics of human movement are described in this section.

Kinematic Assessment

Substantive quantitative evaluation of movement was impossible until appropriate technologies were developed in the 19th and 20th centuries. Photography, developed in the 1830s, permitted the capture of still images from which measurements of position could be made. Nineteenth-century scientists, such as the influential French scientist Etienne-Jules Marey, used photographic techniques (along with other technologies) to capture human motion. Pioneers such as Marey laid the groundwork for the development of cinematography, or motion pictures. Although most people associate motion pictures with movies made for entertainment, it is interesting to note that *scientific* cinematography actually predates entertainment cinematography.

Who invented motion pictures? Historical evidence paints an unclear picture, with various people given credit. Unquestionably, though, two men played an important role. The first was Marey. The other was Eadweard Muybridge, a noted photographer of the late 1800s. Muybridge's photographic collections on human and animal locomotion are classics in the field (figure 6.12).

Muybridge published countless panels of sequential still photographs that depict many kinds of movements. With proper equipment, such image sequences could be projected rapidly in succession and transformed into motion pictures. Muybridge's role in this development has an interesting (see sidebar) and important history. His contributions, while lacking scientific rigor, are enduring. "The richness of his recordings testify to the importance that the language of images, and particularly the language of moving images, was to have in scientific research, documentation, and modern communication" (Tosi, 1992, p. 57).

Cinematography proved an indispensable instrument for the observation, recording, and evaluation of human movement during much of the 20th century. In recent decades, cinematography has been replaced, in many instances, by videographic technologies. Although some laboratories still use film for image capture and movement analysis, the many advantages of videotape (e.g., lower cost, no need to develop film) made videography the dominant medium in recent decades.

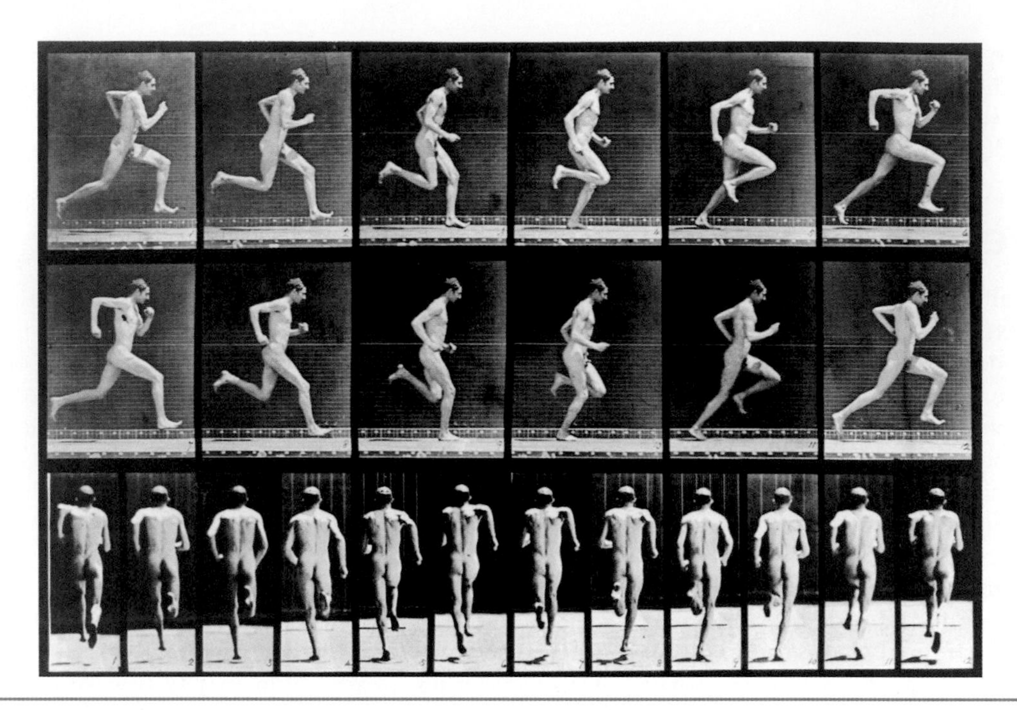

FIGURE 6.12 Sequential photographs by Muybridge of human running.
Reprinted from Library of Congress. LC-USZ62-115103

Sophisticated computerized analysis systems now can provide detailed assessment of human movement. Current systems are based on one of several different technologies. Each technology has its own advantages and limitations, and researchers must determine which system best suits their particular needs. Because some systems are expensive, available resources often dictate the type of system used.

Many systems in the recent past employed videography to perform frame-by-frame analysis of images captured by standard video cameras or by more sophisticated (and expensive) high-speed video cameras. Standard video cameras capture sequential images at a rate of approximately 30 frames per second (fps). Some analysis systems can take these 30 fps images and split them electronically to yield an effective frame rate of 60 fps. This speed is adequate for analyzing many, but not all, human movements. Walking and weightlifting, for example, can be reasonably evaluated at a frame rate of 60 fps.

Faster movements, such as throwing or kicking, or ballistic (i.e., explosive) sports movements (e.g., swinging a golf club or baseball bat) require higher frame rates. To analyze these movements effectively, a movement scientist needs a camera capable of frame rates of 120 fps, 240 fps, or even higher.

Frame-by-frame video images are analyzed with the assistance of a computerized system in a process called **digitization**. Each image is digitized to identify the specific location (coordinates) of anatomical landmarks of interest. The digitized coordinates then can be used to describe the mechanical characteristics (kinematics) of the movements. For example, to describe the lower-extremity movements of a child with cerebral palsy, a clinical researcher might be interested in the movement patterns of the hip, knee, and ankle joints. To quantify these patterns, the locations of joint centers would be digitized and the resulting information used to describe the joint movements.

New technologies have emerged that allow for faster and more sophisticated movement analysis. Optoelectronics, for example, combines the use of electronics and light in multicamera systems that can provide real-time three-dimensional kinematic data. The rapid development of newer technologies such as motion sensors, virtual reality (VR), and wearable strain sensors will take human movement analysis to an even higher level of sophistication and insight.

Applying the Concept

Muybridge and the Stanford Experiment

In the late 1800s, horse racing was both a popular sport and social event. Among its many devotees was Leland Stanford, a wealthy U.S. industrialist. Story has it that Stanford, who was aware of Eadweard Muybridge's work, commissioned him to provide photographic evidence to settle a purported wager as to whether all four of a horse's hooves were off the ground at any point during its running cycle. Muybridge set up a series of cameras on Stanford's farm in Palo Alto, California, and confirmed that horses do indeed have a flight phase, albeit brief, when no hooves are in contact with the ground. As a historical side note, the farm where Muybridge took his photographs later became the site of the now internationally recognized Stanford University.

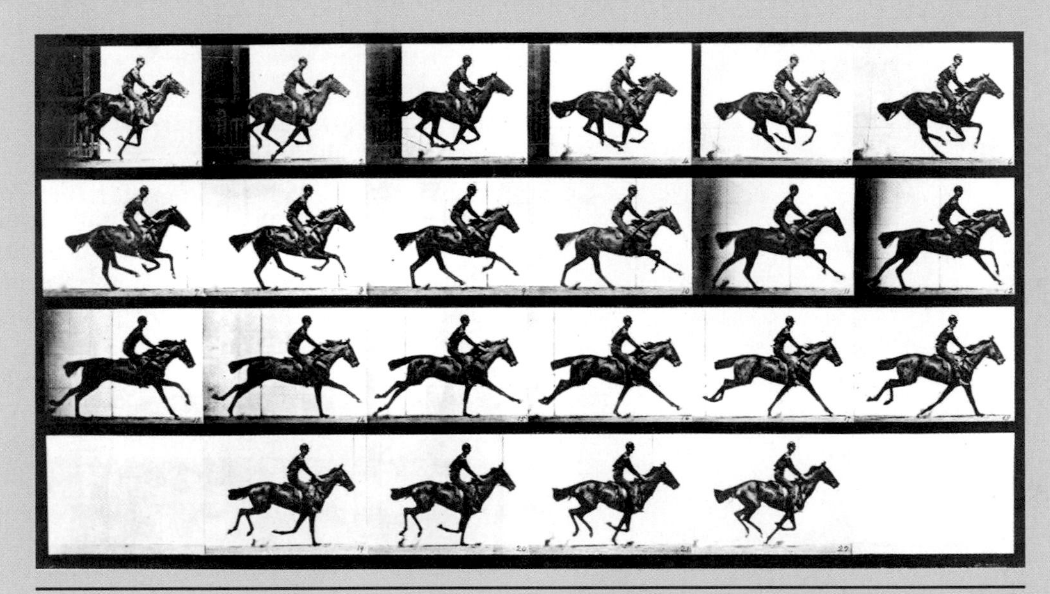

MUYBRIDGE'S PHOTOGRAPHIC evidence showing a galloping horse with a brief period without ground contact (top row).

Reprinted from Library of Congress. LC-USZ62-119473

The need for movement analysis crosses many scientific, industrial, and entertainment areas. Clinicians use motion analysis systems to assess the movement patterns of special populations (e.g., stroke patients) to better identify appropriate treatment interventions and rehabilitation programs, sport scientists use movement analysis to identify the nuances of movement patterns in elite athletes in hopes of finding ways to improve performance and minimize the risk of injury, and ergonomics experts employ motion analysis to assess movements in the workplace to improve worker productivity and efficiency.

In recent years, the most visible application of motion analysis has been in the entertainment industry. Rapid advances in imaging and computer technology have afforded animation artists, cinematographers, and video-game developers the tools to produce remarkably lifelike dynamic images. Experts in these areas use sophisticated motion capture technologies to collect detailed data on the movement patterns of human performers and then use these data to replicate movements in animated form. The astonishing rate of technological advancement in the areas of motion capture and computer animation foretells even more remarkable movement imagery in the future.

FIGURE 6.13 Force platform, or force plate.
Andrius Ramonas

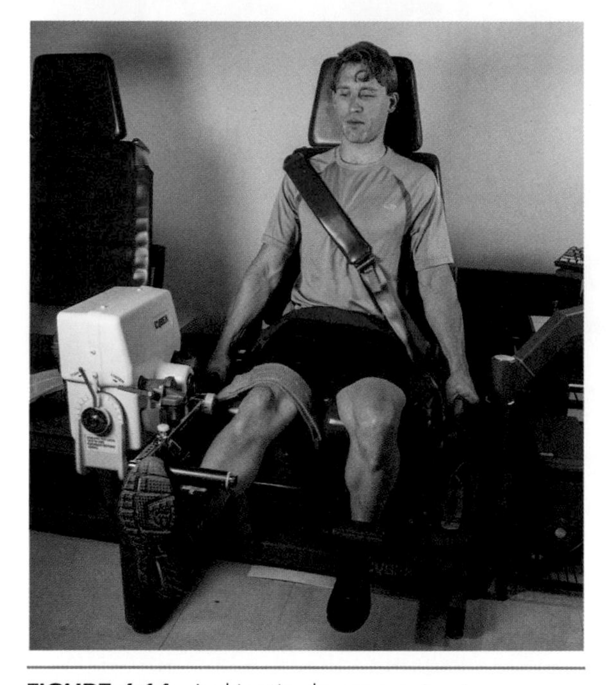

FIGURE 6.14 Isokinetic dynamometer to measure joint strength and speed.

Kinetic Assessment

Descriptive information (kinematics) sometimes is all that is needed to answer the questions or solve the problems of movement analysts. In some cases, however, it is also important to know the causes or mechanisms underlying the observed movements. The force-related (kinetic) variables involved in movement are measured in a variety of ways.

Two of the most commonly used tools to measure force are force plates and isokinetic dynamometers. A force plate (figure 6.13) is a rigid metal platform with embedded sensors that measure the force applied to the plate. Using the principle of Newton's third law of motion, the force plate measures the equal and opposite forces generated between the body (usually through the foot) and the ground. These measured ground reaction forces are useful in understanding the loads the body experiences during certain activities and can be integrated into complex mathematical models to predict the forces and torques (moments) acting at major joints in the body (e.g., ankle, knee, hip) during tasks such as walking, running, and jumping.

Isokinetic dynamometry is widely used in both clinical and research settings (figure 6.14). Dynamometers measure kinetic characteristics of isolated joint motions (e.g., knee flexion and extension) at specific and nominally constant angular velocities. Isokinetic devices can measure parameters such as net joint torque, work, and power and are used to assess muscle strength in controlled, dynamic situations. Isokinetic dynamometry has been widely used over the past quarter century across many areas of study, including rehabilitation and sports medicine.

Electromyography

In chapter 4, we discuss the electrical characteristics of the excitation–contraction coupling process of skeletal muscle and mention how we measure the electrical activity developed by muscle during the force-generating process using electromyography (EMG). Electromyography is used across a wide variety of disciplines ranging from basic research on skeletal muscle mechanics to application in occupational, sport, and clinical settings.

As a clinical tool, EMG is useful in assessing the characteristics of normal voluntary muscle activation and comparing this with the activity patterns of patients with muscle weakness, paralysis, spasticity, or neuromuscular lesions. EMG also is used clinically to measure nerve conduction velocities. Clinical electromyography provides physicians and clinicians with valuable information to assist in diagnosis, treatment, and rehabilitation of neuromuscular disorders.

Electromyography also has application in assessing the coordination (timing and activation level) of muscles during movement tasks. Kinesiological electromyography, as this area is called, employs EMG to measure muscle activity patterns in sport, occupational, and rehabilitation settings to assist in identifying appropriate training programs for performance enhancement.

Electromyography as an area of study traces its roots to the work of Italian anatomist Luigi Galvani (1737-1798), who in the late 18th century observed electrical activity in the muscles of frogs' legs. Advances in electromyographic instrumentation and experimentation, and thereby in our understanding of neuromuscular physiology and mechanics, progressed rapidly in the 19th and 20th centuries. Among the notable scientists engaged in electromyographic research were early pioneers such as Emil Du Bois-Reymond and C.B. Duchenne du Boulogne. Much of what we have learned about the functional aspects of movement control using EMG is summarized in the classic work *Muscles Alive* (Basmajian & DeLuca, 1985). More recent developments are discussed at length in a variety of sources (e.g., Criswell, 2010; Kamen & Gabriel, 2009; Preston & Shapiro, 2012).

Electromyography is one of the principal methods for examining the actions of skeletal muscle in reflex and voluntary movements. EMG essentially is a composite record of momentary electrical events immediately preceding and during active contraction of skeletal muscle. To record the electrical activity of muscle contraction, small electrodes are placed in the muscle (called *indwelling* or *intramuscular electrodes*) or on the skin overlying a superficial muscle or muscle group (*surface electrodes*). Each type has its advantages and disadvantages.

Indwelling electrodes, typically made of fine wires, are inserted directly into the muscle of interest using a hypodermic needle. Advantages include on-site measurement with no time delay or signal attenuation, access to deep muscles, and absence of cross-talk (i.e., electrical "noise" from adjacent muscles). Disadvantages include pain and discomfort, potential for movement alterations due to inserted electrodes, cost, risk of infection due to invasiveness of the procedure, and the requirement of a trained and certified technician.

Surface electrodes, in contrast, are cost-effective, easy to administer, and relatively safe. Drawbacks include limited use for superficial muscles only, risk of cross-talk from nearby muscles, time delay in signal reaching the electrodes, and signal attenuation.

Electrodes typically are placed in or above the muscle belly (i.e., the middle of the muscle). A standard control record normally is made at the outset of an experimental study in the form of a **maximum voluntary contraction** (MVC). The participant is instructed to generate as much force as possible while EMG data is collected. The MVC can be completed *isometrically* against a fixed resistance (IMVC) or *dynamically* against resistance while moving a joint (e.g., using an isokinetic dynamometer, see figure 6.14) in a DMVC. The amount of EMG during prescribed movement tasks is then compared to the MVC. Even though the MVC is termed a maximal contraction, given its voluntary nature, certain movements can elicit more than 100% of the MVC.

Many movement-specific electromyographic studies are presented in subsequent chapters as we explore fundamental human movements, movement dysfunction in clinical practice, and specialized areas of exercise, sport, and dance.

Concluding Comments

Our muscles miraculously control movement with little conscious effort. The nervous system controls these body engines and calls on them when needed to perform all of our daily movement tasks. In a simple lifting task, for example, we do not have to consciously say, "OK, biceps, it's your turn to work. Hey, triceps, not so much there. Take it easy." Our neuromuscular system handles the complex chore all on its own—and usually does so with deceptive ease.

Go to the web study guide to access critical thinking questions for this chapter.

Suggested Readings

Basmajian, J.V., & DeLuca, C. (1985). *Muscles alive* (5th ed.). Baltimore: Williams & Wilkins.

Bernstein, N.A. (1967). *The co-ordination and regulation of movements.* Oxford: Pergamon Press.

Cappozzo, A., Marchetti, M., & Tosi, V. (Eds.). (1992). *Biolocomotion: A century of research using moving pictures.* Rome: Promograph.

Latash, M.L. (2008). *Neurophysiological basis of movement* (2nd ed.). Champaign, IL: Human Kinetics.

Latash, M.L., & Zatsiorsky, V. (2016). *Biomechanics and motor control: Defining central concepts.* Cambridge, MA: Academic Press.

Robertson, D.G.E., Caldwell, G.E., Hamill, J., Kamen, G., & Whittlesey, S.N. (2014). *Research methods in biomechanics* (2nd ed.). Champaign, IL: Human Kinetics.

PART 

Fundamentals of Movements

The three chapters of part III present a summary of fundamental human movements. Chapter 7 (Posture and Balance) examines basic terminology and concepts related to posture and balance, including types of posture and postural control, mechanisms of postural control, and postural alterations and perturbations. It includes a discussion of developmental considerations and postural dysfunction. Chapter 8 (Gait) presents terminology and concepts related to human gait (walking and running), including discussion of the gait cycle, life-span issues, forms of pathological gait, and running-related injuries. This part concludes in chapter 9 (Basic Movement Patterns) with an exploration of general concepts and applications pertinent to the basic movement patterns of jumping, kicking, lifting, throwing, and striking.

7

Posture and Balance

Objectives

After studying this chapter, you will be able to do the following:

- Describe movement concepts related to posture and balance
- Describe standing, sitting, and lying postures
- Explain the mechanisms of postural control
- Describe developmental aspects of balance and balance dysfunction

Movement is essential for all life functions, from motion at the molecular level to whole-body actions. Within the body, movement of blood through the cardiovascular system and air through the respiratory system, for example, serves life-sustaining functions. Fundamental tasks such as eating require movement as well. On a larger scale, one of the most necessary things we do is move from one place to another, whether we are an infant crawling, an adult walking, or an athlete sprinting or jumping.

Why move? We move because it is necessary for our survival. We move because it is essential for performing tasks of daily living. We move because it allows us to experience the joy and exhilaration of dance, physical exercise, and competitive sports. As British author Laurence Sterne notes, "So much of motion, is so much of life, and so much of joy" (1980, p. 345).

This chapter explores the fundamental movement forms of posture and balance. Our description of each movement form in this and subsequent chapters includes assessment of muscle control and takes a life-span approach that includes discussion of developmental aspects and how the movements differ in children, adults, and older individuals. We consider, as well, movement dysfunction due to injury, disease, or congenital factors.

Fundamentals of Posture and Balance

From the first hesitant steps of a 1-year-old child to the remarkable feats of balance seen in circus performers and daredevils, the human body often must struggle to maintain its balance. To do this, people use information from several sensory systems to align their body segments. This positioning, or alignment, of the body is the essence of posture. In much the same way as stacking blocks crookedly makes for an unstable structure prone to falling, crooked body postures can be unstable. To produce a stable posture, the body's segments must be aligned correctly, just as the architect of a skyscraper must begin with a strong foundation and precisely align each successive story one upon another.

Proper posture and balance are essential for the efficient performance of both simple daily tasks and more complex movement patterns. Improper posture and loss of balance can result in poor performance, injury, or even death.

Posture can be defined simply as the alignment, or position, of the body and its parts, or more expansively as "a position or attitude of the body, the relative arrangement of body parts for a specific activity, or a characteristic manner of bearing one's body" (Smith, Weiss, & Lehmkuhl, 1996, p. 401). Clearly, these definitions are not restricted to a single alignment, or posture, but rather suggest an infinite number of possible postures. One might assume, for example, a standing, sitting, or lying posture, or some other distinctive posture appropriate for a particular purpose or task.

To maintain a particular posture or change from one posture to another requires control of the body's alignment. This control of posture is termed balance. **Balance** can be defined as the maintenance of postural stability, or equilibrium, and often is used synonymously with the term **postural control**.

The concepts of posture and balance, although distinct from one another, are critically interdependent. Static postures often are difficult to maintain because the body and its parts continuously respond to forces (e.g., gravity, muscle) that tend to alter body alignment and disrupt the static equilibrium of the system. The body must continuously make subtle adjustments, usually through muscle actions, to maintain the equilibrium necessary for postural stability.

Functions of Posture and Balance

Posture and balance obviously are essential for all tasks we perform. Improper posture or loss of balance can negatively affect performance, decrease movement efficiency, and increase the risk of injury. Proper posture serves three functions: (1) maintenance of body segment alignment in any position: supine, prone, sitting, and standing; (2) anticipation of change to allow engagement in voluntary, goal-directed movements; and (3) reaction to unexpected disturbances, or perturbations, in balance (Cech & Martin, 2011).

Posture should not be considered a static phenomenon. Even while people stand or sit still, there are inevitable small fluctuations in their positions, or posture. Postural positions are maintained within ranges of movement. For a guard at Buckingham Palace, whose intent is to remain perfectly still, these fluctuations are imperceptible; for someone waiting in line to purchase a movie ticket, the movements are considerably greater. Thus, posture from one moment to the next can involve varying amounts of movement, referred to as **postural sway**. Control of postural sway is maintained by muscle action.

Types of Posture

Most of us have been receiving postural advice since childhood. Parents admonish their children to stand up straight and pull their shoulders back. Drill instructors scream at military recruits to stand at attention. Teachers advise their students to sit tall and tell them not to lean back in their chairs. These admonitions try to persuade us to maintain good posture. They also suggest several different types of posture.

Static Postures

Many of our waking hours are spent either standing or sitting. Because these postures typically involve little movement, we refer to them as **static postures**. While resting or sleeping, we usually assume some type of static lying posture. Note that static postures are not completely motionless. Stationary, or static, postures typically involve slight movement, or swaying. Therefore, they sometimes are also referred to as **steady-state postures**.

Standing Posture

The notion of a normal posture might erroneously be interpreted as meaning there is a single best posture. Given the variability in anatomical structure and physiological function (see chapter 1), no single posture is recommended for everyone. The normal posture for individuals depends on many factors, including their body type, joint structure and laxity, and muscular strength. Despite these inherent interindividual differences, certain characteristics are associated with good upright posture. In upright standing, these characteristics include the following:

- Head is held in an erect position.
- Body weight is distributed evenly between the two feet.
- From a frontal view, bilateral structures (e.g., iliac crests, acromion processes) are at the same horizontal level.
- From a sagittal-plane (side) view, the line of gravity passes posterior to the cervical and lumbar vertebrae, anterior to the thoracic vertebrae, posterior to the hip joint, and anterior to the knee and ankle joints (figure 7.1 and table 7.1).
- Appropriate spinal curvatures are evident in the cervical, thoracic, and lumbar regions.

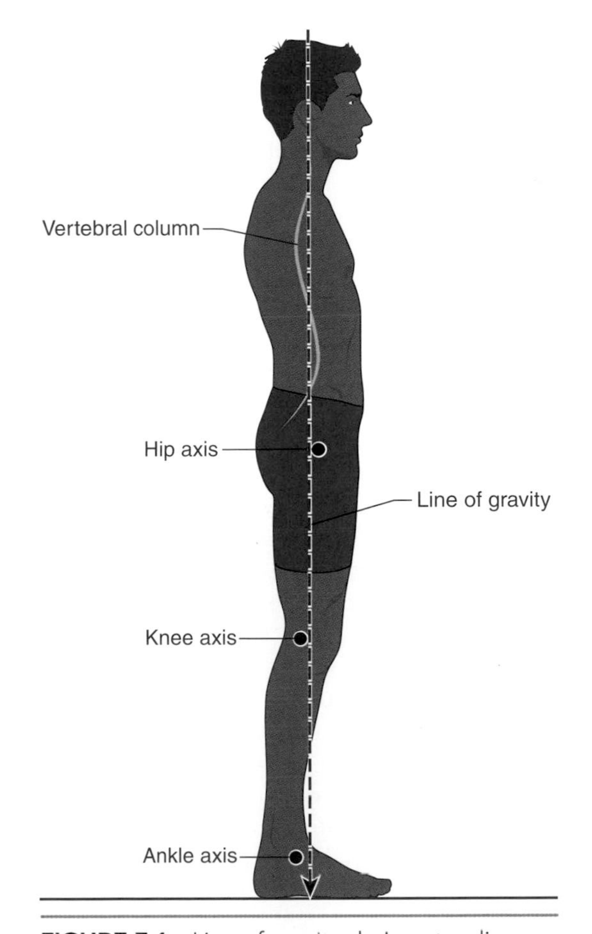

FIGURE 7.1 Line of gravity during standing.

TABLE 7.1 Normal Alignment in the Sagittal Plane

			OPPOSING FORCES	
Joint	**Line of gravity**	**Gravitational moment**	**Passive opposing forces**	**Active opposing forces**
Atlanto-occipital	Anterior Anterior to transverse axis for flexion and extension	Flexion	Ligamentum nuchae Tectorial membrane	Posterior neck muscles
Cervical	Posterior	Extension	Anterior longitudinal ligament	
Thoracic	Anterior	Flexion	Posterior longitudinal ligament Ligamentum flavum Supraspinous ligament	Extensors
Lumbar	Posterior	Extension	Anterior longitudinal ligament	
Sacroiliac	Anterior	Flexion-type motion	Sacrotuberous ligament Sacrospinous ligament Sacroiliac ligament	
Hip	Posterior	Extension	Iliofemoral ligament	Iliopsoas
Knee	Anterior	Extension	Posterior joint capsule	
Ankle	Anterior	Dorsiflexion		Soleus

Adapted from Levangie & Norkin (2011).

This upright posture is uncomfortable if held for an extended period. Most people who must stand for prolonged periods adopt various alternative, and more comfortable, postures (Houglum & Bertoti, 2012). One alternative posture is asymmetric standing, characterized by a weight shift to one leg that is fully extended. By fully extending the knee, the line of gravity passes anterior to the knee joint, creating an extensor moment and reducing the need for quadriceps muscle activity (figure 7.2).

Another alternative posture adopts a wide base of support with both legs fully extended and arms held behind the back or crossed on the chest. A third alternative is the **nilotic stance**, in which the individual stands on one leg with the opposite leg used to brace the standing knee (i.e., like a flamingo).

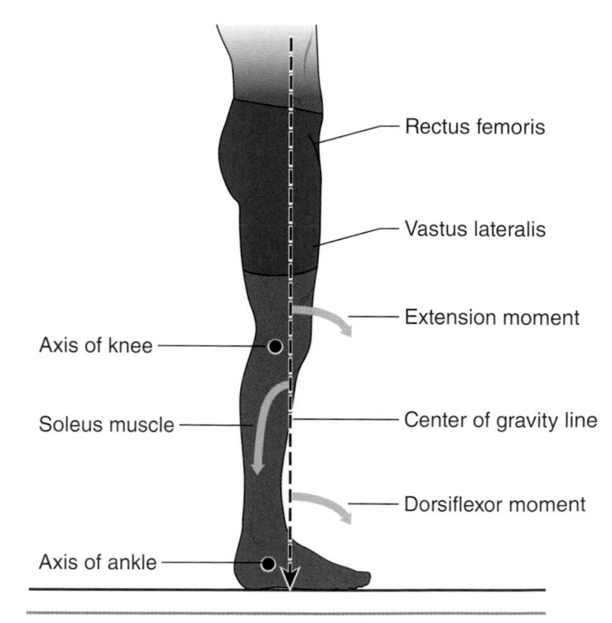

FIGURE 7.2 Effect of fully extending knee on quadriceps activation levels.

Sitting Posture

Many people spend long hours in a sitting, or seated, position, whether at home, work, or school. Correct sitting postures can reduce spinal loading and the chance of injury. Incorrect postures, in contrast, can increase the risk of injury. Historically,

Applying the Concept

Prolonged Standing

Perhaps the most familiar example of standing absolutely still for extended periods of time is provided by the Royal Guards at Buckingham Palace in London, England. They stand motionless for 2 hours per shift. The Guards' performance, impressive as it is, pales in comparison to the record for unaided prolonged standing. That record is held by Akshinthala Seshu Babu (India), who, in 2003, stood motionless for an amazing 30 hours and 12 minutes. In 2015, Babu attempted to break his own record. He stood still for a total of 35 hours, 22 minutes. However, his still posture was interrupted briefly because of an insect bite. Thus, Babu's 2003 record still stands (pun intended).

Prolonged standing (still or not), common in many occupations, has been associated with numerous biomechanical and physiological consequences, including slouching posture, muscular fatigue, sore feet, swollen legs, low back pain, varicose veins, carotid atherosclerosis, joint compression, and neck and shoulder stiffness.

an ideal sitting posture has been characterized by the ischial tuberosities acting as the major base of support, anterior pelvic tilt (which maintains appropriate lumbar curvature), spinal support provided by a slightly inclined seat back, and the feet contacting the floor to share in supporting the body's weight (figure 7.3*a*).

Poor sitting posture, usually characterized by a slouched position (figure 7.3*b*), results in posterior pelvic tilt (which flexes the lumbar spine and reduces the lumbar curvature), increased stretching and weakening of the posterior annulus fibrosis, and increased flexor torque created by an anterior shift in the line of gravity (Neumann, 2016). All these characteristics may increase the chance of lumbar disc injury and low back pain. A slouched posture also affects more than just the lumbar region. Increased lumbar flexion contributes to excessive flexion (kyphosis) in the thoracic spine and a protracted, or thrust-forward, head (see figure 7.3*b*). This head position puts added stress on the vertebrae, muscles, and ligaments of the neck and shoulders.

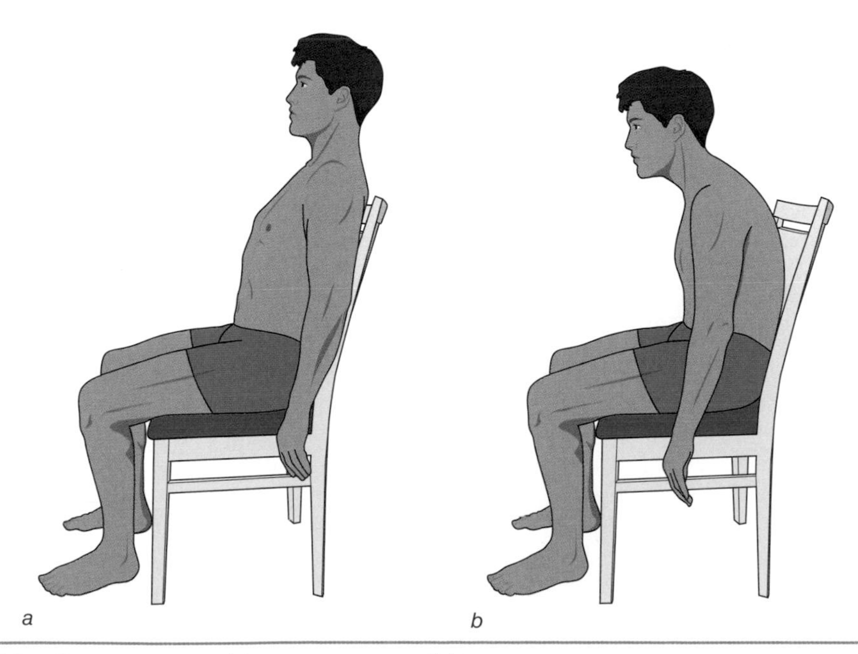

FIGURE 7.3 Sitting posture: *(a)* proper posture, *(b)* improper posture.

Research in Mechanics

Effects of Prolonged Sitting

Certain jobs require workers to sit for extended periods of time. Transit from one place to another can involve extensive sitting. And many people, by choice, sit for long stretches watching television, reading, or doing nothing much at all. While sitting for short periods can, of course, be relaxing, prolonged sitting can have harmful effects. Among the many documented effects are varicose veins, weakened muscles, postural dysfunction, sore neck and shoulders, shortened life expectancy, and increased risk of heart disease, diabetes, and certain cancers (Zhu & Owen, 2017).

The solution? Break up long periods of sitting with intermittent standing and stretching, work while standing instead of sitting, or vary sitting posture. But even that may not be enough. Studies have shown that even regular exercise may not be sufficient to counteract the negative effects of prolonged sitting. For example, in a study of more than 200,000 adults, van der Ploeg and colleagues concluded that "prolonged sitting is a risk factor for all-cause mortality, independent of physical activity. Public health programs should focus on reducing sitting time in addition to increasing physical activity levels" (2012, p. 494).

See the references for the full citations:

Van der Ploeg, Chey, Korda, Banks, & Bauman, 2012.

Zhu & Owen, 2017.

Well-designed chairs can facilitate proper sitting postures, while poorly designed chairs make good postures difficult and may contribute to musculoskeletal disorders such as intervertebral disc degeneration, low back pain, inflexibility, and loss of joint range of motion.

The notion that an ideal sitting posture exists has been challenged. McGill (2016) convincingly argues that one of the greatest risks for low back pain is the prolonged nature of sitting, which increases the risk of disc herniation. Low back health is facilitated by properly using an ergonomic chair to facilitate frequent changes in sitting posture, periodically getting out of the chair and assuming a relaxed standing position, and performing an exercise routine at some time during the workday, preferably *not* in the early morning when the back is more susceptible to injury.

Lying Posture

When resting or sleeping, we normally assume a **recumbent**, or lying, posture because it is the least physiologically demanding. This posture rotates the action of gravity from its longitudinal orientation in standing (figure 7.4*a*) and sitting to a transverse direction relative to the long axis of the body. Basic lying postures include lying facedown (**prone**), lying face up (**supine**), and lying on one side or the other (figure 7.4*b*). Each of these lying postures has advantages and disadvantages.

Surface characteristics play an important role in spinal alignment and how forces are applied to the body. On a very hard surface, only specific body areas (e.g., hip, shoulder) come in contact with the surface (figure 7.5*a*). This creates localized pressure points that can be uncomfortable or even injurious. Sleeping on too soft a surface also can cause problems, including excessive lumbar flexion when in a supine position, exaggerated lumbar extension in a prone posture, and lateral spinal curvatures when side lying (figure 7.5*b*). The surface and head support (e.g., pillow) should be consistent with maintaining proper spinal alignment (figure 7.5*c*).

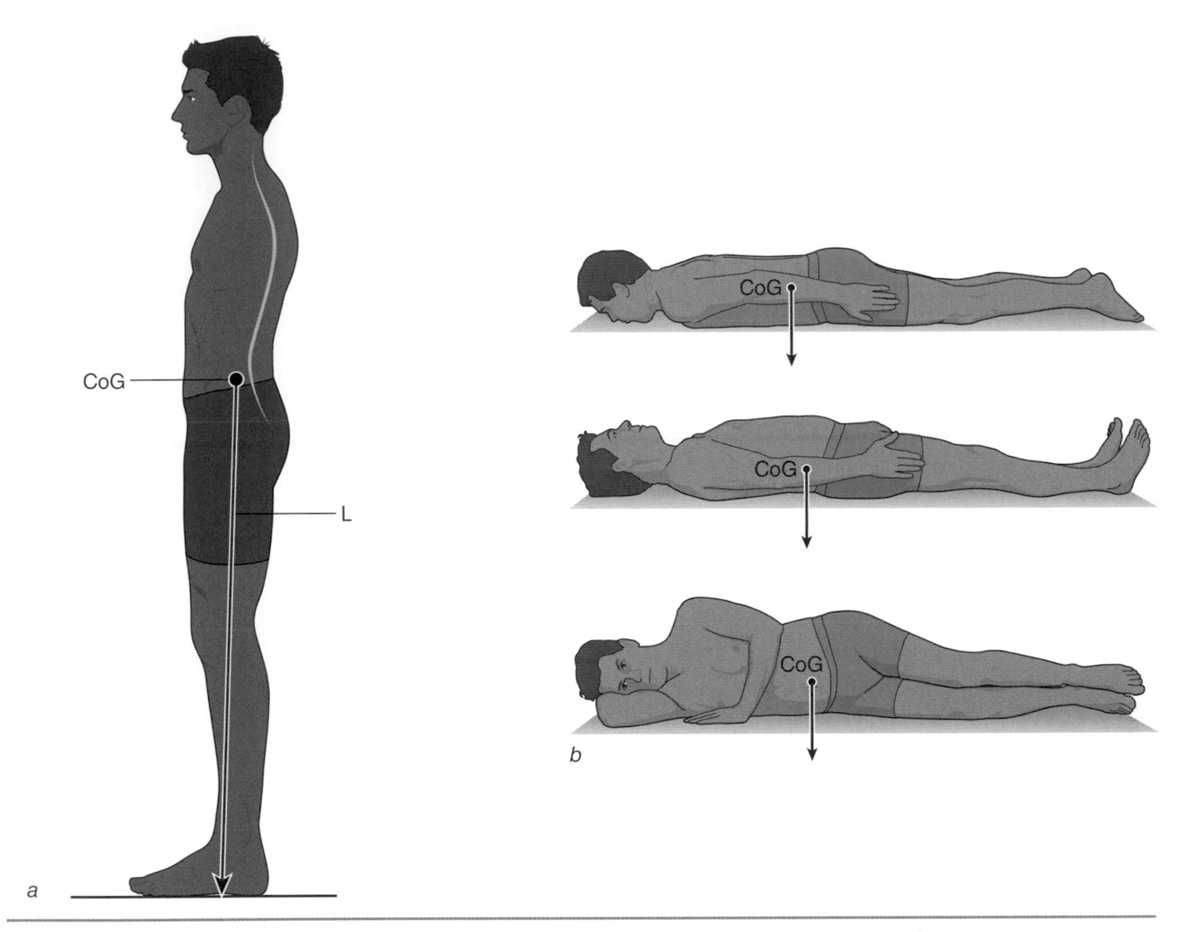

FIGURE 7.4 Line of gravity standing versus lying: *(a)* standing; *(b)* prone, supine, and side postures.

Dynamic Posture

Dynamic posture is the posture of motion, as seen in walking, running, jumping, throwing, and kicking. Each of these movements requires continual changes in the position of the trunk and extremities that must be controlled to maintain the dynamic equilibrium necessary for task completion. Loss of dynamic postural control can result in task failure (e.g., stumbling or falling) and possible injury.

Types of Postural Control

Balance, or postural control, is necessary in dynamic situations and for maintaining static (stationary) positions. There are four types of postural control: static, reactive, anticipatory, and adaptive (Cech & Martin, 2011).

Static postural control involves strategies for keeping the vertical projection of the body's center of gravity (i.e., line of gravity) within the base of support

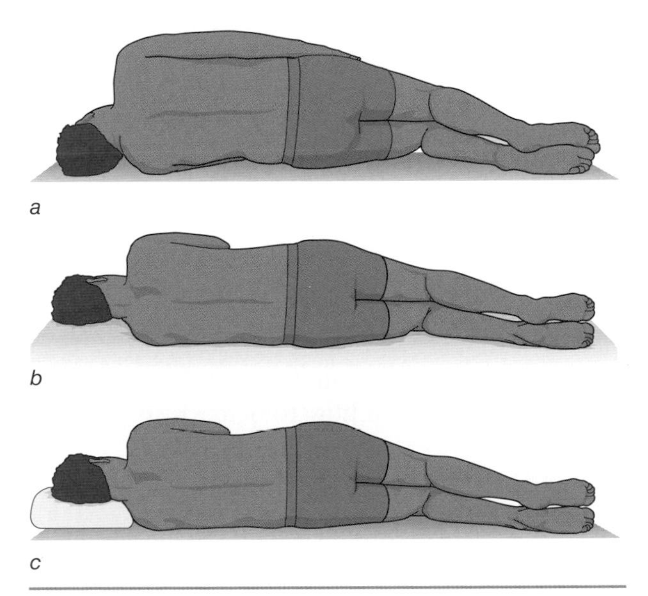

FIGURE 7.5 Lying postures and surfaces: *(a)* hard surface, *(b)* soft surface, *(c)* proper alignment with pillow and surface.

(BOS). In normal upright standing, the body commonly sways slightly from side to side and forward and backward. When the body sways in one direction, the body's line of gravity moves slightly away from the center of the BOS toward its edge. Arresting this movement and moving the line of gravity back toward the center requires muscle action. For example, if the body sways anteriorly (forward), creating a gravitational dorsiflexion moment (torque) at the ankle, the soleus must be recruited to create a plantar flexion countermoment (see chapter 5) to recentralize the line of gravity (figure 7.6). Similarly, a posterior (backward) body lean creates a gravitational plantar flexion moment at the ankle. The tibialis anterior must then be recruited to generate a dorsiflexion countermoment to bring the line of gravity back toward the center of the BOS and reestablish postural equilibrium.

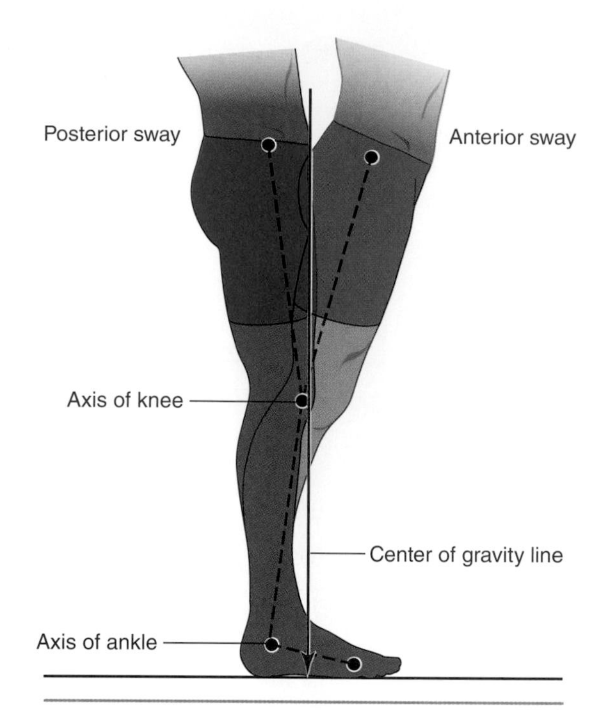

FIGURE 7.6 Postural sway: anterior sway and posterior sway.

When unexpected events (e.g., slipping or tripping) cause the line of gravity to move away from the center of the base of support, **reactive postural control** is necessary for maintaining balance and preventing a fall. If the line of gravity moves away from the BOS center but does not exceed the BOS, the neuromuscular system quickly recruits muscles to regain postural stability through what is termed a *righting response*. In contrast, if the line of gravity moves outside the base of support, the body must employ a different strategy to maintain balance. This strategy typically involves a change in the BOS, such as when a person trips, falls forward, and takes a step to prevent a fall. By taking this step to create a new and larger BOS, the person is exercising reactive postural control.

Anticipatory postural control involves making movements in anticipation of events that will cause postural changes and potential loss of balance. For example, before picking up a heavy suitcase with the right hand, the carrier leans her trunk to the contralateral (left) side to counteract the anticipated load of the suitcase, a rider standing in the aisle of a stationary bus leans forward or grabs the rails in anticipation of the bus's acceleration, and a football player about to collide with an opponent leans in the direction of the impact. In all these examples, the individual makes movements in advance of predicted disruptive events.

Adaptive postural control allows us to modify our movements in response to situational changes. Postural control depends on environmental conditions and task demands, and thus is context dependent. Control of posture is not limited to postural reflexes. Postural responses can adapt to repeated disturbances (perturbations), allowing us to alter our responses and effectively learn how to better respond to the postural disturbance (Enoka, 2015).

Mechanisms of Postural Control

Before the neuromuscular system can initiate corrective postural actions, it needs to receive information from the visual, vestibular, and somatosensory systems. This information, provided by the nervous system, communicates the nature of the physiological, mechanical, and informational perturbations affecting the body and allows the nervous system to recruit the appropriate muscles required to reestablish equilibrium (figure 7.7).

The postural sway inherent to standing is detected by sensory receptors in the eyes (visual), ears (vestibular), skin (tactile), and joints (proprioceptive). Based on the information supplied

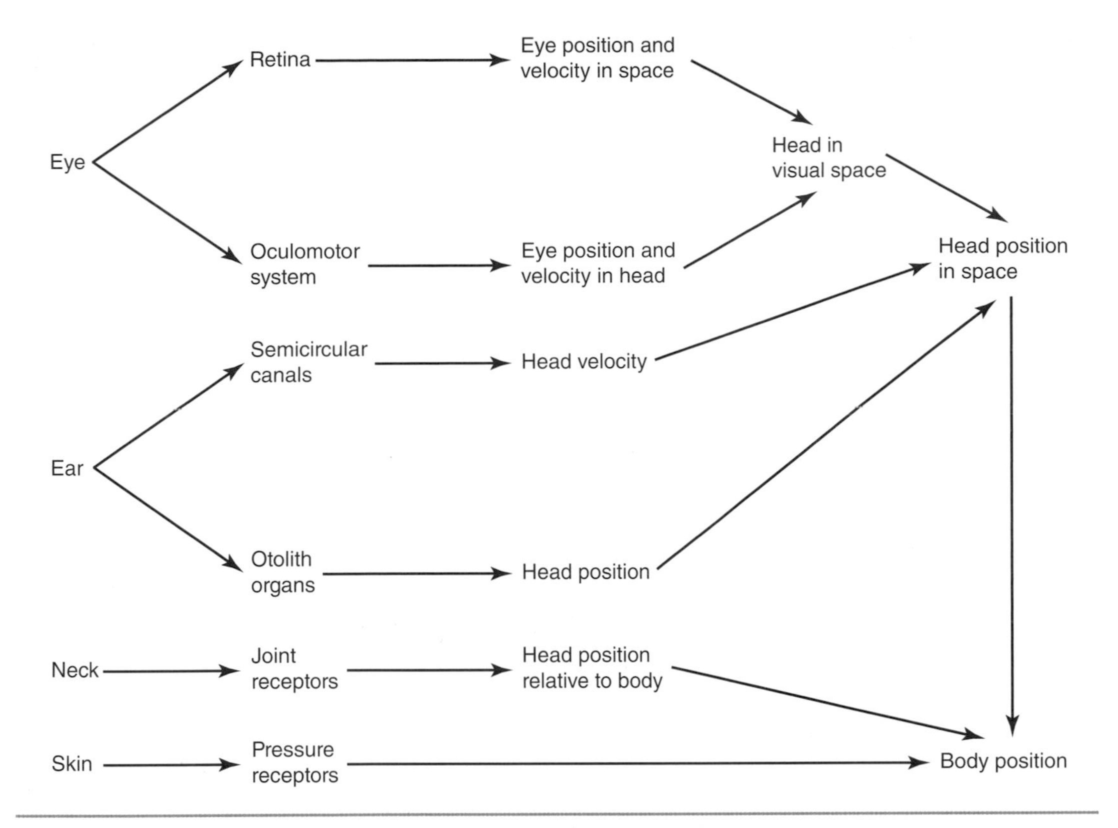

FIGURE 7.7 Nervous system's response to postural sway.

Reprinted from R.H.S. Carpenter, 1984, *Neurophysiology* (New York: Edward Arnold).

by these receptors, the nervous system responds by recruiting muscles to create moments (torques) that counteract the gravitational moments created by the line of gravity sway away from the center of the base of support.

Similarly, in a dynamic task, a gymnast attempting to walk, run, and jump along a narrow balance beam needs continuous sensory feedback to maintain balance and proceed along the beam. Compromised information from any of the sensory systems may result in loss of balance and a fall from the beam.

When someone stumbles or trips while walking, these same sensory systems rapidly provide feedback to the neuromuscular system, which then attempts to recruit the muscles needed to recover balance (e.g., by making a corrective step or reaching for a rail or table for support). Incorrect, incomplete, or delayed sensory information may lead to an unsuccessful recovery and a resulting fall.

Postural Alterations and Perturbations

Deviations from normal postures can be caused by physiological, psychological, environmental, and anatomical factors. A marathoner experiencing physiological fatigue may alter her running posture to compensate for altered muscle activation patterns and joint mechanics; a worker experiencing depression may walk with a lethargic, slumped posture; and an older adult navigating a slippery staircase might adopt a different stepping posture than would a younger person. Anatomically, abnormal posture may be caused by structural defects. For example, a person with a leg-length discrepancy (i.e., one leg longer than the other) may make compensatory postural adjustments, such as a lateral pelvic tilt, that alter joint mechanics.

Altered posture can also be caused by task-dependent disturbances of motion, or **perturbations**. The posture a warehouse worker uses to pick up a dropped pencil, for example, likely differs from the posture he adopts when picking up a heavy box.

Applying the Concept

Pushing the Limits of Balance

Most people pay little attention to balance, except when they lose it. In contrast, a few courageous individuals push the limits of balance. Ashrita Furman, for example, spent a record-setting 5 hours, 7 minutes, and 6 seconds balancing atop an unstable Swiss (stability) ball.

Of balance experts, tightrope walkers are among the most familiar. Tightrope (also tightwire) aerialists walk across a tensioned wire or rope from one site to another. These acrobats maintain their balance by keeping their center of gravity over a very small base of support. Tightrope walkers often carry an implement (e.g., a pole) to help them balance. The pole assists the walker in responding to external torques, such as those created by a gust of wind, by creating an increased moment of inertia, or resistance to rotation.

FAMED TIGHTROPE walker Nik Wallenda crossing Niagara Falls in 2012.
Bert Hoferichter/All Canada Photos/Getty Images

Researchers have used several paradigms to study how people react to perturbations in the environment. One of the most common approaches uses a moving platform to induce horizontal perturbations to standing posture. Investigators then can measure response times, muscle activity patterns, segmental kinematics, and joint forces and torques to assess the mechanisms of postural control. The platforms can be moved in either the anterior–posterior (forward–backward) or lateral (side-to-side) directions, with some also able to change the velocity and inclination of the platform.

The postural responses to the moving platform are task specific and vary with the size of the supporting surface, direction of motion, location and magnitude of the applied force, initial posture at the moment of perturbation, and velocity of perturbation (Levangie & Norkin, 2011).

When the platform moves in the anterior–posterior direction, for example, the subject can adopt joint-specific strategies in response to the perturbation. An *ankle strategy* involves activation of either the plantar flexors (e.g., soleus) or the dorsiflexors (e.g., tibialis anterior) to control ankle motion. If the stationary platform suddenly moves anteriorly (forward), the body tends to fall backward. This creates movement-induced plantar flexion at the ankle. To compensate, the neuromuscular system recruits dorsiflexors to pull the lower leg forward toward the vertical position. If the platform moves posteriorly (backward), the body falls forward, creating movement-induced dorsiflexion at the ankle. In response, plantar flexor muscles are recruited to pull the lower leg backward toward vertical.

The subject could also employ a *hip strategy*. Sudden forward movement of the platform would cause hip extension. Hip flexors (e.g., iliopsoas) would be recruited to maintain an upright posture. Backward platform movement would induce hip flexion, with hip extensors (e.g., gluteus maximus) then required to vertically realign the trunk. If these strategies prove insufficient to reestablish balance, the subject could employ a *stepping strategy* to maintain stability by increasing the base of support.

Developmental Considerations

At both ends of the life span, postural control is less effective than during healthy adulthood. In the early years, infants are still developing their sensorimotor systems and gaining the strength and coordination needed to adopt a bipedal, upright posture. In older adults, disease, strength loss, compromised sensory function, and general motor decline can negatively affect balance and increase the risk of injury.

Posture and Balance in Infants

Postural development in an infant begins with control of the head and proceeds to the trunk. Automatic postural reactions, summarized in table 7.2, develop during the first year of life. Also within the first year of life, an infant reaches postural milestones of sitting and standing and begins developing the balance strategies required for postural control.

Clearly, during this first year, infants respond to environmental cues in what is termed **perception–action coupling**. An infant perceives optical flow (i.e., visual image changes

TABLE 7.2 Postural Reactions in Infants

Reaction	Starting position	Stimulus	Response	Time
Derotative righting	Supine	Turn legs and pelvis to other side	Trunk and head follow rotation	From 4 mo
	Supine	Turn head sideways	Body follows head in rotation	From 4 mo
Labyrinthine righting reflex	Supported upright	Tilt infant	Head moves to stay upright	2-12 mo
Pull-up	Sitting upright held by 1 or 2 hands	Tip infant backward or forward	Arms flex	3-12 mo
Parachute	Held upright	Lower infant toward ground rapidly	Legs extend	From 4 mo
	Held upright	Tilt forward	Arms extend	From 7 mo
	Held upright	Tilt sideways	Arms extend	From 6 mo
	Held upright	Tilt backward	Arms extend	From 9 mo

Reprinted by permission from K.M. Haywood and N. Getchell, *Life span motor development*, 6th edition (Champaign, IL: Human Kinetics, 2014), 107.

due to motion) and coordinates this information to generate postural responses. Infants scale their postural responses to the visual information they receive.

In the first few years of development, an infant reaches numerous posturally related **motor milestones**. According to the Bayley Scales of Infant Development, these include holding the head erect and steady (average age: 1.6 months), sitting with slight support (2.3 months), sitting alone momentarily (5.3 months), rolling from back to front (6.4 months), sitting alone steadily (6.6 months), pulling to stand (8.1 months), standing up by furniture (8.6 months), stepping movements (8.8 months), standing alone (11.0 months), walking alone (11.7 months), walking backward (14.6 months), walking up stairs with help (16.1 months), and jumping off the floor with both feet (23.4 months). The order of these milestones is relatively consistent, with each skill building on earlier ones, but there is a wide age range at which different children reach each milestone. In addition, other scales (e.g., Shirley Sequence) may list somewhat different average ages for each milestone (Haywood & Getchell, 2014).

Balance continually improves through infancy and childhood, but the exact course of the improvement depends on the task. Perception–action coupling must be developed and refined for each new skill, and although the general trend moves toward improved balance, plateaus or even decrements in balance performance may occur along the way (Haywood & Getchell, 2014).

Posture and Balance in Older Adults

Balance impairment in older adults is all too common and a major contributor to the high incidence of falls in the elderly. Falls are the leading cause of fractures in older adults. Fall-related fractures most commonly occur in the radius (from falling on an outstretched arm) and femur (from impact loading at the hip). Hip fractures number more than 300,000 in the United States annually and are a major cause of disability and death. Worldwide, more than 1.6 million hip fractures happen annually, with projections of between 4.5 and 6.3 million annual hip fractures by 2050 (Cooper, Campion, & Melton, 1992; Gullberg, Johnell, & Kanis, 1997).

Many age-related changes compromise postural control ability, including the following (Cech & Martin, 2011; Maki & McIlroy, 1996; Levangie & Norkin, 2011):

- Decline in visual skills (e.g., less able to discern contours and depth cues)
- Visual pathology (e.g., cataracts, macular degeneration)
- Compromised vestibular function (e.g., loss of vestibular hair cells, reduction of nerve fibers in cranial nerve VIII)
- Increased postural sway
- Difficulty using built-in redundancy of the sensory systems
- Slowing of central control mechanisms
- Conflicting sensory inputs
- Slowing of reaction and response times
- Increased reflex latency
- Declines in strength
- Changes in muscle fiber type
- Attention deficits
- Increased sensitivity to absence of visual input
- Slower anticipatory responses
- Diminished control of line of gravity and base of support
- Impaired ability to control compensatory stepping movement
- Psychological state
- Pain

- Decreased joint range of motion
- Decreased balance time and joint torque production
- Anatomical and structural changes (e.g., loss of fluid from intervertebral discs, disc degeneration)
- Increase in flexed, or stooped, posture

Each of these factors contributes to declines in balance, and their composite effect determines the severity of overall decline for a given individual. Some of the changes are inevitable, while others can be ameliorated if interventions are begun soon enough. The first step in arresting the decline or even improving balance in older adults includes awareness and identification of the factors specific to a particular individual. Controllable factors include muscular weakness and imbalance, joint range of motion, psychological state, and control of compensatory movements. The controllable factors can be improved through a combination of education, biofeedback, postural retraining, therapeutic exercise, orthotics, strength training, practice of movement forms (e.g., tai chi) known to enhance postural awareness and balance, and practice of recovery from perturbations.

Research in Mechanics

Better Balance, Fewer Falls

Most falls are related to a loss of balance. Therefore, better balance should result in fewer falls and fall-related injuries (e.g., hip fractures). Substantial research has focused on balance training, especially in older people, with a goal of reducing the incidence of falls.

In general, older individuals have greater fall risk than younger people, due to a variety of factors, including slower reaction time, decreased muscle strength and visual acuity, and slower cognition. Some special populations are at even higher risk. For example, research has shown that older people with type 2 diabetes have a significantly greater falls risk (Maurer et al., 2005; Schwartz et al., 2002). A 2010 study by Morrison and colleagues confirmed the higher falls risk in older people (50-75 years) with type 2 diabetes compared to controls, but also reported that following a 6-week balance training program, the type 2 diabetes group showed improvements in balance, proprioception, lower-limb strength, and reaction time. These improvements decreased falls risk.

Tai Chi Chuan, a slow Chinese movement form, has repeatedly been shown to improve balance in older people (e.g., Gallant et al., 2017; Li et al., 2004; Li et al., 2005; Lin et al., 2006). Tai Chi also has proved beneficial in improving balance and reducing falls risk in special populations, such as older adults with mild cognitive impairment (Sungkarat et al., 2017) and in patients with Parkinson's disease (Fuzhong et al., 2012).

See the references for the full citations:

Fuzhong, Harmer, Fitzgerald, Eckstrom, Stock, Galver, Maddalozzo, & Batya, 2012.

Gallant, Tartaglia, Hardman, & Burke, 2017.

Li, Harmer, Fisher, & McAuley, 2004.

Li, Harmer, Fisher, McAuley, Chaumeton, Eckstrom, & Wilson, 2005.

Lin, Hwang, Wang, Chang, & Wolf, 2006.

Maurer, Burcham, & Cheng, 2005.

Morrison, Colberg, Mariano, Parson, & Vinik, 2010.

Schwartz, Hillier, Sellmeyer, Resnick, Gregg, Ensrud, Schreiner, Margolis, Cauley, Nevitt, Black, & Cummings, 2002.

Sungkarat, Boripuntakul, Chattipakorn, Watcharasaksilp, & Lord, 2017.

Postural Dysfunction

Many factors contribute to poor posture and postural dysfunction: pain, decreased joint range of motion, inflexibility, muscle weakness and imbalances, altered joint biomechanics, joint hypermobility and ligament laxity, altered sensation and proprioception, psychological state, adaptations to the environment, persistent adoption of poor posture (habituation), pregnancy, anatomical defects, fatigue, disease (e.g., muscular dystrophy, osteoporosis), and injury (Neumann, 2016; Everett & Kell, 2010). Poor posture places abnormal loads on anatomical structures and increases the risk of musculoskeletal injury. Physiological functions such as respiration and circulation also can be negatively affected.

As described earlier, the spine's normal curvatures help the body accept compressive loads. Injury, disease, and congenital predisposition can cause deformities of the spinal column (e.g., abnormal structural alignment or alteration of spinal curvatures). These deformities often result in altered force distribution patterns and pathological tissue adaptations that may lead to or exacerbate other musculoskeletal injuries. As described in chapter 3, there are three primary spinal deformities: scoliosis, kyphosis, and lordosis. Classified by their magnitude, location, direction, and cause, these deformities can occur in isolation or in combination.

Concluding Comments

Posture and balance provide the foundation for, and are essential to, all other movement forms. Balance in everyday activities allows us to move effectively without falling or committing other movement errors. Advanced balance skills permit elite performers to execute exceptional, and sometimes miraculous, movements. Loss of balance can result in stumbling and falls. When an infant is learning to walk, balance is the key task to be learned. In older people, loss of balance can lead to falls, injury, incapacitation, and even death.

Go to the web study guide to access critical thinking questions for this chapter.

Suggested Readings

Cech, D.J., & Martin, S.M. (2011). *Functional movement development across the life span* (3rd ed.). Philadelphia: Saunders.

Enoka, R.M. (2015). *Neuromechanics of human movement* (5th ed.). Champaign, IL: Human Kinetics.

Haywood, K.M., & Getchell, N. (2014). *Life span motor development* (6th ed.). Champaign, IL: Human Kinetics.

Houglum, P.A., & Bertoti, D.B. (2012). *Brunnstrom's clinical kinesiology* (6th ed.). Philadelphia: FA Davis.

Kendall, F.P., McCreary, E.K., Provance, P.G., Rodgers, M.M., & Romani, W.A. *Muscles: Testing and function, with posture and pain* (5th ed.). Philadelphia: Lippincott Williams & Wilkins.

Levangie, P.K., & Norkin, C.C. (2011). *Joint structure and function: A comprehensive analysis* (5th ed.). Philadelphia: FA Davis.

McGill, S. (2016). *Low back disorders: Evidence-based prevention and rehabilitation* (3rd ed.). Champaign, IL: Human Kinetics.

Neumann, D.A. (2016). *Kinesiology of the musculoskeletal system: Foundations for rehabilitation* (3rd ed.). St. Louis: Mosby.

Nordin, M., & Frankel, V.H. (2012). *Basic biomechanics of the musculoskeletal system* (4th ed.). Philadelphia: Lippincott Williams & Wilkins.

8

Gait

Objectives

After studying this chapter, you will be able to do the following:

- Define walking and running and describe the walking and running gait cycles
- Describe the components of gait analysis
- Explain the role of lower-extremity muscles in the control of walking and running
- Describe gait and running development across the life span
- Describe examples of pathological gait and running-related injuries

At first glance, the most common forms of gait, walking and running, might appear to be simple movement tasks. Most people perform them with little conscious thought or effort. On closer examination, however, both walking and running are complex tasks that require significant neuromechanical integration.

Walking

U.S. President Franklin D. Roosevelt once said, "There are many ways of going forward, but only one way of standing still." Of all the ways humans have of going forward, walking is by far the most common. Virtually everyone has experience walking and usually gives little thought to this mode of getting from one place to another. Although walking displays certain common characteristics, each of us develops a walking style uniquely our own. Most of us have experienced seeing the silhouette of someone in the distance and recognizing him solely by the way he walks even though we could not see any facial features. What is it about walking that allows us to identify a person this way? This largely involves the mechanics of walking style.

Walking undoubtedly has been the subject of more research than any other movement form. Several excellent books (some listed in the suggested readings at the end of this chapter) are devoted solely or primarily to exploring the intricacies of human walking. Thousands of research articles published in the last century have dealt with various aspects of walking, as have many notable books devoted to the topic. As a result, we arguably know more about walking than any other human movement form. Given the volume of information on walking, we can only scratch the surface in the limited space afforded here. We provide the basics and hope that the interested reader will pursue further study of walking using the many comprehensive resources available.

Terminology

Several terms are important to our understanding of basic movement forms. **Locomotion** is the act or power of moving from place to place. This broad definition obviously includes common movement forms such as walking and running but also embraces less common movement modes such as skipping, crawling, sliding, swimming, and even playing leapfrog or doing cartwheels. All of these movements, and many others, are forms of locomotion because they fulfill the basic requirement of locomotion—getting from one place to another.

Gait refers to a particular form of locomotion. Most commonly, *gait* is used in the context of describing walking gait or running gait. (Note: Some references use *gait* only in reference to walking.)

Walking in humans can be defined as a form of upright, bipedal locomotion in which at least one foot is always in contact with the ground. It is a cyclic activity involving the alternating action of the legs to advance the body forward. At first glance, walking might appear to be a simple movement task. After all, it's something we do every day with little conscious thought or consideration. Voluminous research over recent decades, however, has proved much the opposite; walking entails a complex set of neuromechanical events. The challenges of walking are even more profound for infants, older adults, and individuals affected by disease, injury, or congenital defect.

To study walking gait, we first need to understand several terms commonly found in gait literature. As a person walks, each leg alternates between periods when the foot is in contact with the ground and when it is moving forward through the air (with no foot–ground contact). The period during which the foot contacts the ground is termed the **stance phase**, or stance. When the foot is not in ground contact, the leg is moving forward in what is termed the **swing phase**, or swing. Stance and swing describe the phasing of each leg independently of the other.

The stance phase for a given leg begins when that leg's foot first hits the walking surface. This first contact is called **initial contact**. (Note: Several other terms are found in the literature to describe this first contact, including *heel contact*, *heel strike*, *foot contact*, and *foot strike*. *Initial contact* is preferred here, since it provides the most general description of the event. The other terms may be misleading. *Heel contact*, for example, suggests that the heel is the part of the foot that first contacts the ground. Although usually true, in some conditions, such as in people with paralysis or cerebral palsy, the heel may not be the first contact point on the foot.)

The stance phase continues until the foot leaves the ground in an event termed **toe-off**, or **takeoff**. Toe-off initiates the swing phase, which continues as the leg swings forward and the body advances until the next ipsilateral (i.e., same side) initial contact.

A **gait cycle** refers to the sequential occurrence of a stance and swing phase for a single limb. This period from initial contact of one leg until the next initial contact of the same (ipsilateral) leg is termed a **stride**. Each stride is made up of two steps, with a **step** defined as the period from initial contact of one leg to initial contact of the opposite (contralateral) limb. Strides and steps are measured by length and width as shown in figure 8.1. (Note: The term *stride*, as described here in its common clinical and biomechanical usage, is used differently in track-and-field (athletics) literature. In track-and-field circles, stride refers to the period from ipsilateral contact to contralateral contact, or what we describe here as a step.)

In walking there obviously are periods when only one foot is in contact with the ground, when one leg is in its swing phase and the contralateral leg is solely supporting the entire body. This period is termed **single support**. When both feet are touching the walking surface at the same time (i.e., both legs are in their respective stance phases), the period is termed **double support**.

Another important term is **cadence**, defined as the step rate as measured by steps per minute. A person walking with a cadence of 120 steps per minute, for example, takes 120 steps (60 strides) every 60 seconds, so each step would take 0.5 second.

Walking speed (velocity) is determined by the mathematical product of cadence and step length. If a person has an average step length of 0.7 m and a cadence of 114 steps per minute, her walking speed is 79.8 m/min (0.7 m × 114 steps/min). Obviously, walking speed differs depending on a number of factors, including the person's age, size, and physical condition, as well as the purpose of the walk. A person who is late for an appointment, for example, will choose a faster walking speed than if she is taking a leisurely stroll in the park. Every individual has a free, self-selected, and comfortable walking speed that minimizes energy expenditure.

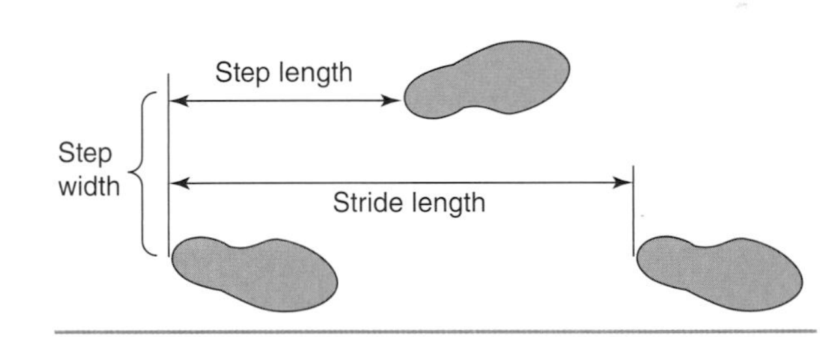

FIGURE 8.1 Step length, stride length, and step width.

Walking Gait Cycle

The gait cycle, as defined earlier, encompasses the stance–swing period of a single leg. The gait cycle is divided into phases, or subcycles. No universally accepted system exists for specifying these phases. We present two widely used systems (figure 8.2) that, although different, share many common elements. The first, a traditional system, identifies critical points in the stance phase (heel strike, foot flat, midstance, heel-off, and toe-off) and divides the swing phase into periods of early swing (acceleration), midswing, and late swing (deceleration).

Another widely adopted system is that proposed by Dr. Jacquelin Perry (Perry & Burnfield, 2010), referred to as the Rancho Los Amigos (RLA) system in recognition of her many decades of pioneering work at the Rancho Los Amigos National Rehabilitation Center in Downey, California. The RLA system divides the gait cycle according to periods (stance and swing), phases of stance (initial contact, loading response, midstance, terminal stance, preswing), and phases of swing (initial swing, midswing or swing-through, terminal swing). Other systems use similar terminology that differs only in some of the details in describing the gait cycle.

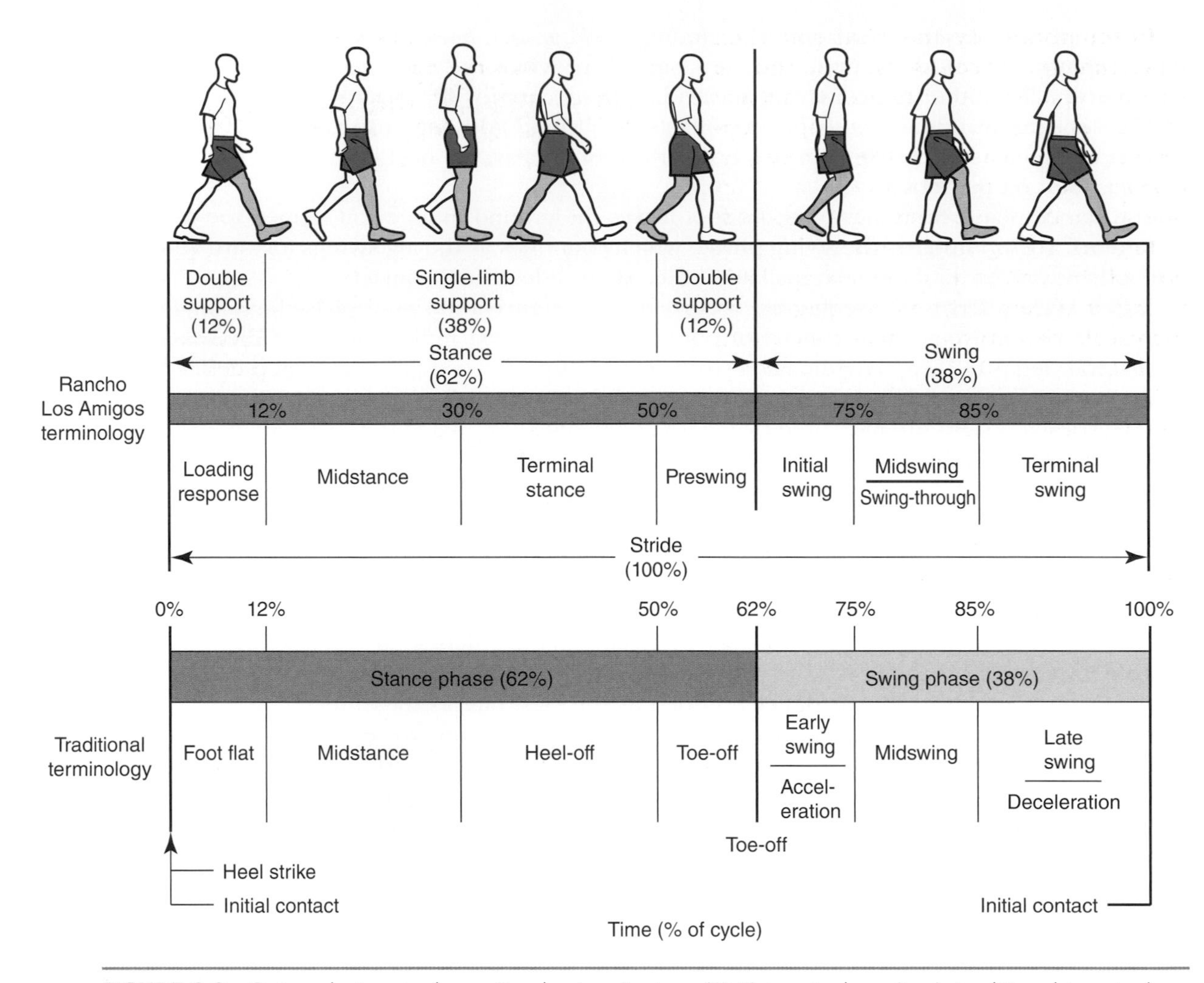

FIGURE 8.2 Gait cycle terminology: Rancho Los Amigos (RLA) terminology (top); traditional terminology (bottom).

Reprinted by permission from J.K. Loudon, R.C. Manske, and M.P. Reiman, *Clinical mechanics and kinesiology* (Champaign, IL: Human Kinetics, 2013), 356.

Gait Analysis

Gait analysis can range from simple measures of timing (**temporal analysis**) to mechanical analysis using sophisticated instrumentation and computer models. The goal of gait analysis is to provide scientific and clinical insights into both normal and pathological walking. Most, if not all, of the variables explored in gait analysis change with walking speed. The characteristics of a slow, leisurely walk differ greatly from those typical of a fast, determined walking gait. Although observational gait analysis has been performed for many centuries, scientific measurement has been limited to the last 100 years or so.

Temporal and Spatial Characteristics

Several timing (temporal) measures of gait are commonly used. Cadence, or step rate, is perhaps the simplest of these measures. The average free walking (i.e., self-selected speed) cadence for adults is about 113 steps per minute. Women typically walk with a higher cadence (117 steps/min) than do men (111 steps/min) to partially compensate for their shorter step length (Perry & Burnfield, 2010).

In temporally symmetrical gait (i.e., equal timing between sides), **step time** is simply the inverse of cadence. A cadence of 114 steps per minute, for example, represents a step time of 0.0088 minutes, or 0.53 seconds. Because two steps make up one stride, the stride time is 1.06 seconds.

In free walking, a person spends about 60% of the gait cycle in stance and 40% in swing (figure 8.3). As walking speed increases, the ratio of stance to swing approaches 50:50. To enhance stability, older adults spend more time in stance and less in swing.

As described earlier, single support is the phase when only one foot contacts the ground; in double support, both feet are in ground contact. Each gait cycle includes two periods of single support and two periods of double support. The relative duration of each of these periods is shown in figure 8.2.

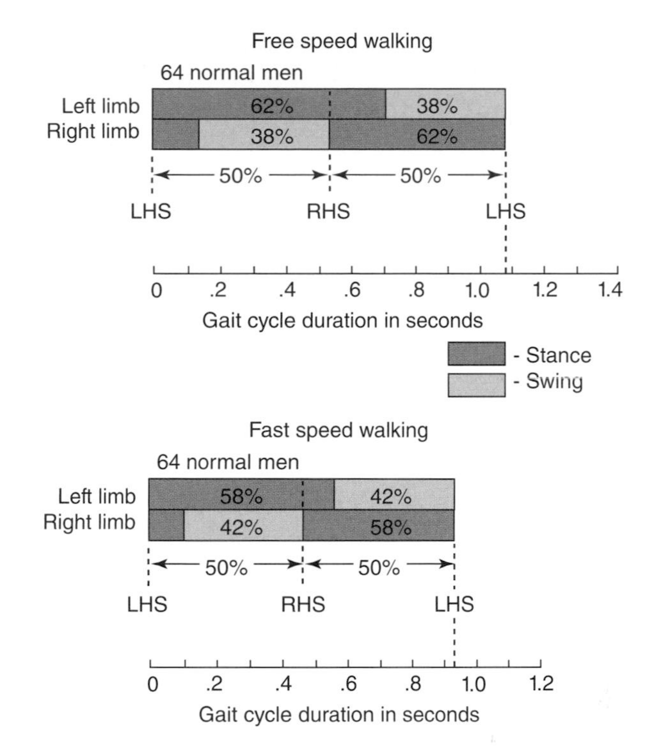

FIGURE 8.3 Stance–swing ratio.

Adapted by permission from M.P. Murray, D.R. Gore, and B.H. Clarkson, "Walking patterns of patients with unilateral hip pain due to osteoarthritis and avascular necrosis," *The Journal of Bone and Joint Surgery* 53, no. 2 (1971): 259-274.

Three common spatial gait measures are step length, stride length, and step width (see figure 8.1). Because of their taller stature, men on average have greater (~14%) step length and stride length than women. Men and women average a stride length of 1.46 m and 1.28 m, respectively (Perry & Burnfield, 2010). Step width averages 7 to 9 cm. In normal gait, the foot also angles outward (laterally) by about 7° from the direction of progression.

Combining temporal (time) and spatial (distance) measures tells us how fast a person is walking. The average adult walks at a speed of about 3 mph (1.34 m/s), with women self-selecting a slightly slower speed than men (Neumann, 2016). Walking speed can vary considerably based on a number of factors, including age, gender, physical condition, environmental conditions, and purpose. (Note: Some sources refer to walking speed, while others describe walking velocity. As described in chapter 5, speed is a scalar measure of how fast a person is moving, while velocity is a vector measure indicating both how fast and in what direction. The difference between speed and velocity in the context of describing gait usually is inconsequential because the direction is forward. Hence, the terms *velocity* and *speed* often are used interchangeably to describe how fast a person walks.)

Gait velocity is a function of stride length and stride rate according to the following equation:

$$\text{velocity} = \text{stride length} \times \text{stride rate} \qquad (8.1)$$

or equivalently (for symmetrical gait)

$$\text{velocity} = \text{step length} \times \text{cadence} \qquad (8.2)$$

For example, using equation 8.2, a person with a cadence of 112 steps per minute and a step length of 0.70 m would have a walking velocity of 78.4 m/min (1.31 m/s or nearly 3 mph).

Muscle Activity and Control of Walking

Proper magnitude and timing in movement of the lower-extremity musculature are essential for normal walking gait. Even subtle muscle deficiencies can markedly affect gait performance. Significantly compromised muscle function, as in paralysis, may render normal gait impossible. In older adults and individuals with injury, pain is often a primary factor in gait alteration. Consider, for example, times you have been injured (e.g., sprained ankle).

Research in Mechanics

Gait Analysis

Gait analysis research has a long and storied history. As long ago as Aristotle, scholars have critically observed human gait. Laboratory research on gait began in 17th century Europe and accelerated from the late 19th and through the 20th century, when technological advancements made measurement of gait kinematics and kinetics possible. Kinematic assessment has predictably become more sophisticated, moving from direct observation to analysis by film, video, optoelectronic, and most recently motion sensor systems. Gait analysis labs have for many years employed electromyographical (EMG) technology to measure muscle activity during gait and force measurement instruments (e.g., force platforms, in-shoe sensors) to measure gait kinetics.

The gait analysis literature is too extensive to review here, but many sources are available that provide in-depth discussion of both biomechanical and clinical gait analysis. Among these are review articles, such as Sutherland's trilogy (2001, 2002, 2005) on the evolution of clinical gait analysis, which provides historical detail on the development of gait analysis and its pioneers.

Numerous books have documented the history, development, results, and application of gait analysis research, including seminal works by Verne Inman and colleagues (1981) and Jacquelin Perry (1992).

See the references for the full citations:

Inman, Ralston, & Todd, 1981.

Perry, 1992.

Sutherland, 2001.

Sutherland, 2002.

Sutherland, 2005.

The pain from the injury likely caused you to alter your muscle activation patterns and gait mechanics. You probably tired more quickly than normal because altered gait is physiologically less efficient than self-selected normal gait, and you may also have felt discomfort or pain in other body regions (e.g., contralateral knee or hip) resulting from the mechanical changes in how you walked.

In normal gait, muscles of the hip, knee, and ankle are primarily responsible for controlling joint motion. Excellent detailed analysis of muscle action during gait can be found in a number of sources (e.g., Levangie & Norkin, 2011; Neumann, 2016; Perry & Burnfield, 2010; Rose & Gamble, 2005).

Highlights of muscle activity during phases of the gait cycle (using the system described by Perry & Burnfield, 2010) include the following:

- *Initial contact and loading response.* Hip abductors (upper fibers of the gluteus maximus, gluteus medius, gluteus minimus, tensor fasciae latae) eccentrically prevent excessive lateral pelvic tilt, knee extensors (quadriceps: vastus intermedius, vastus lateralis, vastus medialis, rectus femoris) eccentrically prevent excessive knee flexion, and ankle dorsiflexors (tibialis anterior, extensor hallucis longus, extensor digitorum longus, peroneus tertius) eccentrically control ankle plantar flexion. These actions work to accomplish weight acceptance, pelvic stabilization, and deceleration of the body.

- *Midstance.* Plantar flexors (soleus, gastrocnemius, tibialis posterior, flexor digitorum longus, flexor hallucis longus, peroneus longus, peroneus brevis) eccentrically control tibial advancement over the foot. The gastrocnemius (with two-joint action at the ankle and knee) also helps stabilize the knee.

- *Terminal stance.* Plantar flexors act concentrically to assist with push-off.
- *Preswing and initial swing.* Hip flexors (iliacus, psoas, rectus femoris, gracilis, sartorius) initiate hip flexion early in the swing phase.
- *Initial swing and midswing.* Ankle dorsiflexors (tibialis anterior, extensor hallucis longus, extensor digitorum longus, peroneus tertius) act concentrically to dorsiflex the foot at the ankle to guarantee toe clearance.
- *Late midswing and terminal swing.* Ankle dorsiflexors eccentrically control ankle plantar flexion; hamstrings (semitendinosus, semimembranosus, long and short heads of biceps femoris) eccentrically control knee extension in preparation for initial contact.
- Any disturbance in the sequencing or level of muscle activation may alter gait mechanics and create the need for compensatory muscle action. Several examples of altered, or pathological, gait are described later in this chapter.

Although we have focused on the lower-extremity muscles, the arms and trunk also play an essential role in gait mechanics. The arms, for example, typically work in alternating fashion with the legs to provide a counterbalancing effect. When walking, the left arm moves synchronously with the right leg, while the right arm and left leg move in concert with one another. If you doubt the importance of the arms in gait, try walking briskly or running with your hands straight at your sides or placed on top of your head. You'll instantly feel the discomfort and uncoordinated motion this change in normal mechanics causes.

Life-Span Perspective

Newborn infants are essentially immobile. Developmental changes during the first year allow infants to take their first halting steps, and by 3 to 4 years of age, children usually demonstrate a mature walking gait pattern. The ability to alter walking speed, change pace and direction, and react to hazards continues to develop into adulthood and is maintained well into one's 50s and beyond. Numerous age-related changes make individuals in the 60-plus age range more susceptible to declines in gait function.

Infants and Children

The road from infant immobility to upright, bipedal walking gait is marked by a number of motor milestones. To complete these progressively more demanding tasks, an infant must develop the needed strength and neuromuscular coordination, and the environment must provide a flat, firm surface with sufficient friction to allow locomotion. The first forms of infant locomotion typically are crawling and creeping, which emerge from 7 to 10 months of age. **Crawling** is progressing on hands and stomach; **creeping** is moving on hands and knees. The normal developmental progression is (1) crawling with the chest and stomach on the floor, (2) symmetrical leg movements that create low creeping with the stomach off the floor, (3) back and forth rocking in a high creep position, and (4) creeping with alternating action of the arms and legs (Haywood & Getchell, 2014).

Before the end of their first year, most infants have developed sufficiently to stand without support and take their first steps. By 2 years of age, most of the characteristics of mature walking are in place. Over the next several years, children refine these characteristics through increased strength, coordination, and joint range of motion. By about 5 years, children exhibit a mature gait pattern that shows most, if not all, of the characteristics of proficient walking, which include the following (Haywood & Getchell, 2014):

- Absolute increase in stride length (resulting from increased leg length and greater force application and leg extension at push-off)
- Change from flat-foot planting to heel-to-forefoot pattern, resulting from increased range of motion
- Reduced out-toeing

- Narrowed base of support laterally to emphasize anterior–posterior force application
- Walking pattern that includes full knee extension at initial contact (heel strike), followed by slight knee flexion through midstance and full extension again during push-off
- Pelvis rotation to allow full range of leg motion and reciprocal movement of the upper and lower body segments
- Improved balance and reduction in forward trunk inclination
- Coordinated arm and leg movements so that the opposite arm and leg move forward and backward in unison

From this point on, only subtle gait changes are seen through the rest of childhood and into the adult years.

Older Adults

Older adults show a number of predictable changes in their walking gait. Although some older adults maintain gait patterns similar to younger adults well into their 60s and 70s, most individuals in this age range begin to exhibit changes that include slower walking speed; shorter step length and relative swing time; longer relative stance time; decreased maximum walking speed, joint range of motion, and cadence; and increased base of support, double limb support, and use of visual scanning (Bohannon, 1997; Judge, Ounpuu, & Davis, 1996; Ostrosky, VanSwearingen, Burdett, & Gee, 1994). Chronic conditions such as heart disease, arthritis, osteoporosis, or pain may exacerbate these changes or lead to their earlier onset.

The consequences of age-related changes in gait, coupled with declines in reaction time, strength, endurance, and visual acuity, greatly increase the chance of accidents such as tripping, slipping, and falling. Injuries associated with falls in older adults are a serious and growing public health problem. Considerable research is directed at characterizing the mechanics of falls and devising strategies to reduce their incidence. The human toll and monetary costs (in the billions of dollars) associated with the more than 1.6 million fall-related hip fractures annually worldwide (Johnell & Kanis, 2006) are but two examples of how gait changes in older adults are integrally related to a major public health issue.

Pathological Gait

Countless conditions can contribute to dysfunctional, or pathological, gait, including disease, injury, paralysis, anatomical abnormalities, and congenital defects. Comprehensive examination of pathological gait is beyond the scope of this book, but we present an overview of three examples that demonstrate how anatomical, physiological, and environmental conditions can alter the gait pattern, increase the loads placed on anatomical structures, stress physiological systems, and increase the chance of injury.

The first example, **Trendelenburg gait**, results from weakness or paralysis of the hip abductors (e.g., gluteus medius, gluteus minimus). In normal walking, during early to midstance of a given leg, the pelvis tends to drop to the contralateral side. This drop is controlled by eccentric action of the hip abductors. Compromised hip abductor action allows for the excessive pelvic drop characteristic of Trendelenburg gait. To compensate for this pelvic drop, subjects typically lean their trunks toward the side of the support leg (figure 8.4).

The second example involves gait changes characteristic of people with anterior cruciate ligament (ACL) deficiencies. The ACL ligament in the knee restricts anterior translation of the tibia relative to a fixed femur or, conversely, limits posterior movement of the femur relative to a fixed tibia. The ACL, one of the most commonly injured ligaments in the body, suffers damage when the knee experiences extreme valgus-rotation loads or is violently hyperextended.

An ACL-deficient knee results in altered gait mechanics and various compensatory adaptations. Individuals with ACL deficiency commonly adopt a **quadriceps avoidance** gait pattern. The quadriceps muscle group, acting through the knee extensor mechanism, tends to pull the tibia anteriorly. One of the primary functions of an intact ACL is to restrict this anterior tibial translation relative to the femur. In the case of a partially or completely torn ACL, the quadriceps avoidance strategy results in decreased activation of the quadriceps and an accompanying decrease in knee joint moment. This reduces the anteriorly directed forces on the tibia and effectively provides compensatory stabilization.

The third example involves gait pathologies in children with **cerebral palsy** (CP). Cerebral palsy is a nonprogressive condition of muscle dysfunction and paralysis caused by brain injury at or near the time of birth. CP affects more than half a million Americans. There are many forms of cerebral palsy, the most common being spastic diplegia. In **spastic diplegia**, the child's posture and gait are characterized by abnormally flexed, adducted, and internally rotated hips; hyperflexed knees; and **equinus** of the foot and ankle.

In addition to the mechanical deviations characteristic of CP gait, walking exacts a much greater physiological demand on children with CP. One study found 63% greater oxygen consumption (relative $\dot{V}O_2$) in children with spastic CP compared with controls. This increase was largely caused by inefficient energy transfer between and within adjacent body segments during the gait cycle (Unnithan, Dowling, Frost, & Bar-Or, 1999).

Treatment approaches for CP-related gait pathologies include neuromuscular training, orthotics (e.g., ankle–foot orthotics, or AFOs), drugs, and surgery.

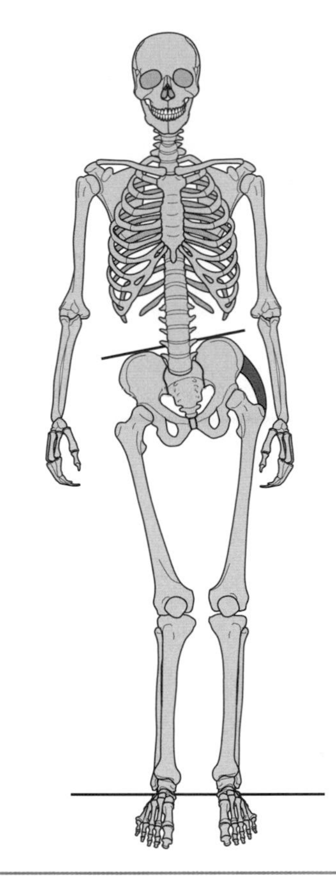

FIGURE 8.4 Trendelenburg gait. Note the ipsilateral pelvic drop.

Running

Why do we run? Usually, to move faster. Whether we are late for class or an important meeting, taking advantage of the increased metabolic demands of running to increase our level of physical conditioning, or sprinting to win a race, we adopt a running gait to move faster.

The transition from walking to running in adults typically happens at about 2 m/s (4.5 mph). In most cases, running is faster than walking, but speed is *not* the distinguishing characteristic between these two gait modes. Race walkers, for example, can reach walking speeds of more than 12 kph (7.45 mph; Murray, Guten, Mollinger, & Gardner, 1983), while a slow jogger runs at a much slower speed. Instead of speed, the defining difference between walking and running is that walking has periods when both feet are in contact with the ground (which running does not) and **running** has a period of nonsupport when both legs are in the air (which walking does not).

The definition of running is simple. The execution of running is not. As with walking, running requires a complex interaction of many body systems. As even casual observation of runners makes clear, we do not all run in the same way. Superb athletes run with remarkable beauty and effectiveness, a person with injury or disease will run much more slowly, and infants and older adults run much differently than do young adults.

Each person's running style depends on many factors, including sex, size, strength, balance, anatomical structure, level of conditioning, and skill. These factors result in a wide range of running styles and considerable variability among individuals. Although many of the aspects of running discussed in this section may seem to imply a uniform running style, always keep in mind that each of us is different, and we run in different ways.

Applying the Concept

Prosthetic Feet and Gait

Amputation of a foot may be required for many reasons, including severe accident-related trauma, compromised blood flow, diabetes-related complications, chronic and persistent infections, tumors, severe burns or frostbite, and nonhealing wounds. Following amputation, there are many types of prosthetic feet to choose from:

- Solid ankle cushioned heel (SACH) and elastic keel prostheses are made of neoprene or urethane foam molded to model a human foot. A keel is an energy-storing component in a prosthesis that bends the foot upward when weight is applied to the foot. The SACH and elastic keel designs have no hinged parts and therefore are durable and relatively inexpensive, and require little maintenance.
- Single-axis and multiaxis feet allow for foot movement. A single-axis design includes a uniplanar ankle joint that permits sagittal plane (up-and-down) foot motion. A multiaxis design allows for sagittal-plane and frontal-plane (side-to-side) motion.
- Dynamic response foot prostheses accommodate more active lifestyles by storing and releasing energy during the stance phases of walking and running, and provide a more natural feel. The energy is stored by a deformable keel.
- Microprocessor-controlled (MPC) foot prostheses are relatively new. They use computer-controlled sensors to process information from the person's limb and the external environment to adjust prosthetic response to a variety of conditions.

Running Gait Cycle

The running gait cycle differs from that of walking. As explained earlier, walking is characterized by alternating periods of single support, when one of the legs is in its swing phase while the other singly supports the body's weight, and brief periods of double support, during which both feet are in contact with the ground.

Running, in contrast, does not have a period of double support. Rather it has alternating periods of single support, separated by a double **float phase**, or **flight phase**, when both feet are airborne. The phases of the running gait cycle are shown in figure 8.5.

Temporal and Spatial Characteristics of Running Gait

Phase timing and stride lengths during running change as speed increases. As gait velocity increases from slow to fast walking and then to running, the duration of the stance phase decreases, while the duration of the swing phase changes little. With increases in speed, the stance–swing ratio changes from the 60:40 of normal walking to 20:80 in sprinting at 9.0 m/s (20 mph; see figure 8.6).

As with walking, running speed is also calculated as the product of stride length and stride rate (see equation 8.1). Initial increases in speed are made primarily by lengthening the stride. Velocity increases at higher speeds are achieved by quickening the stride rate, largely because there is an anatomical limit to stride length.

Muscle Activity and Control

Given the kinematic differences between walking and running, it is not surprising that electromyographic (EMG) patterns between gait modes differ as well. For running, at the hip, the gluteus maximus is active at the end of the swing phase, just before foot contact, in order to eccentrically slow hip flexion. Similarly, the hamstring group (semitendinosus,

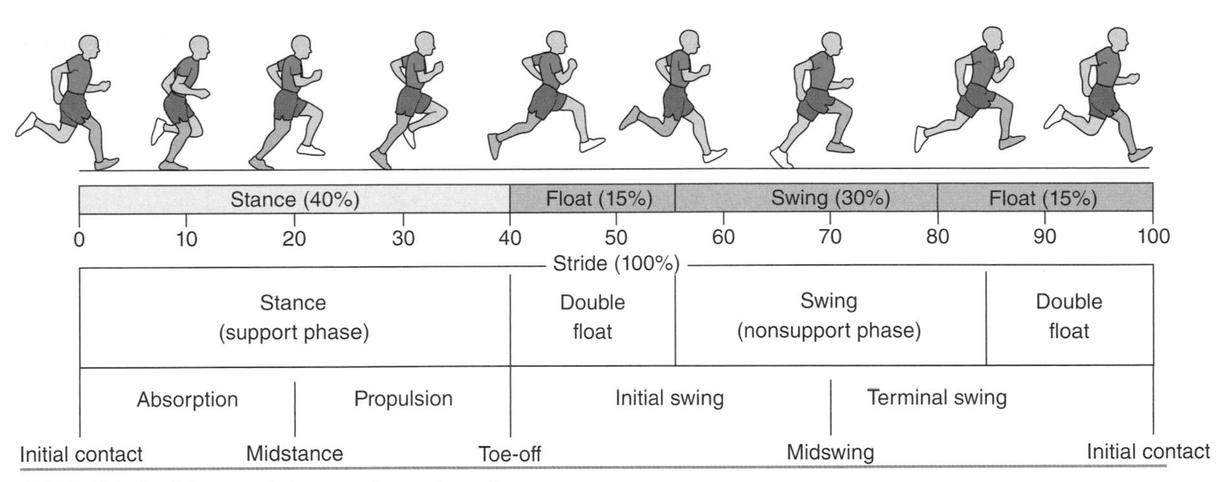

FIGURE 8.5 Phases of the running gait cycle.

Reprinted by permission from P.A. Houglum, *Therapeutic exercise for musculoskeletal injuries,* 4th ed. (Champaign, IL: Human Kinetics, 2016), 331.

semimembranosus, biceps femoris) is active in late swing to slow knee extension before contact.

During early stance, the gluteus maximus, quadriceps group (vastus medialis, vastus lateralis, vastus intermedius, rectus femoris), and gastrocnemius actively generate extensor moments at the hip, knee, and ankle, respectively. These extensor moments provide support and resist the tendency of the ground reaction force to flex the lower-extremity joints.

The plantar flexors, especially the gastrocnemius, are active in late stance to produce the propulsive force necessary for effective push-off into the swing phase. The hamstrings are active to produce the knee flexion during swing. In the swing phase, the rectus femoris is active to aid with both hip flexion and knee extension.

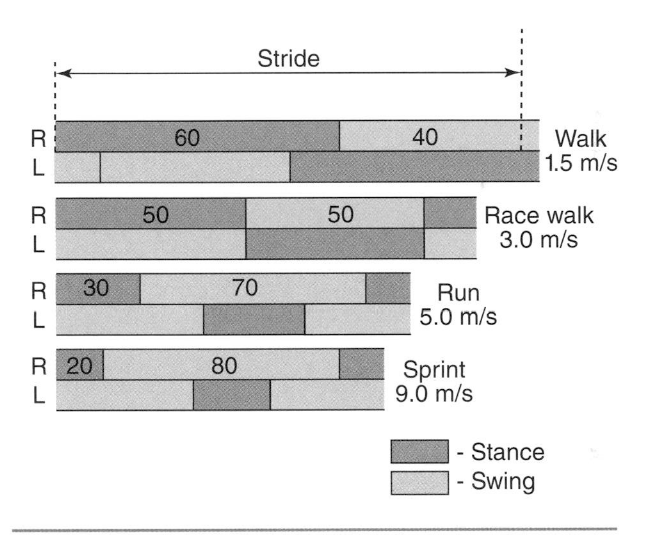

FIGURE 8.6 Stance–swing ratio changes from walking to sprinting.

Adapted by permission from C.L. Vaughan, "Biomechanics of running gait," *CRC Critical Reviews in Biomedical Engineering* 12 (1984): 6.

The tibialis anterior shows activity through much of the running gait cycle. During swing, the tibialis anterior acts concentrically to dorsiflex the ankle, while in early stance, it is coactive with the plantar flexors to stabilize the foot and ankle.

Increases in running speed are accompanied by greater EMG peak activity and overall levels of activation, shorter *absolute* periods of activity (due to shorter gait cycle duration), but higher *relative* duration as a percent of the gait cycle (Mero & Komi, 1986).

Life-Span Perspective

In comparison with walking, running has a shorter life-span perspective. This is simply because infants lack the strength to propel themselves into the air and thus cannot begin to run until well into their second year or later, and most older adults no longer run, either by choice or because of limitations imposed by injury, disease, or declines in strength and endurance. As a result, most of the research literature on running focuses on children and younger adults.

Research in Mechanics

Barefoot Versus Shod Running

Anthropologists estimate that humans have been wearing shoes for the past 30 to 40,000 years. Before that, humans obviously walked and ran barefoot. Footwear has developed considerably over the millennia, beginning with animal skins and evolving into the high-tech footwear of today. However, shoe development didn't eliminate barefoot gait. One notable example of barefoot running is Abebe Bikila, an Ethiopian distance runner, who won the 26.2 mile (42.2 km) Olympic Games marathon in Rome (1960) without wearing shoes.

The running and jogging boom in the 1970s and '80s resulted in extensive running shoe design and development. One of the primary features of running shoes then and now is a broad and cushioned heel. This design feature provides shock absorption and allows runners to comfortably impact the ground heel first. Runners adopting this technique are known as heel-strikers. Barefoot runners don't use a heel-strike technique, since the uncushioned impact on a bare heel would be too high. Barefoot runners use a technique in which initial ground contact is on the forefoot (forefoot strikers) or middle of the foot (midfoot strikers).

In more recent years, there has been a movement by some back to unshod (i.e., barefoot) gait or, since about 2004, to wearing of so-called *minimalist footwear*. Minimalist footwear comes in a variety of designs, but essentially is a compromise between wearing traditional running shoes and going barefoot. This footwear provides limited cushioning but some degree of protection for the foot.

Numerous research studies have documented differences in running kinematics and kinetics between barefoot and shod runners and between traditional and minimalist footwear (e.g., Bonacci et al., 2013; Hollander et al., 2015; Mullen et al., 2013; Shih et al., 2013; Squadrone et al., 2015). To date, the evidence is equivocal as to whether one type of footwear condition is better than another. One thing that is consistent in the literature, though, is advice that anyone considering making a change (e.g., from traditional footwear to either minimalist shoes or barefoot) should do so gradually to avoid injury.

See the references for the full citations:

Bonacci, Saunders, Hicks, Rantalainen, Vicenzino, & Spratford, 2013.

Hollander, Argubi-Wollesen, Reer, & Zech, 2015.

Mullen & Roby, 2013.

Shih, Lin, & Shiang, 2013.

Squadrone, Rodano, Hamill, & Preatoni, 2015.

Infants and Children

Until infants develop sufficient muscle strength to produce a flight phase, they are limited to walking gait. When muscular development allows force production great enough to push the body off the ground, the infant begins development of a running gait. This typically occurs during an infant's second year but can vary from one infant to another.

Early running is characterized by short steps and brief flight periods. As strength and balance improve, running gait matures and becomes more like that of an adult, with increases in stride length, joint range of motion, and trunk rotation, along with a narrower base of support, less lateral limb movement, and coordinated action between the arms and legs. Although these changes begin early in childhood, complete maturation of running gait may not occur until children reach their teenage years.

Applying the Concept

How Fast?

How fast can humans run? The winners of the 100 m dash in the Olympic Games generally are recognized as the world's fastest man and woman. The winning times at the 2016 Olympics in Rio de Janeiro, Brazil, were 9.81 seconds by Usain Bolt (Jamaica) in the men's competition and 10.71 seconds by Elaine Thompson (Jamaica) on the women's side. These winning times are just ticks slower than the world records as of early 2018: 9.58 seconds for men (Usain Bolt, Jamaica, set in 2009) and 10.49 seconds for women (Florence Griffith-Joyner, United States, set in 1988). By the time you read this information, these remarkable sprint times may have been eclipsed.

The peak speed in a 100 m sprint is not reached until about 60 m into the race. From that point on, the sprinter's speed typically decreases a bit. The winner, therefore, is not necessarily the fastest runner at a given moment but rather the one who can maintain speed (i.e., hold on) through the finish line.

The highest measured human peak running speed is just under 28 mph (just over 12 m/s). Compare this with the fastest land animal, the cheetah, whose top speed can exceed 70 mph. One of the slowest is the snail, who pokes along at 0.03 mph.

Cech and Martin (2011) describe the following developmental levels of running:

▸ Level 1

- Upper extremities: Arms are held high to assist with balance control but otherwise are not active.
- Lower extremities: Feet are flat; minimal flight, with swing leg slightly abducted.

▸ Level 2

- Upper extremities: Arms begin to swing as trunk rotation counterbalances pelvic rotation; arms may appear to flail.
- Lower extremities: Feet remain flat and may support knee flexion more during weight transfer; longer flight phase.

▸ Level 3

- Upper extremities: Arm swing increases in response to trunk rotation.
- Lower extremities: Heel contact made at foot strike; swing leg is in the sagittal plane; support leg reaches full extension at toe-off.

▸ Level 4

- Upper extremities: Arm swing becomes independent of trunk rotation; arms move in opposition to one another and contralateral to leg swing.
- Lower extremities: Similar to level 3.

Each progressive level depends on the developmental status of the child. Increases in size and strength, coupled with neuromuscular system development, allow the child to adopt more complex movement patterns and eventually reach a mature running gait.

Older Adults

While most adults reduce or stop running as they get older, some older adults continue to run well into their later years. Remarkable feats of running prowess are not uncommon.

American Bill Galbrecht, for example, completed marathons on each of the seven continents between 1997 and 1999, finishing the last race at age 71 (Cunningham, 2002). By that age, most people have given up running in favor of less-demanding fitness activities such as walking or swimming.

Those who do continue to run in their later years may show some changes in their temporal and spatial characteristics, often due to declines in strength, balance, and joint range of motion. Older adults typically jog and run more slowly, show less knee flexion during swing, and have shorter stride lengths than do younger runners.

Running Injuries

Running imposes high forces on the body that can lead to a variety of injuries. Because forces are felt first by the feet, and then progressively at the ankle, knee, and hip, running injuries most often afflict the lower extremities. The response to injury varies. Most injured runners modify their running (e.g., lower their distance) or stop running altogether until the injury heals sufficiently. Some determined runners, however, continue to run while injured. This can worsen the injury or lead to another injury, as when altered running mechanics in response to foot pain causes a compensatory, or secondary, injury at the knee or hip. Highly competitive athletes, such as professionals whose livelihood depends on their continued running, often try to play through an injury and train or compete while in pain. They do so at their own peril.

Many different injuries are associated with running. Injuries can be *acute* (e.g., twisting an ankle while running on an uneven surface; straining a hamstring muscle during fast running or sprinting) but more commonly are *chronic*, caused by the repetitive ground reaction forces applied to the foot and transmitted up through the joints of the lower extremities. Among the most common chronic running injuries are knee pain (e.g., chondromalacia patella), calcaneal (Achilles) tendinitis, plantar fasciitis, stress fractures (usually of the tibia or metatarsals), iliotibial band syndrome, and so-called shin splints (see sidebar).

Applying the Concept

Shin Splints—A Case of Vague Identity

"Of the many catchall terms used in the medical literature, perhaps none can match *shin splints* when it comes to nonspecificity, lack of consensus on meaning, and continuing misunderstanding and confusion" (Whiting & Zernicke, 2008, p. 188). O'Donoghue astutely notes,

> As with many names in common use, there is considerable and often heated argument as to what is actually meant by the term. As is usual in these circumstances, the term "shin splints" is a wastebasket one including many different conditions. The authors of various articles on the subject are inclined to state very definitely that it is caused by one particular thing to the exclusion of all others, which causes great confusion. (1984, p. 591)

Although many continue to use the term *shin splints*, clarity would be served by instead using terms that are clinically specific.

Altered running gait can be caused by physical pathologies or by other factors, whimsically suggested by the following poem:

> **The centipede was happy quite**
> **Until a toad in fun**
> **Said, "Pray, which leg goes after which?"**
> **That worked her mind to such a pitch**
> **She lay distracted in a ditch**
> **Considering how to run.**
>
> Attributed to Katherine Crater (1841-1874)

Concluding Comments

This chapter makes clear that fundamental movements of walking and running are complex phenomena requiring sophisticated integration of our body systems with the external environment. Common developmental sequences are seen in learning to walk and run. Despite these sequences, each of us develops unique gait patterns. Our individual patterns evolve throughout our lifetimes as a consequence of age-related changes in balance, muscular strength and coordination, cognition, joint range of motion, and visual acuity.

Go to the web study guide to access critical thinking questions for this chapter.

Suggested Readings

Cech, D.J., & Martin, S.M. (2011). *Functional movement development across the life span* (3rd ed.). Philadelphia: Saunders.

Enoka, R.M. (2015). *Neuromechanics of human movement* (5th ed.). Champaign, IL: Human Kinetics.

Haywood, K.M., & Getchell, N. (2014). *Life span motor development* (6th ed.). Champaign, IL: Human Kinetics.

Houglum, P.A., & Bertoti, D.B. (2011). *Brunnstrom's clinical kinesiology* (6th ed.). Philadelphia: FA Davis.

Levangie, P.K., & Norkin, C.C. (2011). *Joint structure and function: A comprehensive analysis* (5th ed.). Philadelphia: FA Davis.

Levine, D. (2012). *Whittle's gait analysis* (5th ed.). London: Churchill Livingstone.

Neumann, D.A. (2016). *Kinesiology of the musculoskeletal system: Foundations for rehabilitation* (3rd ed.). St. Louis: Mosby.

Nordin, M., & Frankel, V.H. (2012). *Basic biomechanics of the musculoskeletal system* (4th ed.). Philadelphia: Lippincott Williams & Wilkins.

Perry, J., & Burnfield, J. (2010). *Gait analysis: Normal and pathological function* (2nd ed.). Thorofare, NJ: Slack.

Rose, J., & Gamble, J.G. (2005). *Human walking* (3rd ed.). Philadelphia: Lippincott Williams & Wilkins.

9

Basic Movement Patterns

Objectives

After studying this chapter, you will be able to do the following:

- Describe the movement characteristics and muscle control of the fundamental skills of jumping, kicking, lifting, throwing, and striking
- Explain how these movements are affected by age and skill
- Give examples of injuries common to these movement skills that might restrict or preclude movement

In the previous chapter, we explored the fundamental movement patterns of walking and running gait. In this chapter, we examine several other common movement patterns, including jumping, kicking, lifting, throwing, and striking.

Jumping

Citius, Altius, Fortius, the motto of the International Olympic Committee (IOC), means "swifter, higher, stronger." To reach the second goal of going higher, one must jump.

Jumping means to spring free from the ground or other base by the muscular action of feet and legs. This definition provides a general description of the jumping action but does not distinguish between different ways of launching and landing. To make these distinctions, *jumping* applies to when individuals propel themselves from the ground with one or both feet and then land on both feet. *Hopping* involves propelling from one foot and landing on the same foot. *Leaping* describes the movement when individuals propel from one foot and land on the other foot (Haywood & Getchell, 2014).

Even though these terms are descriptive and specific, they still do not include all forms of jumping. In athletic competition, for example, high jumpers leave the ground from one foot and land in the pit on their backs. They clearly jump, but their actions do not fit into any of the standard definitions.

Types of Jumping

Jumping comes in many forms. Children at play jump out of sheer joy. Athletes jump to grab a rebound in basketball or catch a pass in American football. Ballet dancers jump when performing a grand jeté. Physical education students do jumping jacks. Boxers jump rope. The list goes on and on.

Jumping also is used to test lower-extremity power output (e.g., vertical jump test) and provides performance challenges to see how high (e.g., high jump) and how far (e.g., long jump, triple jump) one can jump. Each jump type has a specific goal and therefore requires a unique set of movements and pattern of muscle involvement.

With so many different types of jumps, it is infeasible to analyze here the joint motions and muscle control of them all. Thus, we describe here only a basic standing vertical jump with a two-foot takeoff and landing. The fundamental patterns described here are modified for other jump types, but many of the basic concepts, such as preparatory leg and arm action (i.e., countermovements), apply to most jump types.

A standing vertical jump can be divided into four phases: preparatory, propulsive, flight, and landing (figure 9.1). The jump begins from a normal standing position. During the preparatory (down) phase, the hip and knee joints flex, the ankles dorsiflex, and the arms swing back into hyperextension. In the propulsive (up) phase, the hips and knees extend, the ankles plantar flex, and the arms swing forward in flexion. The flight phase

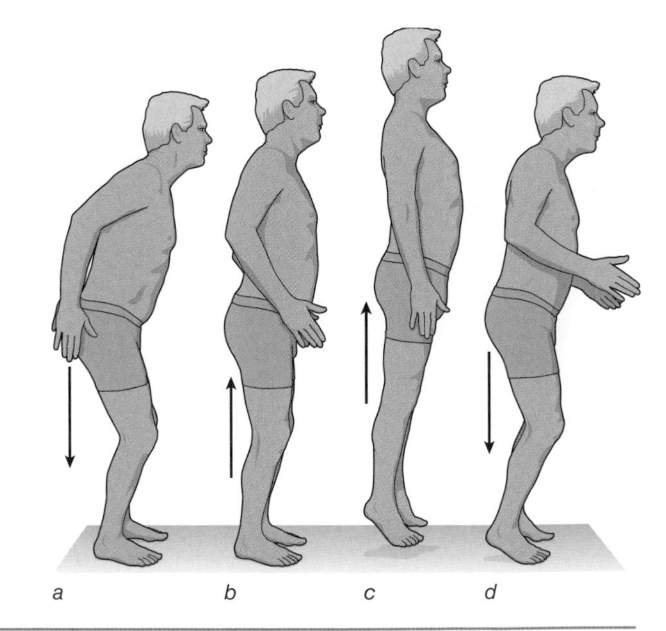

FIGURE 9.1 Phases of a standing vertical jump: *(a)* preparatory, *(b)* propulsive, *(c)* flight, *(d)* landing.

begins at takeoff when the toes leave the ground. Throughout the flight phase, the body assumes a relatively upright posture that is maintained until landing. Following flight, first ground contact begins the landing phase, during which the hips and knees flex, with ankle dorsiflexion and extension of the arms, as the body absorbs the forces of landing.

The proper timing of joint motions is critical for a successful and proficient jump. During the propulsive phase, for example, there is a rapid proximal-to-distal sequencing of maximum angular velocity at the hip, knee, and ankle joints, with very small delays between adjacent segments (Chiu et al., 2014; Hudson, 1986). This sequencing is necessary for the effective transfer of energy, from one segment to the next, required for optimal jumping performance. Alterations in this sequencing, such as when a jumper is fatigued, can alter the mechanics of the jump and result in a lower jump height.

Muscle Activity and Control

We know from our discussion in chapter 4 that actively stretching a muscle before its active shortening enhances the muscle's force capability. This eccentric–concentric stretch–shortening cycle is used to good advantage in jumping. More than a century ago, researchers demonstrated that a jumper performing a countermovement jump (i.e., with a preparatory squat immediately before the upward propulsive phase) could jump higher than when executing a static (or squat) jump (i.e., a jump begun from a squat position with no preparatory downward movement). One of the primary reasons for the better performance is that muscles during the preparatory phase act eccentrically immediately before their concentric action during the propulsive phase.

More specifically, hip and knee extensors (e.g., gluteus maximus at the hip and the vasti muscles at the knee) and ankle plantar flexors (e.g., gastrocnemius) act eccentrically during the preparatory phase to control flexion of the lower-extremity joints. When the hip, knee, and ankle joints reverse from flexion (during the preparatory phase) to extension (propulsive phase), the muscle actions also reverse from eccentric to concentric. This creates a classic stretch–shortening cycle for all the involved muscles, thereby facilitating force enhancement and thus a higher jump. And just as we saw a proximal-to-distal sequential pattern in the joint angular velocities, we see a similar pattern in the maximal activation of the hip, knee, and ankle extensors as well.

During the early flight phase, the lower-extremity muscles show markedly lower activity. Toward the end of the flight phase and before ground contact, the hip and knee extensors and ankle plantar flexors show anticipatory activity in preparation for landing. This preactivation is necessary to stiffen the muscles and better prepare them to eccentrically accommodate the high ground reaction forces at impact and early in the landing phase. In the case of an immediate second, or repeated, jump, this eccentric loading allows a smooth transition into the next stretch–shortening cycle for the subsequent jump.

The arms play an important role in successful jumping. They provide balance throughout the jump and add to the energy and momentum in propelling the body upward. Biomechanical studies (e.g., Luhtanen & Komi, 1978; Feltner et al., 2004; Blache & Monteil, 2013) show that arm swing significantly increases jump height and vertical velocity at takeoff. During the preparatory phase, the arms swing backward into hyperextension through concentric action of the glenohumeral (shoulder) extensors (e.g., posterior deltoid). In the propulsive phase, the arms swing forward through concentric action of the shoulder flexors (e.g., anterior deltoid, pectoralis major).

Jumping is an essential element in many sports and dance forms. In some, the jump itself is the primary focus. For example, in athletic field events such as the long jump, triple jump, and high jump, the goal is to jump as far or as high as possible (figure 9.2*a*). Success in these activities requires considerable power, speed, and coordination. In other activities, jumping plays an important role but is not the sole focus of the activity. Success in ballet, basketball, and volleyball, for example, is enhanced by jumping skills (figure 9.2*b*).

FIGURE 9.2 *(a)* Jumping in athletics (track and field—long jump) and *(b)* jumping in ballet.

(b) Peter Mueller

Life-Span Perspective

The life span of jumping is shorter than that of walking or running. The explosive strength necessary to propel the body through the air does not develop until the end of a child's second year. At the other end of the life span, older adults, with rare exceptions, abandon jumping as a movement form.

Infants and Children

As children learn to jump, they typically move from simple jumping patterns to more complex ones. Initially, they jump from two feet and land on two feet. Next, they may jump down from one foot to two feet, then jump down from two feet to one foot. This is followed by jumping forward from two feet to two feet, and then to running and jumping forward from one foot to two feet. Children later develop the ability to jump over objects and perform more complex jumping patterns that may involve body rotations and more use of the upper extremities. The development of jumping skills begins at about age 2 years, with many of the basic jump types in place by the end of a child's fourth year. Many children do not master jumping skills until much later, with some showing poor jumping technique even in their teens.

Developmental progress in jumping can be assessed by the age at which a child performs a particular jump, the height or distance of a jump, and the jumping form or technique. Proficient jumping is characterized by a preparatory crouch that prestretches the muscles of propulsion, allowing them to generate maximal force as the lower-extremity joints fully extend at takeoff, and a backward arm swing followed by a vigorous forward arm swing to initiate takeoff.

When jumping for maximum height, proficient jumpers direct force downward to the ground, extend the body throughout the flight phase, keep their trunks relatively upright throughout the jump, and flex the hips, knees, and ankles on landing to absorb the ground reaction forces. In jumping for maximum horizontal distance, jumpers lean their trunks forward, direct force downward and backward with the heels leaving the ground before knee extension during push-off, flex the knees during flight, swing the lower legs forward for a two-foot landing, flex at the hips to assume a jackknife position at landing, and flex the hips, knees, and ankles on landing to absorb the contact forces (Haywood & Getchell, 2014).

Research has well established that the stretch–shortening cycle (SSC) aids performance in adults by enhancing muscles' ability to generate force. But do children make use of the SSC? If so, to what extent? To explore this question, Harrison and Gaffney (2001) compared vertical jump performance of countermovement and static jumps in children and adults. Using vertical velocity at takeoff as the criterion measure, they found that children do indeed make use of the SSC, but performance was more variable in children than in adults, suggesting that children may perform countermovement jumps in a nonoptimal way.

Older Adults

Older adults typically cease jumping even earlier than they might abandon running. Many factors contribute to this tendency, including declines in muscle strength and power, compromised balance, increased risk of injury, and fear of falling. As a result, there is scant

Applying the Concept

Movement Evolution Versus Revolution

Changes in human movement patterns typically evolve gradually. On rare occasions, however, a single person dramatically changes, or revolutionizes, a movement pattern. Such is the case of Dick Fosbury. Prior to Fosbury, high jumpers used either a scissors technique, in which the jumper faces the bar and alternates lifting one leg over the bar, quickly followed by the other leg, or a straddling or roll technique, in which the jumper sequentially lifts one leg over the bar, followed by a rolling action of the body, with face down, and then clearance by the trailing leg.

While in high school, Fosbury developed a technique, later dubbed the "Fosbury Flop," which revolutionized high jumping. In the Fosbury Flop technique, the jumper quickly approaches the bar in a diagonal arc, plants his foot, and arches his body over the bar. In doing so, the jumper's center of gravity may actually pass under the bar (see figure 9.3).

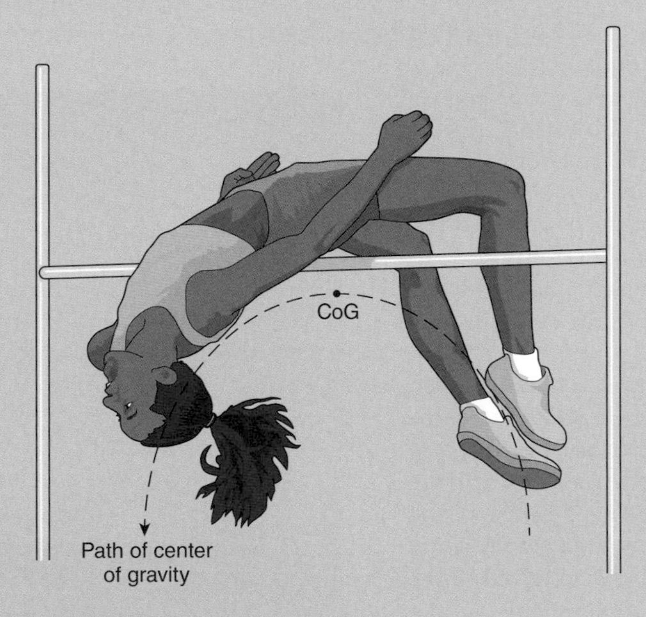

FIGURE 9.3 The center of gravity in the Fosbury flop.

Fosbury's success, highlighted by a gold medal at the 1968 Summer Olympics in Mexico City, cemented his technique as a de facto standard. More than 50 years after its development, the Fosbury Flop remains the dominant technique for high jumpers worldwide.

research describing the characteristics of jumping in older adults. We know little about how the jumping patterns of older individuals compare with those of younger adults.

Jumping Injuries

Most jumping-related injuries involve the knee. The knee joint complex forms the critical middle link in the kinetic chain of the lower extremity. In this role, its loading and motion characteristics dictate effective limb function. The most important component of the knee complex is the **knee extensor mechanism** (KEM), consisting of the quadriceps muscle group, the patellofemoral joint, and the tendon group connecting these elements. The patella serves as the central structure in the KEM. In that role, the patella acts as a pivot to enhance the mechanical advantage of the quadriceps during knee flexion and extension.

In jumping, the KEM is essential for effective performance. Injuries affecting the KEM compromise jumping ability. These injuries include patellar maltracking (described in chapter 3), quadriceps tendinitis (at the superior pole of the patella), patellar tendinitis (at the inferior pole), chondromalacia patella (softening degeneration of the articular cartilage on the back side of the patella), and Osgood-Schlatter disease (inflammation of the bone at the tibial tuberosity where the patellar tendon attaches to the tibia).

Kicking

Kicking uses a proximal-to-distal sequencing to perform the movement. Kicking is performed by the lower extremity, which strikes an object (e.g., a ball) that is not initially in contact with the body. The kicking leg is non–weight bearing throughout the movement as it swings freely; the contralateral leg bears all of the body's weight in supporting the kicking motion.

The final contact in kicking is made by the foot, but not without contributions from the upper extremities and trunk.

Kicking is an essential movement in sports (e.g., soccer [also fútbol], American football), various dance forms, and some martial arts (figure 9.4). The plane of action and goal of the kick may differ from one task to another, but there is a common element of a proximal-to-distal segmental progression in swinging the leg forward.

FIGURE 9.4 Example of kicking.

The kicking motion can be divided into phases. The kick usually begins with some type of approach, wherein the kicker moves forward toward the object to be kicked. The purpose of the approach phase is to build momentum that can ultimately be transferred to the object. A pre-impact phase follows the approach. This phase begins when the nonkicking leg contacts the ground to provide support for the kick. The support leg blocks the forward motion of the body and helps initiate the thigh swing of the kicking leg. The lower leg (shank) of the kicking leg continues to flex while the thigh begins its forward swing. Immediately before impact, the thigh slows down (decelerates) rapidly, and the shank reverses its direction and extends quickly toward contact.

During the impact phase, the foot is briefly in contact with the object (e.g., ball). The contact time between the foot and the object typically lasts for about 0.1 second or less. After impact, the leg continues into a follow-through phase, during which the leg slows down and the kicker regains balance in preparation for the next movement task.

Muscle Activity and Control

The hip joint muscles play a prominent role in any kicking movement. They are responsible for decelerating thigh hyperextension during the approach and reversing its motion to flexion as impact nears. This reversal from eccentric to concentric action allows the hip flexors to make use of the stretch–shortening cycle and produce more forceful action.

Applying the Concept

A Unique Kicker

Tom Dempsey is a former American football placekicker who played professionally in the National Football League (NFL) for a variety of teams from 1969 to 1979. While many placekickers of that era were transitioning to a soccer-style sweeping kicking motion, Dempsey maintained the traditional straight-toe style in which the kicking leg was swung straight forward in the sagittal plane.

On November 8, 1970, Dempsey broke the NFL field goal record with a 63 yd (58 m) game-winning kick. Dempsey's achievement is all the more remarkable given that he was born without toes on his right foot or fingers on his right hand. Dempsey wore a modified kicking shoe with a flattened front.

TOM DEMPSEY kicking a field goal in 1971.
Focus on Sport/Getty Images

At the time, Dempsey's shoe created controversy, since some contended that the special shoe afforded him a competitive advantage. Despite the fact that subsequent analysis concluded that Dempsey was not advantaged, the NFL instituted Rule 5, Section 4, Article 3(g) in 1977, which states (in part) that "kicking shoes must not be modified . . . and any shoe that is worn by a player with an artificial limb on his kicking leg must have a kicking surface that conforms to that of a normal kicking shoe." This rule has become known colloquially as the Tom Dempsey rule.

Hip muscles also are indirectly involved in knee extension through the whiplike proximal-to-distal sequencing as momentum is transferred from the thigh to the shank and foot. Interestingly, the knee extensors play little part in knee extension during the latter phase of the kick. In a study of soccer kicking, Robertson and Mosher (1985) did not find any knee extensor activity just before ball contact. Instead, they noted knee flexor torques just before contact. These torques may be the body's mechanism to prevent violent hyperextension at the end of knee extension, thereby reducing the chance of a hyperextension knee injury. The authors also suggest that knee extension during the preimpact phase may be too rapid for the knee extensor muscles to keep up with because of limitations dictated by the extensor muscles' force–velocity properties.

Life-Span Perspective

Kicking is a skill used by children and young adults, most often in sports such as soccer and American football, the martial arts, and dancing. Most of the research on kicking, therefore, focuses on younger populations. Older people usually do not participate in activities involving kicking. Thus, little, if any, research information details the kicking profile of older individuals.

Proficient kicking entails a preparatory windup, sequential segmental movement, a full range of motion at the swinging hip, trunk rotation to improve range of motion, backward body lean at contact, and oppositional use of the arms for balance. Young children initially kick with a simple leg push and exhibit none of the characteristics of skilled kickers. Unskilled kickers typically do not exhibit an approach or preparatory step and often kick with a bent knee. As skill develops, kickers begin to show proficiency, with proper foot placement, increased range of motion, full extension of the kicking leg at contact, trunk rotation, and arm opposition (Haywood & Getchell, 2014).

Children in the initial learning stages find it difficult to control kicking direction. Developmentally, focusing on kicking form rather than aim is advisable. Once proper form is achieved, further practice will improve aiming ability. Trying to teach proper aiming strategies too early in the learning process may result in tentative kicking motions without progression toward characteristics of proficient kicking (e.g., full range of motion).

Kicking Injuries

Kicking injuries can result from the mechanics of the kick itself or from contact during the kick (e.g., from an opposing player crashing into the kicker). Given its explosive (ballistic) nature, kicking can cause so-called deceleration injuries, when muscles work eccentrically to slow down a rapidly extending joint such as the knee. If the decelerating muscles (e.g., hamstrings) are overwhelmed by the forces of joint extension, injury may result.

In contact sports such as American football and soccer, injuries may occur when the kicker is hit by another player. In football, the punter is vulnerable to rushing opponents who are intent on blocking the kick. The supporting (nonkicking) leg is exposed to contact when the kicking leg is extended and the hip flexed.

In soccer, an opponent may hit the kicker and cause injury. One of the most common soccer injuries is knee damage resulting from the violent impact of an opposing defender. Most susceptible to injury are the knee's medial collateral ligament, medial meniscus, and anterior cruciate ligament.

Lifting

Lifting involves grasping an object and moving it to another, often higher, location. Many occupations require lifting as an essential job component. Construction workers, for example, must lift materials such as lumber. Nurses lift patients from their beds. Physical therapists provide lifting support for patients who cannot perform a task by themselves. Athletes lift weights to improve their strength (figure 9.5).

Lifting Techniques

The technique used to lift an object depends on many factors, including characteristics of the object (e.g., weight, size, shape), lifter (e.g., height, weight, strength, range of motion), environment (e.g., surface conditions), and task (e.g., distance moved, lifting speed). Some lifts are restricted to the sagittal plane, while others involve some rotation, or twisting, of the trunk.

Fundamental lifting postures include the squat and the stoop (figure 9.6). Some lifts are performed using a style in between a stoop and a squat. Traditionally, the squat lift has been recommended over the stoop. This advice, however, is based on the classic adage to lift with your legs and keep your back straight, which may be overly simplistic. In practice, the stoop lift is used more often than the squat, especially for lifting light objects.

Perhaps the most important factor when performing either lift is to maintain a neutral lumbar spine (i.e., avoid excessive spinal flexion). Keeping the load close to the body reduces the compressive and shear loads on the spine.

FIGURE 9.5 Example of lifting.

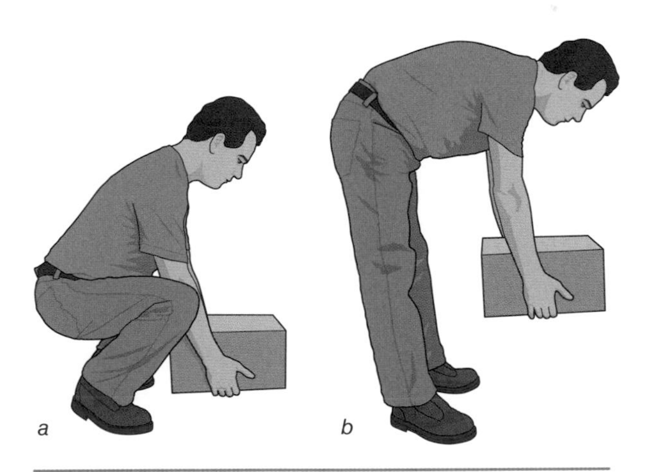

FIGURE 9.6 Lifting postures: *(a)* squat; *(b)* stoop.

Reprinted by permission from J. Watkins, *Structure and function of the musculoskeletal system* (Champaign, IL: Human Kinetics. 1999, p. 155).

Muscle Activity and Control

To lift an object from the ground to a higher level generally involves concentric action of muscles at major joints, along with isometric muscle action for joint stabilization. Controlled lowering of an object (e.g., moving a box from a tabletop to the floor) requires eccentric muscle action. Although lifting heavy objects requires high muscle strength, many lifting tasks involve repeated lifts with lower loads and thus emphasize muscle endurance over muscle strength.

In a traditional squat lift (see figure 9.6*a*), the lifter primarily uses the major joints of the lower extremity, producing hip and knee extension. At the hip, the gluteus maximus provides the primary hip extensor component. At the knee, the quadriceps muscles (vastus medialis, vastus lateralis, vastus intermedius, rectus femoris) produce knee extension. The degree of arm muscle involvement depends on the task. In sagittal-plane lifts above the head, glenohumeral flexors act concentrically to swing the arms forward and upward. The triceps brachii gets involved when elbow extension is required. The stoop lift (see figure 9.6*b*) involves less leg muscle action and more contribution by the trunk extensors (e.g., erector spinae).

Life-Span Perspective

Most of the research on lifting focuses on healthy adults. Far fewer studies involve lifting in children and older people. Many of the differences in lifting technique and capability across the life span are dictated by muscular strength and endurance, joint range of motion, and balance.

Infants and Children

Infants and children are limited in what they can lift because of their small stature, short limbs, and limited strength. As they physically mature, their capacity for lifting increases. Because of their greater flexibility and relative body dimensions, children use the squatting technique more often than do adults.

Anticipatory postural adjustments (APAs) are required to perform lifting tasks. Research shows that children are still developing their APA ability at 3 to 4 years, demonstrating inconsistent and immature kinematic and electromyographic patterns in a bimanual load-lifting task (Schmitz, Martin, & Assaiante, 1999). The children in this study also exhibited high intraindividual variability that is expected to decline with age and task mastery.

Older Adults

Lifting ability declines with age, largely because of decreases in muscular strength, range of motion, and balance. Consistent with these declines are changes in lifting strategies based on strength capacity. For example, Puniello, McGibbon, and Krebs (2001) identified three lifting strategies adopted by 91 functionally limited older adults. Subjects with relatively strong hip and knee extensors adopted a strategy dominated by the legs. Those with relatively strong knee extensors but weak hip extensors favored a mixed strategy characterized by initially using the back to lift, followed by the legs. The third approach, a back-dominant strategy, was used by subjects with weak hip and knee extensors. The authors concluded that older adults self-select a lifting strategy based on their hip and knee extensor strength.

Lifting and Low Back Disorders

Low back pain is one of the costliest musculoskeletal disorders in industrial societies. Up to 80% of the population will suffer from low back pain in their lifetimes, and many cases of low back pain result from lifting, especially manual material-handling tasks. The exact cause of back pain often proves elusive. As noted by McGill (2002, p. 118), "Clearly, [low back disorder] causality is often extremely complex with all sorts of factors interacting." Possible mechanisms of low back injury include lifting excessive loads, too many repetitions, and cumulative exposure.

Applying the Concept

Absolute Versus Relative Lifting Strength

A fundamental question related to human movement performance is, What are the limits? In terms of lifting, the limits are measured by world records. Let's compare, for example, the limits in the squat movement, in which lifters balance a barbell across their shoulders, squat down until their thighs are level with the ground, and then rise to a standing position.

In terms of *absolute* weight lifted, consider the 2017 men's record holder Blaine Sumner, who squatted 1,113 lb (505 kg) in March of that year. The women's record holder (47 kg class), Wei-Ling Chen from Taiwan, squatted 462 lb (210 kg). Both lifting records are exceptional. In *relative* terms, Sumner lifted three times his body weight in his record squat. Chen, who weighs a mere 103 lb (47 kg), lifted an astounding 4.5 times her body weight!

McGill (2002) suggests a two-pronged approach for people with low back pain to reduce discomfort and improve function. First, remove the stressors (e.g., heavy loads) that cause or exacerbate damage. Second, engage in activities aimed at building healthy supportive tissues (e.g., muscles, ligaments).

The pain associated with low back dysfunction can arise from many sources, including chemical irritation of tissue due to biochemical events associated with inflammation; stretching of connective tissues such as ligaments, the periosteum, tendons, and the joint capsule; compression of spinal nerves; herniation of intervertebral discs; and local muscle spasms (Whiting & Zernicke, 2008).

Although we will never completely eliminate lifting injuries, we can reduce the risk of lifting-related low back pain through prudent lifting strategies and physical training to improve strength, flexibility, and balance. Lifting technique, injuries, and safety recommendations are discussed more fully in chapter 13 (Ergonomics Applications).

Throwing

Throwing is as old as humankind. In prehistoric times, hunters threw rocks and spears at animals in hope of securing food for survival. Through the millennia, throwing has been an essential combat skill. Early on, people used rocks and primitive weapons and more recently they have employed destructive implements such as hand grenades. Many contemporary sports include throwing as an essential skill. These include softball and baseball, American football, basketball, and several events in athletics (i.e., track and field) such as the shot put, discus, and javelin. In noncompetitive situations, throwing sometimes provides nothing more than a pleasant diversion, as when a thrower tries to skip rocks across the still surface of a mountain lake.

Despite the wide range of venues and goals, all throws are similar in that they involve using the upper extremity to launch a handheld object (**projectile**) through the air. The study of projectile motion is called **ballistics**, and throwing is one of several ballistic skills in which force is imparted to an object to project it through the air. Other ballistic skills include kicking and striking.

Throws are categorized according to upper-extremity limb segment motion and the method of imparting force to the projectile. Classifications are *overarm throws*, *underarm throws*, *push throws*, and *pull throws* (figure 9.7). Overarm throwing is used, for example, by baseball pitchers and javelin throwers. Softball pitchers employ an underarm throwing motion to deliver the ball to the plate. Shot-putters use a push throw to project the shot, while discus and hammer throwers employ a pull throw to project their respective implements.

FIGURE 9.7 Types of throws: *(a)* overarm, *(b)* underarm, *(c)* push, *(d)* pull.

Throwing Principles

Throwing depends on a number of principles, including the transfer of momentum in a proximal-to-distal manner to an object held in the hand. As a result, the object is thrust, or propelled, into the air. The proper sequencing of limb segment motion presents the neuromuscular system with a challenging muscular control problem. In executing a throw, the body makes good use of the stretch–shortening cycle to enhance force production and throwing distance.

Throwing and Projectile Motion

Projectiles move through the air under the influence of only gravity and air resistance along a path called the **trajectory**. The trajectory is determined by three factors: release height (above the ground), release speed (how fast the object is thrown), and release angle (relative to the horizontal) as shown in figure 9.8. All the thrower's actions before release are intended to produce the proper combination of height, speed, and angle and thereby achieve the throwing goal.

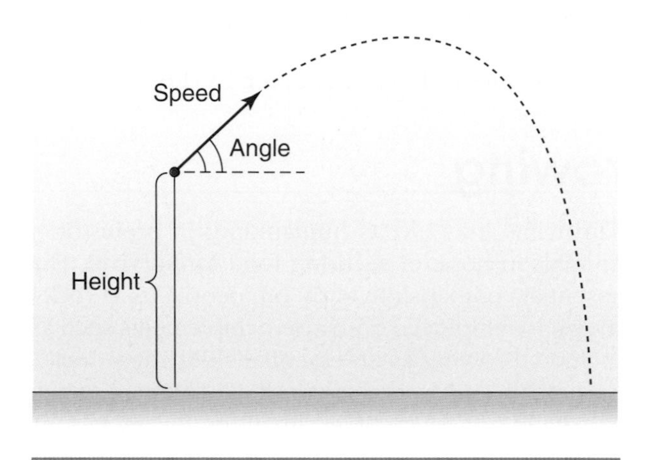

FIGURE 9.8 Initial conditions at release determine trajectory of a projectile: height, speed, and angle of projection.

Goals of Throwing

Each throw has a unique goal. Some throwing tasks, such as the shot put and javelin, seek to maximize throwing distance. For distance throws, the most important factor is release speed because speed is more important for distance than are height and angle of release. The goal of other throws may be to maximize height. Still others may have accuracy (e.g., dart throw) or maximization of throwing speed as their goal.

In some cases, a throw may combine goals. An American football quarterback, for example, in attempting to pass the ball to his receiver, must use the right combination of height (to clear the outstretched arms of oncoming defensive linemen), distance (to clear the defensive back), and ball speed (to get the ball to the receiver before another defender can intercept it).

Throwing Phases

The throwing pattern often is divided into phases to facilitate analysis. Each phase has defined beginning and end points, as well as specific biomechanical functions that contribute to the success of the throw.

One general phasing scheme describes three throwing phases: preparation, action, and recovery (Bartlett, 2000). The primary functions of the preparation phase are to (1) put the body in a favorable position for execution of the throw, (2) maximize the range of movement, (3) allow for larger body segments to initiate the throw, (4) actively stretch the agonist muscles to make use of the stretch–shortening cycle, (5) place the muscles at an advantageous length on their respective length–tension curves, and (6) store elastic energy to be used during the action phase.

During the action phase, skillful throwers use sequential muscle actions to execute the throw, beginning with muscles of larger segments, followed rapidly by muscles of the smaller, more distal segments. In most throws, there is proximal-to-distal muscle action and transfer of momentum and kinetic energy. The exact pattern of muscle action and mechanical transfer depends on the goal of the throw.

The primary purpose of the recovery phase is to slow down, or decelerate, the body and its limb segments through eccentric muscle action. This places the body in a favorable balanced position and reduces the chance of injury.

These general phases often are modified, or subdivided, in describing the throwing motion of a particular sport or type of throw. In baseball, for example, the pitching motion typically is divided into five phases: windup, cocking, acceleration, deceleration, and follow-through (figure 9.9). A sixth phase, stride, is sometimes included between windup and cocking. In context of the general scheme just presented, windup and cocking would constitute preparation, acceleration would correspond with action, and deceleration and follow-through would combine for recovery. Further details of throwing in baseball and softball are presented in chapter 11 (Sport and Dance Applications).

FIGURE 9.9 Phases of baseball pitching: *(a-e)* windup, *(f-h)* arm cocking and acceleration, *(i-k)* arm deceleration and follow-through.

Muscle Activity and Control

Muscle control of throwing varies depending on the type of throw being made. In general, muscles of the shoulder girdle and elbow are primarily involved, with secondary involvement of other upper-extremity muscles and musculature of the trunk and lower extremities. Our focus here is on muscles of the shoulder during a typical overhead throwing motion. Detailed discussion of muscle action in throws specific to baseball, softball, and American football is found in chapter 11 (Sport and Dance Applications).

During the windup phase, there is minimal shoulder muscle activity. As the arm is abducted, extended, and externally rotated in the cocking phase, the supraspinatus, posterior deltoid, infraspinatus, and teres minor are active. During the late cocking phase, the internal rotators (e.g., subscapularis, latissimus dorsi, anterior deltoid, and pectoralis major) are active eccentrically to slow down and control glenohumeral external rotation. As the throw transitions from the cocking to acceleration phase, the glenohumeral internal rotators reverse from eccentric to concentric action (i.e., a stretch–shortening cycle) to internally rotate the glenohumeral joint. Also, during the cocking phase, the triceps brachii may be active concentrically to begin elbow extension, but elbow control is quickly replaced by eccentric action of the elbow flexors (e.g., biceps brachii) to slow down elbow extension. During the deceleration phase, most of the glenohumeral and elbow muscles act eccentrically to slow down motion of the arm.

Life-Span Perspective

Many mechanical changes occur in the development of throwing from the first tentative motions of a 2-year-old to the intimidating power of a professional baseball pitcher's fastball. Improved strength, range of motion, and intersegmental coordination all contribute to improved throwing performance.

Infants and Children

When children first throw, they typically use only arm action by reaching back and then extending the elbow. As development progresses, they add trunk action and stepping and use more complex arm action (Roberton & Halverson, 1984).

Trunk action, when present in early throwers, usually involves forward trunk flexion accompanying hip flexion during the throw. Later, the child adds pelvic and trunk rotation, initially as a coupled block motion in which the pelvis and trunk rotate together as a single unit. As development continues, the child uncouples the pelvis and upper trunk and begins the forward throwing motion with pelvic rotation (while the trunk remains rotated away from the throwing direction), followed by forward trunk rotation.

Initially, children throw from a standing position with no foot movement. They then throw with an ipsilateral step (i.e., step forward with the foot on the same side as the throwing arm). As development progresses, the child learns to take a short contralateral step and later makes a longer contralateral step.

Throwing development also involves more complex arm action, including a preparatory backswing, alignment of the humerus perpendicular to the trunk, humeral lag (i.e., forward movement of the humerus temporally lags behind trunk and shoulder movement), and forearm lag (i.e., the forearm and hand remain behind until the trunk is facing front).

All these developmental changes allow the young thrower to generate more force and throw the object farther. Developmental pathways share common aspects across all throwers, as just described, but also exhibit individual characteristics unique to each child (Langendorfer & Roberton, 2002).

Older Adults

Mirroring the age-related performance declines in many movement skills, the throwing performance of older adults is similar in terms of velocity to that of children in middle elementary school (Williams, Haywood, & VanSant, 1991). One of the most notable changes is a decline in force production. Even though force production and velocity decline, only small declines in movement form appear evident. Williams, Haywood, and VanSant (1998), in a 7-year longitudinal study of throwing in older adults, report only small declines in movement form and suggest that elderly participants in their study coordinated their movements in a similar way to younger throwers but controlled them differently. Individual cases in the study showed slower movement speeds and decreased range of motion.

Throwing Injuries

In many ways, the human arm is not well designed structurally for repeated, vigorous overarm throwing. The high number of injuries seen in athletes who throw hard and often (e.g., baseball pitchers, water polo players, javelin throwers) suggests there are limits to the body's ability to withstand the forces of repeated throws. Injuries most commonly happen to the upper extremity but can also afflict the trunk, spine, and lower extremities.

Common upper-extremity throwing-related injuries include impingement syndrome, rotator cuff tears, and medial epicondylitis. **Impingement syndrome** of the shoulder results from repeated arm abduction in which suprahumeral structures (most notably the supraspinatus tendon and the subacromial bursa) are forcibly pressed against the anterior surface of the acromion and the coracoacromial ligament. Overarm throwers are particularly susceptible to impingement syndrome because of the repeated abduction motions inherent to their throwing.

Research in Mechanics

Tommy John Surgery

In 1974, Los Angeles Dodgers pitcher Tommy John was en route to one of his best seasons ever. However, in July, John ruptured the ulnar collateral ligament (UCL) in his pitching arm, and his career appeared to be over. He reportedly asked team physician Dr. Frank Jobe to "make up something" to salvage his pitching arm. And Dr. Jobe did just that. In what has become known as "Tommy John surgery," Dr. Jobe reconstructed John's UCL using a tendon graft from John's nonpitching arm.

At the time, no one knew what the outcome would be. John missed the 1975 baseball season and returned in 1976 to test his repaired arm. He passed the test with flying colors. Postsurgery, Tommy John pitched for an additional 13 years and amassed 164 wins. These wins, combined with 124 preinjury victories, left John with 288 career triumphs when he finally retired in 1989. Postinjury, John was a four-time All-Star (1968, 1978, 1979, 1980) and finished second in the Cy Young Award voting in 1977 and 1979.

Dr. Jobe continued to perform and refine his UCL reconstruction techniques (Jobe et al., 1986) and share his procedure with other professional colleagues. Jobe and colleagues reported good-to-excellent results in 80% of patients (Conway et al., 1992).

The most notable result, however, remains that of Tommy John. John's 288 career wins may be forgotten, but his name will not. The name Tommy John will always be associated with the pioneering surgery that saved his pitching career. As John said, "I'd never have been able to win 288 games without the surgery. We're going to be linked forever."

As for John's perspective of the surgery and his baseball career? "You know what I'm most proud of? ... I pitched 13 years after the procedure and I never missed a start. I had not one iota of trouble. I'd like people to remember that about me, too."

See the references for the full citations:

Conway, Jobe, Glousman, & Pink, 1992.

Jobe, Stark, & Lombardo, 1986.

Rupture of musculotendinous structures in the rotator cuff (supraspinatus, subscapularis, infraspinatus, teres minor) is typically the final result of a chain of events that begins with minor inflammation and progresses with continued overuse to advanced inflammation, microtearing of tissue, and finally partial or complete rupture. Compromised tissue integrity and muscle fatigue contribute to altered movement mechanics, and these modified movements further stress the involved tissues and hasten their eventual failure. The supraspinatus is the most commonly injured muscle in the rotator cuff group. Less frequently, other cuff muscles suffer damage. Supraspinatus injury, in particular, is associated with repeated and often violent overhead movement patterns such as throwing.

One of the most common throwing-related injuries at the elbow is **medial epicondylitis**. The wrist flexors in the forearm share a common proximal attachment at the medial epicondyle of the humerus. Eccentric action of these flexors in controlling wrist extension, together with violent valgus-extension loading during the end of the cocking phase, places considerable forces on the medial epicondyle. Repeated loading can lead to tissue damage on the medial aspect of the elbow. While the term *medial epicondylitis* implies an inflammatory condition, it may be misleading, since there is little evidence of inflammatory cells at the site of damaged tendinous attachments on the medial epicondyle.

On rare occasions, vigorous throwing can cause bone fractures. Humeral fractures, for example, have been documented as a result of throwing objects as varied as baseballs, javelins, and hand grenades. Various theories have been proposed to explain throwing-related fractures, including factors of antagonistic muscle action, violent uncoordinated muscle action, poor throwing mechanics, excessive torsional forces, and fatigue. Branch, Partin, Chamberland, Emeterio, and Sabetelle (1992) identify additional risk factors in a report on a series of 12 spontaneous humeral fractures in baseball players (average age 36): age, prolonged absence from pitching activity, lack of a regular exercise program, and precursory arm pain.

Although fairly common, throwing injuries are not inevitable. Correct throwing mechanics, proper physical conditioning, and moderation in throwing volume can greatly reduce the risk of throwing-related injuries to the upper extremity.

Striking

Striking is a hitting action, or dealing a blow to a person or object by use of a body part (e.g., hand, foot) or an implement (e.g., tennis racket, softball bat). Striking movements involving the hand include hitting an opponent during a boxing match and spiking a volleyball, while the foot may be involved in striking motions in any of the many martial arts. Striking actions typically are fast, short-duration, high-impact-force events.

Striking movements (figure 9.10) can be classified as *sidearm striking*, in which the striking happens at or just below shoulder level (e.g., baseball or softball swing), *overhead striking*, where the action occurs above shoulder level (e.g., tennis serve, volleyball spike), or *underhand striking*, in which the striking is below waist level (e.g., golf swing; Haywood & Getchell, 2014).

Further details of striking actions in tennis, baseball, softball, volleyball, and golf are presented in chapter 11 (Sport and Dance Applications).

Muscle Activity and Control

Given the wide variety of striking movements, few generalizations can be made related to muscle activity and control. Since most striking movements are ballistic (i.e., explosive) in nature, they often involve a stretch–shortening cycle in which eccentric muscle action, to slow and eventually stop joint motion, is immediately followed by concentric action of the same muscles. Muscle activity and control for specific sports are considered in detail in chapter 11 (Sport and Dance Applications).

FIGURE 9.10 Types of striking movements: *(a)* sidearm, *(b)* overhead, *(c)* underhand.

Life-Span Perspective

Striking activities can span a lifetime, beginning when a child learns to strike a plastic ball with a bat and continuing into advanced age with sports such as tennis and golf.

Infants and Children

The developmental stages involved in learning striking movements are similar to those already described for throwing and kicking. In sidearm striking in tennis, for example, the child's first arm movement typically is a vertical chopping action. With practice, the child gradually changes her swing from a vertical chop to an oblique swing, and eventually to a more horizontal-plane action using arm swing only. The swing then progresses to a stage where the tennis racket lags behind the child's trunk rotation, but moves ahead of the trunk when the child is facing forward. Finally, the child's racket lags behind the body throughout the striking movement when the body is front facing (Haywood & Getchell, 2014).

Foot action development during striking movements is similar to that found in throwing. Initially, the child remains in his initial foot position and takes no step. In the next stage, the child steps with the ipsilateral foot (i.e., the same side as the striking arm). Next, the child takes a small contralateral step (i.e., step with the foot on the opposite side as the striking arm). Finally, the child takes a longer contralateral step prior to striking (Roberton & Halverson, 1984).

Developmentally mature and proficient sidearm striking is characterized by stepping into the hit and using differentiated trunk rotation for enhanced force, fully swinging range of motion, arm extension just prior to contact, and a sequential pattern of "backswing and forward step, pelvic rotation, spinal rotation and swing, arm extension, contact, and follow-through" (Haywood & Getchell, 2014, p. 176).

Older Adults

Some adults maintain striking-related activities, particularly tennis and golf, well into their older years. One notable example is Gus Andreone, who took his first golf lesson in 1934 and made a hole in one in 2014 at the age of 103! Age-related declines in strength and flexibility can lead to reduced striking force and power production, but declines in tasks involving accuracy (e.g., golf putting) may be less affected by advancing age.

Applying the Concept

Range of Striking Strategies in Golf

In golf, players must master a wide range of strokes. Off the tee on long holes, golfers try to hit the ball as far as possible. In championship competition, top recorded drives are as much as 450 yd (411 m). In contrast, a short-shot to the green may be between 50 and 100 yd (46-91 m). And an all-important putt, which may determine a tournament champion, may be as short as a few feet in length. The movement strategies involved in a long drive off the tee are substantially different from the sensitive touch and judgment required for a short putt.

Striking Injuries

Striking-related injuries are of two general types. *Acute injuries* result from a single, usually high-level, force application. A striking martial arts kick or punch, for example, might cause a contusion (bruise), joint dislocation, or bone fracture. *Chronic injuries* from repeated force applications are quite common in striking sports such as tennis and golf. These injuries may result from overuse, poor technique, or both.

Specific injuries to the elbow have taken on sport-related monikers. Lateral epicondylitis often is referred to as *tennis elbow*. Caution is warranted in using this term, however, since it might be misinterpreted to mean that lateral epicondylitis happens only to those who play tennis, or that all tennis players acquire lateral epicondylitis. Neither of these suppositions is true. In fact, only a small percentage of those diagnosed with lateral epicondylitis attribute it to tennis (Peterson & Renström, 2001). Nonetheless, lateral epicondylitis is prevalent in tennis players, with 40% to 50% of players experiencing this injury at some time during their years of playing tennis.

Medial epicondylitis (discussed earlier in relation to throwing) sometimes is referred to as *golfer's elbow*. Though much less common than lateral epicondylitis, medial epicondylitis can afflict the elbow of golfers through repeated strokes that involve the wrist flexor muscles of the forearm, which have a common attachment on the medial epicondyle of the humerus.

In both tennis and golf, elbow and shoulder injuries in particular may be exacerbated by faulty stroke mechanics, off-center ball contact, grip tightness, and racket or club vibration.

Other common chronic injuries involving striking movements include glenohumeral impingement syndrome and biceps tendinitis in volleyball spiking, rotator cuff tendinitis and impingement syndrome in tennis serving and overhead strokes, and a variety of wrist, back, elbow, and knee injuries in golf.

Concluding Comments

Athletes and performers combine basic skills, such as jumping, kicking, lifting, throwing, and striking, into movement forms unique to their sport or activity. A softball player, for example, employs running and throwing as an integral part of her performance, just as soccer players combine running, kicking, and throwing to play their sport and ballet dancers include elements of running, jumping, kicking, and lifting in their performances. Integration of basic skills forms the foundation for successful performance in all activities.

Go to the web study guide to access critical thinking questions for this chapter.

Suggested Readings

Cech, D.J, & Martin, S.M. (2012). *Functional movement development across the life span* (3rd ed.). Philadelphia: Saunders.

Enoka, R.M. (2015). *Neuromechanics of human movement* (5th ed.). Champaign, IL: Human Kinetics.

Haywood, K.M., & Getchell, N. (2014). *Life span motor development* (6th ed.). Champaign, IL: Human Kinetics.

Houglum, P.A., & Bertoti, D.B. (2011). *Brunnstrom's clinical kinesiology* (6th ed.). Philadelphia: FA Davis.

Levangie, P.K., & Norkin, C.C. (2011). *Joint structure and function: A comprehensive analysis* (5th ed.). Philadelphia: FA Davis.

McGill, S. (2015). *Low back disorders: Evidence-based prevention and rehabilitation* (3rd ed.). Champaign, IL: Human Kinetics.

Neumann, D.A. (2016). *Kinesiology of the musculoskeletal system: Foundations for rehabilitation.* St. Louis: Mosby.

PART IV

Movement Applications

The four chapters of part IV explore movement applications across a variety of areas. Chapter 10 (Strength and Conditioning Applications) presents general principles of physical conditioning, with emphasis on resistance training and related movement analysis and muscle actions. Chapter 11 (Sport and Dance Applications) presents general concepts, applications, and muscle involvement in the sports of baseball and softball, basketball, cycling, American football, golf, soccer, swimming, tennis, and volleyball, as well as basic movements in dance. Chapter 12 (Clinical Applications) examines movement-related concepts of musculoskeletal injury, injury prevention, and rehabilitation. This chapter also explores issues related to the fields of prosthetics and orthotics. Chapter 13 (Ergonomics Applications) presents concepts related to ergonomics, the study of how humans interact with their immediate environment, with specific discussion of human–machine interface, lifting mechanics, and overuse conditions.

10

Strength and Conditioning Applications

Objectives

After studying this chapter, you will be able to do the following:

- Explain general concepts of strength and conditioning
- Describe the interdisciplinary nature of strength and conditioning
- Explain strength and conditioning principles of frequency, intensity, time, type, volume, and progression
- Discuss different types of strength and conditioning, including anaerobic, aerobic, endurance, strength, and power
- Describe the factors to be considered in designing a resistance training program
- Analyze resistance training exercises in terms of muscle action and contraction type

The theoretical and practical aspects of strength and conditioning encompass a variety of scientific disciplines, including human anatomy, biomechanics, physiology, endocrinology, nutrition, and psychology (Haff & Triplett, 2015). Our focus in this chapter is on anatomical and biomechanical applications.

In designing a strength and conditioning program, one primary consideration is the individual's goals. The program for a sedentary person beginning to exercise or for people looking to improve their general fitness and well-being will be quite different from that of elite athletes seeking to maximize their performance and compete at the highest level. The goals for a child may differ from those of an older person. Similarly, performance conditioning goals will vary from those for rehabilitation or reconditioning following injury.

Goals may involve improvement in (1) muscular strength, power, or endurance, (2) posture and balance, (3) flexibility, (4) speed and agility, or (5) body image (e.g., bodybuilding), or some combination of these areas.

General Principles

One of the fundamental principles of strength and conditioning is specificity. As described in chapter 1, body tissues and systems can adapt, either structurally or physiologically, in a way that is specific to the mechanical or metabolic demands placed on them. This characteristic of adaptability is described by the **SAID** (specific adaptation to imposed demands) **principle**. In addition to specificity, strength and conditioning programs should follow the FITT-VP principle, as outlined by the American College of Sports Medicine (2017). FITT-VP is an acronym for frequency, intensity, time, type, volume, and progression.

- Frequency refers to how often an exercise is performed.
- Intensity specifies how hard a person works (e.g., moderate or intense).
- Time addresses the duration of an exercise or session.
- Type refers to exercise mode, or what kind of exercises are performed.
- Volume is the combination of frequency, intensity, and time, and typically is calculated on a per-day or per-week basis. The same volume can be achieved by high intensity for a short period of time or by moderate intensity over a longer time span.
- Progression refers to the gradual increase in the other components as individuals improve or progress toward their goals.

One final principle relates to the need for a person to exercise at increasing levels of difficulty in order to maintain progress. This principle is termed **overload**. If the body is not continually challenged by new levels of stimulation, progress will slow and eventually plateau. In resistance training, for example, overload can be achieved in a number of ways, including increasing the amount of weight lifted, the speed of the lift, or the number of repetitions.

Types of Strength and Conditioning Programs

Given the many and varied goals of strength and conditioning programs, there is a commensurate number of program types. For example, anaerobic programs consist of intermittent bouts of high-intensity exercise, such as resistance training, plyometric training, and interval training. Anaerobic training can improve muscular strength and power and movement explosiveness.

Aerobic programs, in contrast, involve development of the body's aerobic capacity through improvements in cardiovascular and respiratory function, including increased muscular endurance, capillary density, mitochondrial density, enzyme activity, metabolic energy stores, ligament and tendon strength, and decreased body fat percentage.

Research in Mechanics

Power and Velocity

Power production is essential to rapid, ballistic movements (e.g., vertical jumping, throwing, striking). In concentric action, muscle force is inversely related to velocity (see chapter 4). Mechanical power can be calculated as the product of force times velocity (P = F × v). The curve of the force–velocity relationship (figure 4.8) can be used to draw a theoretical power–velocity curve (figure 10.1).

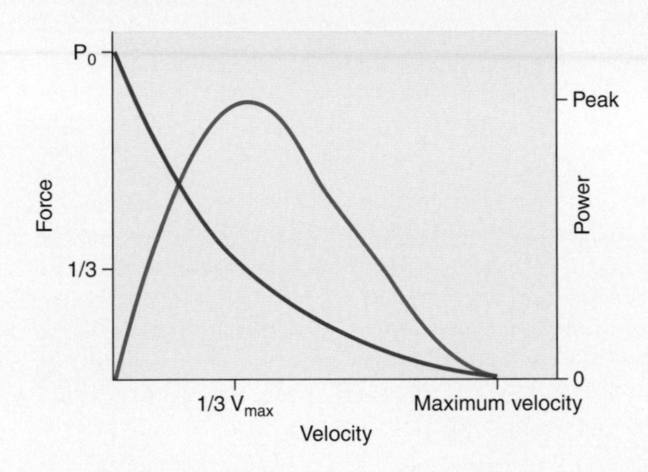

FIGURE 10.1 Force–velocity and power–velocity curves.

The nature of the power–velocity curve raises the question, what resistance (force) should be used to maximize power output? Or, conversely, at what velocity should one move for maximal power? The answer to these questions is of great practical significance for athletes training to maximize power output. Early research (e.g., Perrine & Edgerton, 1978) showed that maximum power does not happen at maximum force or maximum velocity, but rather at submaximal values for each variable such that the mathematical product of the two (i.e., force × velocity) is maximized. The literature varies considerably in specifying exact levels for each variable, but values in the 30% to 70% range are common.

See the references for the full citation:

Perrine & Edgerton, 1978.

Resistance Training

Seven design variables have been identified for development of a resistance training program: (1) needs analysis (including sport evaluation and athlete assessment), (2) exercise selection, (3) training frequency, (4) exercise order, (5) training load and repetitions, (6) volume, and (7) rest periods (Sheppard & Triplett, 2015).

In designing any type of **resistance training** program, exercises should be selected that target specific body regions, muscle groups, or specific muscles. To do this, we analyze exercise movements as described in chapter 6. The muscle control formula in that chapter facilitates movement analysis by outlining a step-by-step procedure for identifying the movements at relevant joints, the type of muscle action (i.e., concentric, isometric, eccentric), and the muscles being trained. Although the formula enables us to determine the principal

Applying the Concept

Resistance Training for Youth

The long-held belief that kids shouldn't lift weights because it will stunt their growth has been disproven. A growing body of research has shown that resistance training can provide unique benefits for children and adolescents. The National Strength and Conditioning Association's position statement on youth resistance training concludes that properly designed and supervised resistance training programs are relatively safe and can enhance muscular strength and power, improve cardiovascular risk profiles, improve motor skill performance, increase resistance to sport-related injuries, improve psychosocial well-being, and promote and develop exercise habits in youth (Faigenbaum et al., 2009). Similar benefits can accrue through other modes of conditioning as well to promote long-term athletic development (Lloyd et al., 2016).

The American College of Sports Medicine states,

> The goal of youth strength training should be to improve the musculoskeletal strength and general fitness of children and adolescents while exposing them to a variety of safe, effective and fun training methods. Adult strength training guidelines and training philosophies should not be imposed on youngsters who are anatomically, physiologically or psychologically less mature. Strength training should be one part of a well-rounded fitness program that also includes endurance, flexibility, agility and skill-building exercises. (Faigenbaum & Micheli, 2017, p. 1).

muscles being used, it does not tell anything about the level of muscle recruitment because that depends on many factors, including joint range of motion, technique, movement speed, and the external resistance being lifted.

Table 10.1 lists some of the most effective and popular resistance training exercises used to train different regions of the body. The exercises can be performed with dumbbells, barbells, or machines or simply by using one's own body weight. Most of the exercises, however, require external weights. Table 10.1 also includes the principal concentric joint movements needed to perform each exercise.

Movement Analysis and Muscle Actions

By using the muscle control formula presented in chapter 6 and the information in table 10.1, we explore the primary muscle groups used when performing the following exemplar exercises:

Wide-grip lat pull-down

Seated front barbell press

Seated pulley row

Squat

The following examples list the muscles being trained by groups (e.g., shoulder abductors, elbow flexors, knee extensors). For each of these four exercises and for all the exercises analyzed in table 10.1, challenge yourself to identify the specific muscles being used. Check your answers by reviewing tables 4.3 through 4.7.

Wide-Grip Lat Pull-Down

The concentric phase of the wide-grip lat pull-down (figure 10.2) consists of elbow flexion, glenohumeral adduction, and primarily downward rotation of the shoulder girdle. The

TABLE 10.1 Popular Resistance Exercises That Effectively Train the Major Regions of the Body

Muscle group	Exercise	Concentric joint movement
Forearm	Wrist curls Reverse wrist curls	Wrist flexion Wrist extension
Upper arm (anterior)	Standing dumbbell curls Hammer dumbbell curls Barbell curls Reverse curls Standing EZ-bar curls Concentration dumbbell curls Preacher curls	Elbow flexion
Upper arm (posterior)	Triceps extension Triceps kickbacks Triceps press-downs Triceps dips	Elbow extension
Shoulder	Front press Dumbbell press	Shoulder girdle upward rotation, glenohumeral abduction, and elbow extension
	Standing dumbbell side laterals	Shoulder girdle upward rotation (only partial movement) and glenohumeral abduction
	Bent-over lateral raises Pec deck rear deltoid laterals	Glenohumeral horizontal abduction (extension) and shoulder girdle adduction
	Front raises	Glenohumeral flexion
Chest	Flat bench press Push-ups	Glenohumeral horizontal adduction (flexion), elbow extension, and shoulder girdle abduction
	Incline bench press	Glenohumeral horizontal adduction (flexion), elbow extension, shoulder girdle abduction, and glenohumeral flexion
	Parallel bar dips	Glenohumeral flexion, elbow extension, and shoulder girdle depression and abduction
	Flat dumbbell flys Cable crossover flys	Glenohumeral horizontal adduction (flexion), shoulder girdle abduction
Back	Close-grip lat pull-downs Chin-ups	Shoulder girdle depression, glenohumeral extension, and elbow flexion
	Wide-grip lat pull-downs	Shoulder girdle downward rotation, glenohumeral adduction, and elbow flexion
	Seated pulley rows	Vertebral column extension, shoulder girdle adduction, glenohumeral extension, glenohumeral horizontal abduction (extension), and elbow flexion
	Back extensions	Vertebral column extension and hyperextension
Leg	Wide-stance squats	Hip extension, hip adduction, and knee extension
	Narrow-stance squats Leg press Lunges	Hip extension and knee extension
	Leg extensions	Knee extension
	Leg curls	Knee flexion
	Standing calf raises	Ankle plantar flexion
Abdomen	Crunches Reverse crunches Stability ball crunches	Vertebral column flexion
	Bicycle maneuver	Vertebral column flexion and vertebral column rotation

Note: Each exercise includes the principal joint movements needed to perform the concentric phase of the motion.

Applying the Concept

What's in a Name?

Some names used to identify resistance training exercises clearly describe the exercise. Exercises such as the barbell bicep curl, bent-over row, wrist curl, squat, and leg (knee) extension are easily visualized. Other terms are less clear. Variations of the basic curl exercise, for example, include concentration curls, EZ-bar curls (using a bar bent to position the forearms in a semisupinated position), the hammer curl with radioulnar joints in midposition (palms facing inward), and the preacher curl with upper arms isolated on an inclined surface. Other exercises with less-than-obvious names include the lat pull-down, named based on primary involvement of the latissimus dorsi muscles, the Romanian deadlift, and the good morning exercise.

RESISTANCE EXERCISE names: *(a)* barbell biceps curl, *(b)* concentration curl, *(c)* EZ-bar curl, *(d)* hammer curl, *(e)* preacher curl, *(f)* lat pull-down, *(g)* Romanian deadlift, and *(h)* good morning exercise.

principal muscles being trained, therefore, are the elbow flexors, glenohumeral adductors, and shoulder girdle downward rotators. Using tables 4.4 and 4.5, we see that the following muscles are being trained:

- **Elbow flexors**
  - Biceps brachii
  - Brachialis
  - Brachioradialis
- **Shoulder adductors**
  - Latissimus dorsi
  - Teres major
  - Pectoralis major, sternal portion
- **Shoulder girdle downward rotators**
  - Rhomboids
  - Levator scapulae
  - Pectoralis minor

FIGURE 10.2 Wide-grip lat pull-down.

Keep in mind that the muscles that *produce* the concentric motions are the same muscles that *control* the eccentric motions. In the return to the starting position, elbow extension is controlled eccentrically by the elbow flexors, shoulder abduction is controlled by eccentric action of the shoulder adductors, and shoulder girdle upward rotation is controlled eccentrically by the shoulder girdle downward rotators.

Seated Front Barbell Press

The concentric phase of the front barbell press (figure 10.3) consists of elbow extension, shoulder abduction, and shoulder girdle upward rotation. The principal muscles being trained, therefore, are the elbow extensors, shoulder joint abductors, and shoulder girdle upward rotators. Using tables 4.4 and 4.5, we see that the following muscles are being trained:

- **Elbow extensors**
  - Triceps brachii
  - Anconeus
- **Shoulder abductors**
  - Middle deltoid
  - Supraspinatus
  - Anterior deltoid
  - Biceps brachii, long head
- **Shoulder girdle upward rotators**
  - Trapezius
  - Serratus anterior

FIGURE 10.3 Seated front barbell press.

As the weight is lowered to the starting position, elbow flexion is controlled eccentrically by the elbow extensors, shoulder adduction is controlled by eccentric action of the shoulder abductors, and shoulder girdle downward rotation is controlled eccentrically by the shoulder girdle upward rotators.

Seated Pulley Row

The concentric phase of the seated pulley row (figure 10.4) consists of elbow flexion, shoulder extension, shoulder girdle adduction, and vertebral column extension. The principal muscles being trained, therefore, are the elbow flexors, shoulder extensors, shoulder girdle adductors, and vertebral column extensors. Using tables 4.3 through 4.5, we see that the following muscles are being trained:

▸ Elbow flexors

- Biceps brachii
- Brachialis
- Brachioradialis

▸ Shoulder extensors

- Latissimus dorsi
- Teres major
- Pectoralis major, sternal portion
- Posterior deltoid
- Triceps brachii, long head

▸ Shoulder girdle adductors

- Rhomboids
- Middle trapezius

▸ Vertebral column extensors

- Erector spinae

In the return to the starting position, elbow extension is controlled eccentrically by the elbow flexors, shoulder flexion is controlled eccentrically by the shoulder extensors, shoulder girdle abduction is controlled eccentrically by the shoulder girdle adductors, and vertebral column flexion is controlled eccentrically by the spinal extensors.

FIGURE 10.4 Seated pulley row.

Squat

In contrast to the previous three examples, each of which began with its concentric phase and returned using eccentric actions, the squat begins with an eccentric phase during descent, followed by concentric muscle action during the ascent (figure 10.5). Because the concentric phase of the squat consists of hip extension and knee extension, the principal muscles being trained are the hip and knee extensors. Using table 4.6, we can see that the following muscles are being trained:

- **Hip extensors**
  - Gluteus maximus
  - Semimembranosus
  - Semitendinosus
  - Biceps femoris, long head
  - Adductor magnus, posterior fibers
- **Knee extensors**
  - Vastus lateralis
  - Vastus intermedius
  - Vastus medialis
  - Rectus femoris

FIGURE 10.5 Squat.

In the downward phase of the squat, hip flexion is controlled eccentrically by the hip extensors, with knee flexion controlled eccentrically by the knee extensors.

Variations in Technique

How do variations in technique affect muscle recruitment? For example, what principal differences exist between the seven biceps curl exercises listed for the upper arm in table 10.1? Because elbow flexion is the concentric motion for each exercise, the elbow flexors are the principal muscles used. The position of the forearm, however, does affect recruitment of the biceps brachii and brachioradialis. For example, hammer curls and EZ-bar curls maximize

Applying the Concept

Squat: How Deep?

The squat exercise is a fundamental movement frequently used in resistance training programs. Many variables affect squat performance, including foot placement, load, speed, and depth. Of these variables, squat depth is perhaps the most controversial. From a standing position, with the knees fully extended (i.e., at zero degrees of flexion), the lifter can perform a partial squat with knee flexion of ~45° to 70°, 90° squat, parallel squat with thighs parallel with the ground, or a deep, full squat with knee flexion of 120° or more.

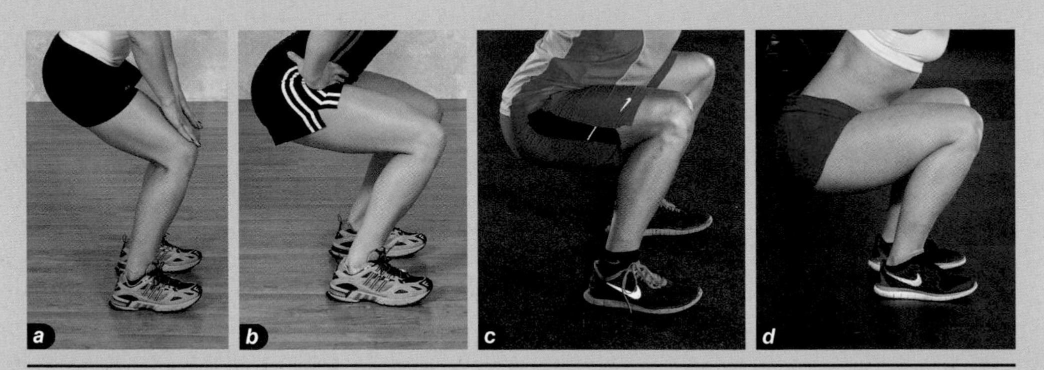

SQUAT DEPTHS: *(a)* partial, *(b)* 90°, *(c)* parallel, *(d)* deep.

Researchers have measured the relative activity of the quadriceps and hamstring muscle groups and estimated compressive loads and shearing forces at the tibiofemoral (knee) and patellofemoral joints, as well as tensile forces in the anterior and posterior cruciate ligaments, during squatting movements. Results and conclusions vary. From an injury-risk perspective, some authors have concluded that deep squats (i.e., below parallel thigh) may increase certain injury risks (e.g., Escamilla, 2001), while others have found the contrary (e.g., Schoenfeld, 2010). The topic of the efficacy and injury potential of deep squats remains equivocal. In any case, lifters should squat only as deep as proper technique will allow.

While much of the research on squat depth has focused on forces and injury risk at the knee, other body regions such as the hip joint and low back deserve consideration based on high loads. For example, compressive forces at the lumbar spine (L3/L4) during a half squat, with barbell loads between 0.8 and 1.6 times body weight, have been estimated as varying between 6 and 10 times body weight (Cappozzo et al., 1985). Average compressive loads on L4/L5 have been estimated as high as 17,192 N (~3,863 lb; 1,752 kg) in national-level powerlifters (Cholewicki et al., 1991).

the contribution from the brachioradialis because the brachioradialis moves the forearm to midposition (between supination and pronation). Supinated curls and EZ-bar curls target the biceps brachii because this muscle is also the strongest supinator of the forearm. In contrast, the brachialis is the prime mover for reverse curls because its insertion on the ulna means its line of pull, and therefore its length, is unaffected by forearm pronation or supination. Although the biceps brachii is recruited in a reverse curl, it is at an anatomical disadvantage to maximize its force contribution. This explains why you can't pronate (reverse) curl as much weight as you can during a hammer, EZ-bar, or supinated curl.

Concentration curls and preacher curls help isolate the elbow flexors because the upper arm is supported, thereby reducing the contribution from the shoulder flexors. Next time you perform a standing curl, be conscious of the contraction produced by the anterior deltoid and the clavicular portion of the pectoralis major. Why are these muscles active? As soon as you start to flex the elbow, the weight wants to drive the shoulder joint into hyperextension to balance the load against your body. In other words, the shoulder flexors contract to counteract the extensor torque produced by the weight.

What is the principal difference between a flat and an incline bench press, and why can you lift more weight when performing a flat bench press? For a flat bench press, the primary concentric shoulder joint motion is horizontal adduction, but with an incline bench press, the concentric motion becomes a combination of horizontal adduction and flexion. Because the sternal portion of the pectoralis major is a shoulder joint extensor (once the arm is flexed in front of the trunk), it will start to drop out (i.e., become less active) as the incline increases. Using the incline bench, therefore, targets the clavicular portion of the pectoralis major because it functions as both a horizontal adductor and flexor of the shoulder. Notice that as the incline steepens to 90°, you switch from performing a bench press to performing a shoulder press. The steeper the incline, therefore, the more you lose the sternal portion of the pectoralis major and the less weight you can lift.

What is the principal difference between a narrow- and wide-stance squat, and why can you lift more weight when using a wide stance? In a narrow-stance squat, the principal concentric hip joint movement is extension. In the wide-stance squat, the concentric hip motion is a combination of extension and adduction. Using this technique, therefore, calls on the hip adductors to work with the hip extensors to control the weight.

Sometimes variations in exercise technique recruit the same principal muscles but for different reasons. For example, both the close-grip and wide-grip lat pull-down recruit the latissimus dorsi, teres major, and sternal portion of the pectoralis major. In the close-grip lat pull-down, these three muscles are recruited as shoulder joint extensors. In the wide-grip lat pull-down, however, they function as shoulder joint adductors.

Although the examples just discussed are specific to resistance training, the basic steps outlined in chapter 6 can be used to analyze any movement. In addition, the exercises in table 10.1, although by no means comprehensive, represent fundamental movements used across all sports and activities and help illustrate the principles of joint movement and muscle function.

Concluding Comments

Our brief discussion of strength and conditioning principles and applications has emphasized the interdisciplinary nature of the area, the applicability to a variety of populations (ranging from youth to older people, beginners to experts, and healthy to injured), and different types of strength and conditioning programs and factors to be considered in their design. Using concepts of functional anatomy and biomechanics, we can analyze a variety of resistance training exercises.

Go to the web study guide to access critical thinking questions for this chapter.

Suggested Readings

Alvar, B.A., Sell, K., & Deuster, P.A. (Eds.). (2017). *NSCA's essentials of tactical strength and conditioning*. Champaign, IL: Human Kinetics.

American College of Sports Medicine. (2017). *ACSM's guidelines for exercise testing and prescription* (10th ed.). Philadelphia: Lippincott Williams & Wilkins.

Chandler, T.J., & Brown, L.E. (Eds.). (2012). *Conditioning for strength and human performance* (2nd ed.). Philadelphia: Lippincott Williams & Wilkins.

Chu, D.A., & Myer, G.D. (2013). *Plyometrics*. Champaign, IL: Human Kinetics.

Coburn, J.W., & Malek, M.H. (Eds.). (2011). *NSCA's essentials of personal training* (2nd ed.). Champaign, IL: Human Kinetics.

Delavier, F. (2010). *Strength training anatomy* (3rd ed.). Champaign, IL: Human Kinetics.

Haff, G.G., & Triplett, N.T. (Eds.). (2015). *Essentials of strength and conditioning* (4th ed.). Champaign, IL: Human Kinetics.

National Strength & Conditioning Association. (2016). *Strength training* (2nd ed.). Champaign, IL: Human Kinetics.

Ratamess, N. (Ed.) (2011). *ACSM's foundation of strength training and conditioning*. Philadelphia: Lippincott Williams & Wilkins.

Zatsiorsky, V., & Kraemer, W. (2006). *Science and practice of strength training* (2nd ed.). Champaign, IL: Human Kinetics.

11 Sport and Dance Applications

Objectives

After studying this chapter, you will be able to do the following:

- Describe and explain basic movement patterns, muscular involvement, and mechanics for the following sports:
    - Baseball and softball
    - Basketball
    - Cycling
    - American football
    - Golf
    - Soccer
    - Swimming
    - Tennis
    - Volleyball
- Describe and explain basic movement patterns, muscular involvement, and mechanics for various forms of dance

In this chapter, we look at specific sport and dance movements. Entire books are devoted to the functional anatomy and biomechanics of sports (e.g., Bartlett, 2014; Blazevich, 2017; pioneering volume by Hay, 1993; McGinnis, 2013) and dance (e.g., Clippinger, 2016). Given that volume of information, we clearly cannot discuss all sports here. We therefore present selected sports—baseball and softball, basketball, cycling, American football, golf, soccer, swimming, tennis, and volleyball—and some of their movement forms to show how the principles of applied anatomy can be used to analyze movements specific to a given sport or exercise activity, highlighting muscle involvement and movement mechanics for each. In addition, we consider movements and muscle involvement in dance.

Baseball and Softball

Baseball and softball are bat-and-ball sports played by two opposing teams that consist of nine players per team. Each team takes alternating turns on offense, trying to score runs, and on defense, trying to prevent the other team from scoring. Batters try to hit a ball thrown by the opposing team's pitcher and advance around a series of four bases (first, second, and third bases and home plate). A player rounding all four bases scores a run for the player's team. A primary difference between baseball and softball is the manner in which the pitcher delivers the ball to the hitter. In baseball, pitchers use an overhead throwing motion, while softball pitchers use an underhand, windmill motion. Variants of softball include slow-pitch, fast-pitch, and modified.

General Concepts

The primary movement forms involved in baseball and softball are running, throwing, and striking, with occasional jumping (to try to catch a ball). Offensive players, after hitting a pitched ball, run to first base. If successful in getting to first base before being thrown out by a defensive player, the runner attempts to run around the bases, with the ultimate goal of reaching home plate (and thereby scoring a run). Defensive players run when trying to field or catch a ball hit by the batter.

Baseball pitchers throw a variety of pitch types (e.g., fastball, curveball, slider, changeup, knuckleball), as do softball pitchers (e.g., fastball, changeup, drop pitch, curveball, rise ball), from a raised pitching mound toward home plate, a distance in baseball of 60.5 ft (18.4 m) and in softball of 46 ft (14 m) for men and 43 ft (13.1 m) for women. Defensive infielders (first base, second base, third base, shortstop) and outfielders (left fielder, center fielder, right fielder) throw the ball across a variety of conditions and distances. Striking in baseball consists of a batter swinging a bat (wood or aluminum) in an attempt to hit a pitched ball.

Applications

Most of the research on muscle activity in baseball and softball has focused on pitching. Electromyographic studies, using indwelling electrodes, have documented the activity of numerous shoulder, elbow, and forearm muscles during baseball pitching (e.g., Gowan et al., 1987; Jobe et al., 1984; Sisto et al., 1987). During the windup and early cocking phases (figure 11.1), minimal shoulder and elbow muscle activity occurs. In the late cocking phase, when the arm approaches maximal external rotation, moderate activity is detected in the biceps brachii. The cocking phase ends with the pectoralis major and latissimus dorsi acting eccentrically to slow, and eventually stop, external rotation at the shoulder. Also active during the cocking phase are the supraspinatus, infraspinatus, teres minor, deltoid, and trapezius. These muscles position the shoulder and elbow for delivery of the pitch. During the acceleration phase, the biceps brachii shows little activity, with propulsion produced by the pectoralis major, serratus anterior, subscapularis, latissimus dorsi, and triceps brachii.

Applying the Concept

Baseball's Sporting Cousin—Cricket

Based on participation and fan support, cricket is among the most popular sports worldwide. Cricket's governing body is the International Cricket Council (ICC), founded in 1909 by charter members England, Australia, and South Africa. The International Women's Cricket Council, founded in 1958, merged with the ICC in 2005. The ICC currently has 105 member nations. Twelve of the member nations have full membership and therefore are entitled to play at the highest level, known as Test cricket. In addition to the three founding nations, Test cricket is played by Bangladesh, India, New Zealand, Pakistan, Sri Lanka, West Indies, and Zimbabwe, along with 2017 additions Afghanistan and Ireland.

Many of the movement elements in baseball and softball are found in their sporting cousin, cricket. With a fan base second only to soccer, cricket has strong international appeal. Cricket, like baseball and softball, is a bat-and-ball sport. The baseball and softball players who deliver the ball (pitcher), catch the ball (catcher), and hit the ball (batter) have cricket counterparts known as the bowler, wicketkeeper, and striker batsman, respectively. While the rules of baseball and softball are, of course, different from the laws of cricket, and the implements (bat and ball) have different dimensions, some of the players' movement patterns are similar between the sports. Other movement patterns are distinct. A baseball pitcher, for example, throws the ball with one foot touching a rectangular pad called the rubber, while a cricket bowler takes a running head start before releasing the ball.

In their review of the biomechanics of bowling in cricket, Bartlett and colleagues (1996) began by stating that "cricket has not been well served by biomechanical research" (p. 403). At that time, there was a dearth of research on cricket, with particularly little exploration of muscle action involved in the bowling motion. In the decades since Bartlett's review, more research on cricket has been forthcoming, but it is likely still not commensurate with the popularity of the sport worldwide.

Ahamed and others (2014) presented a comprehensive review and research comparing the muscle action of cricket bowlers and other overhead-throwing athletes. In reviewing studies from 1990 to 2011, they found only four that examined the electromyography of muscles during cricket bowling. A few recent studies (e.g., Hazari, Warsi, & Agouris, 2016) have advanced our understanding of muscle action in cricket, but more work is needed in this area.

CRICKET BOWLER delivering the ball to a batsman, backed by a wicketkeeper.
Lance Bellers/fotolia.com

FIGURE 11.1 Baseball pitch.

The deceleration and follow-through phases are characterized by eccentric muscle action that decelerates the arm. The biceps brachii, for example, acts eccentrically to slow elbow extension, while glenohumeral external rotators (e.g., infraspinatus, teres minor) actively lengthen to slow the shoulder's internal rotation.

Noteworthy differences are evident in the EMG patterns of professional pitchers compared with less-skilled amateur pitchers. Professional pitchers achieved higher throwing velocities, in part due to stronger shoulder muscle activity (e.g., pectoralis major, serratus anterior, subscapularis, latissimus dorsi). Differences were also evident during the later throwing phases, with amateur pitchers making greater use of the rotator cuff muscles and biceps brachii than professionals (Gowan et al., 1987).

The softball windmill pitch commonly is divided into five phases: (1) windup with first ball motion forward to 6 o'clock position (arm straight down), (2) from 6 o'clock to 3 o'clock, (3) from 3 o'clock to 12 o'clock (arm overhead), (4) 12 o'clock to 9 o'clock, and (5) 9 o'clock to ball release (figure 11.2). After ball release, the arm slows down in the follow-through phase.

Maffet and colleagues (1997) studied intramuscular EMG activity of eight shoulder girdle muscles in 10 female collegiate-level softball pitchers. Among their findings were that (1) the supraspinatus was maximally active during phase 2 in aligning the humeral head with the glenoid fossa, (2) the posterior deltoid and teres minor were maximally active in phase 3 to maintain arm elevation and externally rotate the humerus, (3) the pectoralis major accelerated the arm in phases 4 and 5, (4) the serratus anterior positioned the scapula to maintain glenohumeral congruency, and (5) the subscapularis worked as an internal rotator and protected the anterior joint capsule.

FIGURE 11.2 Softball pitch.

Research in Mechanics

Softball Windmill Pitching Mechanics and Injury

Several research studies have examined the biomechanics of the softball windmill pitch and injury risk in fast-pitch softball pitchers. A common misconception persists that underhand windmill pitchers experience lower forces and injury risk than overhead baseball pitchers. Barrentine and others (1998) reported high forces and torques at the shoulder and elbow during the delivery phase of the windmill pitch, including peak compressive forces at the elbow and shoulder of 70% to 98% body weight and shoulder extension and abduction torques of 9% to 10% of body weight times height. The authors concluded that "demand on the biceps labrum complex to simultaneously resist glenohumeral distraction and produce elbow flexion makes this structure susceptible to overuse injury" (p. 405).

Werner and colleagues studied the biomechanics of windmill softball pitching in youth (2005) and elite softball pitchers at the 1996 Olympic Games (2006). The authors reported average shoulder distraction forces of 80% of body weight for Olympic pitchers and 94% of body weight for youth pitchers. In both studies, excessive distraction forces in softball pitchers were comparable to those found in baseball pitchers, suggesting that windmill softball pitchers are at risk for overuse injuries.

See the references for the full citations:

Barrentine, Fleisig, Whiteside, Escamilla, & Andrews, 1998.

Werner, Guido, McNeice, Richardson, Delude, & Stewart, 2005.

Werner, Jones, Guido, & Brunet, 2006.

Oliver and others (2011) reported high gluteus maximus activity and consistent gluteus medius activity, especially in phase 3 when the pitcher was in single support. The biceps brachii was most active during phase 4, while the scapular stabilizers had their highest activity in phase 2. Triceps brachii activity was consistently high during the entire pitching motion. High biceps activity during phases 4 and 5 also was reported by Rojas and colleagues (2009).

Batting involves a precisely coordinated rotational motion that concludes with the bat crossing home plate in an attempt to hit a pitched ball (figure 11.3). Hitting a pitched ball is considered one of the most difficult neuromotor tasks in all of sports. Peak baseball pitching speeds typically are in the 90 to 95 mph (144.8-152.9 kph) range, with occasional pitches topping 100 mph (160.9 kph). Softball pitches can reach 75 mph (120.7 kph) for women and 85 mph (136.8 kph) for men. Given the shorter distance from the pitching mound to home plate in softball (as compared to baseball), hitting a fast-pitch softball arguably is as difficult as, if not more difficult than, hitting a fastball baseball pitch.

FIGURE 11.3 Baseball batting swing.

While the mechanics of batting are similar in softball and baseball, the limited research on muscle activity in batting has focused on baseball. Shaffer and others (1993) examined the activity patterns of 12 muscles of the lower extremity, trunk, and upper extremity during a batting swing. They reported high hamstring and gluteal muscle activity during the preswing and early swing phases, but diminished activity in later phases of the swing. The vastus medialis showed peak activity throughout the swing phases and follow-through. The erector spinae and abdominal obliques were active during the swing phases and follow-through, while the supraspinatus and serratus anterior demonstrated relatively low muscle activity throughout. The authors concluded that batting requires sequential and coordinated muscle activity, starting with the hip, followed by the trunk, and finally with the arms. Most of the power in the swing begins at the hip, with high muscle activity in the trunk musculature throughout the swing.

Nakata and colleagues (2013) reported on EMG activity of the lower limbs in baseball batting in skilled and novice batters. Based on EMG measurements of eight muscles (rectus femoris, biceps femoris, tibialis anterior, and medial gastrocnemius on both legs), the authors found that skilled batters showed higher levels of muscle activity, with their timing for shifting, stepping, and landing occurring earlier. The authors suggested that preparations for the swing are made earlier in skilled batters, who recruit lower extremity muscles more effectively than novice batters do.

Basketball

Basketball is an internationally popular sport involving two teams of five players each who play on a rectangular court with baskets at each end of the court. These baskets have 18 in. (46 cm) circular metal rims with a net hanging from the rim, and they are elevated 10 ft (3.05 m) above the floor. The rims are attached to a rectangular backboard. Players on offense attempt to shoot the ball through the rim. Defensive players try to prevent the offensive players from shooting and scoring. A moving player in possession of the ball must dribble (bounce) the ball using only one hand.

FIGURE 11.4 Basketball cut (change of direction).
Zach Bolinger/Icon Sportswire via Getty Images

General Concepts

Among the movement skills in basketball are running, stopping, cutting (i.e., change of direction), dribbling, passing, shooting, and rebounding.

Applications

Given the nature and movement requirements of the game, basketball players physiologically have high aerobic and anaerobic power demands (de Araujo et al., 2013; Narazaki et al., 2009; Pojskić et al., 2015). Basketball is a fast-paced game involving short bouts of running, quick stops, and rapid changes of direction, along with jumping, passing, rebounding, and shooting.

In executing a change of direction (known as a *cut*; figure 11.4), the player

Applying the Concept

Basketball—Special Populations

Basketball is well suited for, and popular with, people with special needs, including those with physical and intellectual disabilities. Wheelchair basketball (WCB), in particular, ranks as one of the most popular sports for those with spinal cord injury. While the wheelchair provides on-court mobility, the lack of lower-extremity function in WCB players requires them to make effective use of their trunk and upper-extremity musculature to pass, shoot, dribble, rebound the ball, and propel their wheelchairs.

The International Wheelchair Basketball Federation (IWBF, 2014) classifies players into groups ranging from 1.0 (lowest functioning) to 4.5 (highest functioning) in 0.5 increments, based on the amount of a player's trunk movement and control. To maintain competitive balance (Gil-Agudo et al., 2010), each five-player team cannot have a team classification total of more than 14.

Research on muscular involvement in WCB has highlighted the role of shoulder internal rotation, elbow extension, wrist flexion–extension range of motion, and wrist flexion–extension strength (Wang et al., 2005); respiratory muscle strength (Moreno et al., 2012); and trunk muscle strength (Santos et al., 2017; Yildirim et al. 2010) in effective performance.

WHEELCHAIR BASKETBALL.

uses eccentric action of the ankle plantar flexors (to control dorsiflexion), knee extensors (to control flexion), and hip extensors (to control hip flexion), followed immediately by concentric action of the same muscles to powerfully propel the player's body in a different direction.

When dribbling, the player uses one hand to push the ball to the floor using the fingers on the dribbling hand (figure 11.5). The pushing motion is created by simultaneous action of the elbow extensors and wrist and finger flexors. After the ball bounces from the floor and returns to the fingers, the elbow flexes and the wrist and fingers extend slightly before the next pushing phase.

Passing involves pushing the ball into the air toward a teammate. Basic basketball passes include the chest pass (figure 11.6*a*), bounce pass (figure 11.6*b*), and overhead pass (figure 11.6*c*). In terms of muscle activity, pushing the ball in the chest pass and bounce pass involves bilateral action of the glenohumeral flexors, elbow extensors, and wrist flexors. The overhead pass is performed by action of the glenohumeral extensors, elbow extensors, and wrist flexors.

Rebounding involves players trying to secure possession of a missed shot. To do this, players need to be in the proper position, jump explosively, and reach for the ball with one or both hands (figure 11.7). Jumping for a rebound requires powerful action of the ankle plantar flexors, knee extensors, and hip extensors and typically involves a stretch–shortening cycle of these muscles, with eccentric muscle action at the hip, knee, and ankle muscles immediately followed by concentric action as the player propels upward.

FIGURE 11.5 Basketball dribbling.

FIGURE 11.6 Basketball passing: *(a)* chest pass, *(b)* bounce pass, *(c)* overhead pass.

FIGURE 11.7 Basketball rebounding.

Shooting is one of the most fundamental and important basketball skills. Among the shot types are the layup (figure 11.8*a*), jump shot (figure 11.8*b*), free throw (figure 11.8*c*), and dunk (figure 11.8*d*). The jump shot and free throw have been the subject of most shooting research. Research has focused on a variety of measures, including ball trajectory, release velocity, release angle, release height, movement phases, anthropometrics (i.e., physical characteristics), experience, basket height, ball size and weight, fatigue, shooting distance, visual field, and presence of an opponent (Okazaki et al., 2015). Given the importance of shooting as an essential basketball skill, surprisingly little research exists on EMG activity in muscles involved in basketball shooting. In the jump shot, upper-extremity movements are produced by coordinated action of the shooting arm's glenohumeral flexors, elbow extensors, and wrist and finger flexors.

FIGURE 11.8 Basketball shooting: *(a)* layup, *(b)* jump shot, *(c)* free throw, *(d)* dunk.
(a) Leon Bennett/Getty Images

Cycling

Cycling is one of the most popular exercise and sport activities. Millions of cyclists pedal on a regular basis for health benefits and recreation and many others are involved in competitive cycling. The sport's popularity is evident in the remarkable worldwide interest in cycling's premier event, the grueling Tour de France. The three-week tour, held each July in France, is one of the most demanding of all athletic events. Cycling also provides effective exercise for people with injuries or musculoskeletal conditions (e.g., osteoarthritis) that make walking and running painful. In certain populations (e.g., stroke patients or those with spinal cord injury), cycling also can provide rehabilitative benefits.

General Concepts

Despite its many forms (e.g., unicycle, bicycle, tricycle, recumbent bike, human powered vehicles) and purposes (e.g., transportation, recreation, racing [road and sprint], mountain biking, exercise, rehabilitation), cycling fundamentally involves the application of force, provided by the lower extremities, to pedals and a crank and chain system to rotate the cycle's wheels.

Applying the Concept

Clinical Aspects of Cycling Biomechanics

Cycling is a commonly used modality in musculoskeletal rehabilitation. Among the clinically relevant points to consider in cycling are that (1) patients can learn to redistribute forces applied through the foot–pedal interface to improve performance, (2) when cycling, reactive loads are produced at all locations where the body contacts the bike, (3) in special populations, particular attention is required when securing the foot to the pedal, (4) clipless pedals allow for normal foot rotation and reduce torques acting at the knee, (5) changes in seat height can affect joint range of motion and the level of shear force at the knee, and (6) changes in cadence and resistance alter loads on muscles, ligaments, and bones (Gregor et al., 2011).

Pedaling Cycle

Just as many sport skills (e.g., rowing, throwing, running, swimming) are divided into phases for ease of analysis, so is pedaling. The **pedaling cycle** normally is divided into two phases: a power phase that drives the bike forward and a recovery phase. The power phase starts when the crank arm (the portion of the crank from the crank axis to the pedal) is at top dead center (TDC), or 0°, and ends when the crank arm reaches bottom dead center (BDC), or 180°. Although most of the useful force is applied to the cranks during the power phase, it is possible to apply force to rotate the cranks during portions of the recovery phase. Even for elite cyclists, however, the quantity of force used to propel the bike forward during the recovery phase is small compared with the force produced during the power phase.

Propulsion and Force Application at the Pedals

Because the force applied to the pedals turns the cranks and propels the bike forward, it is important to understand how pedal forces are measured and what they tell us. To record the force produced by the cyclist's legs during the pedaling cycle, researchers use instrumented force pedals to measure the normal and tangential forces acting on the pedal. The **normal force** component acts perpendicular to the pedal surface, whereas the **tangential force** component acts in the anterior–posterior direction along the surface of the pedal (figure 11.9). By measuring the pedal angle relative to the crank arm, the forces acting on the pedal can then be resolved into an effective, or perpendicular, component and an ineffective, or tangential, component acting down the crank arm. As the name implies, the effective component produces the torque needed to turn the cranks. In contrast, the ineffective component acts parallel to the crank and, as a result, produces no useful external work in propelling the bike.

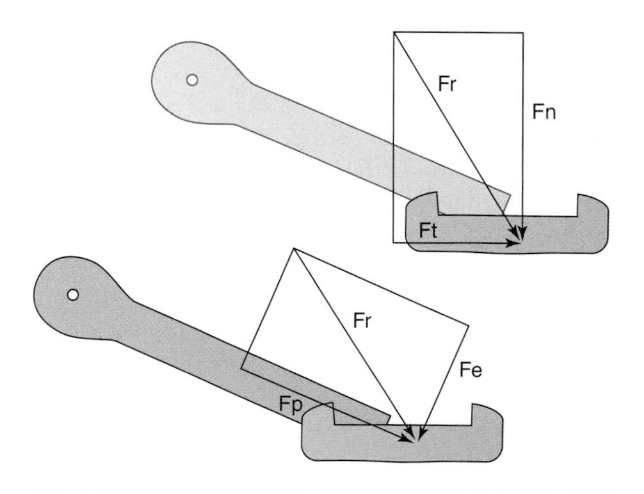

FIGURE 11.9 Application of pedal forces: resultant force (Fr), normal force (Fn), tangential force (Ft), force parallel to crank arm (Fp), and effective force (Fe).

Adapted by permission from P.R. Cavanah and D.J. Sanderson, The biomechanics of cycling: Studies of the pedaling mechanics of elite pursuit riders, in *Science of cycling*, edited by E.R. Burke, (Champaign, IL: Human Kinetics, 1986), 103.

A clock diagram is a useful way to visualize the force applied to the pedal throughout the pedaling cycle (figure 11.10). For experienced and elite

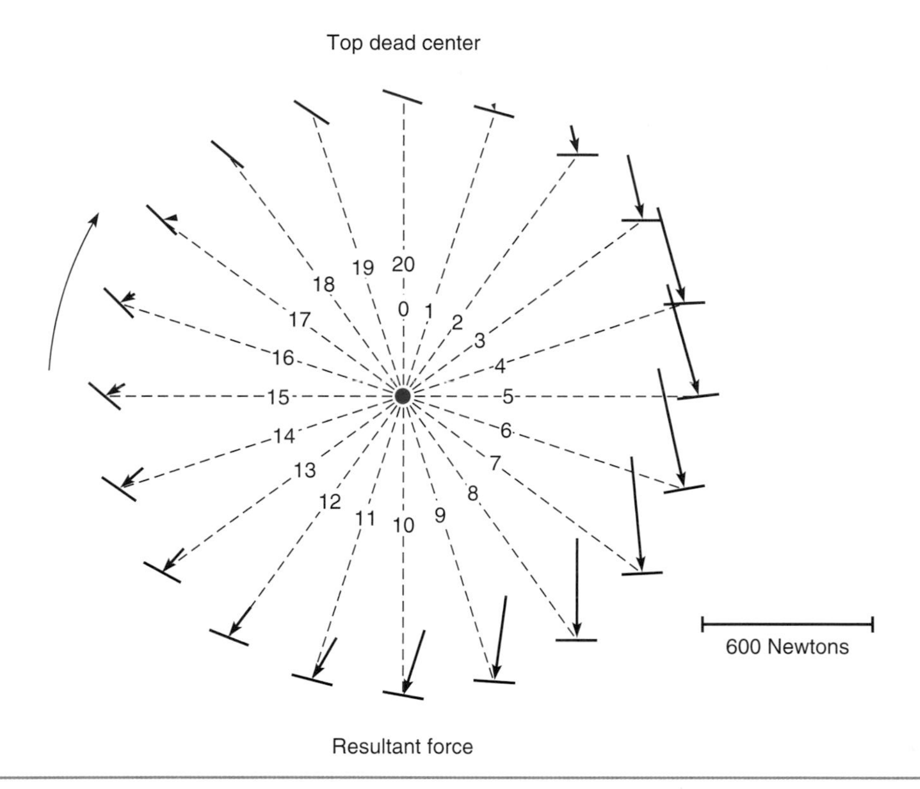

FIGURE 11.10 Force applied to the pedal during one complete pedaling cycle.

Adapted by permission from P.R. Cavanah and D.J. Sanderson, The biomechanics of cycling: Studies of the pedaling mechanics of elite pursuit riders, in *Science of cycling*, edited by E.R. Burke, (Champaign, IL: Human Kinetics, 1986), 105.

riders, the maximum normal force, and therefore crank torque, usually occurs at approximately 100° after TDC (at approximately 4 on the clock). The clock diagram shows both the magnitude and direction of the applied force at the pedal. Because the muscle actions during cycling are primarily concentric, each muscle's activation pattern helps explain the duration and timing of the forces acting on the crank throughout the pedaling cycle.

Cadence

Most competitive cyclists use a pedaling rate, or *cadence*, between 70 and 100 revolutions per minute (rpm). Although track cyclists may reach pedaling rates as high as 160 rpm for brief sprints, pedaling rates above 110 rpm are not recommended for road-racing cyclists unless needed for short sprints.

For metabolic efficiency, a slow pedaling cadence should be avoided because of the large energy expenditure needed to maintain prolonged muscle contractions. Although the amount of force increases as the contraction speed decreases, the increased blood vessel occlusion and metabolic energy needed to maintain the contraction could eventually neutralize any advantage of the enhanced force production. If the pedaling rate is too high, little external work can be performed, and energy will be wasted not only in the form of heat but also in overcoming the internal resistance of the muscle.

High cadences and power output elicit an increased reliance on the recruitment of fast-twitch fibers and thus a greater dependence on anaerobic pathways for energy production. The increased blood flow, along with decreased blood vessel occlusion and muscle stress (i.e., less force produced per pedaling cycle) at the higher cadences, however, may help compensate for the greater metabolic cost. In addition, high cadences may produce less metabolic stress on elite cyclists because of their high aerobic capacity. Because the most efficient cadence increases with power output, cadences between 70 and 100 rpm are not excessive for elite cyclists when matched with their high power output during training and racing. The cyclist's optimal cadence, therefore, should produce the most sustainable power for the duration of the event without accumulation of metabolites (e.g., lactic acid) that might impair performance.

Body Position and Seat Height

Although the position of the trunk and arms is critical for aerodynamic efficiency, particularly during cycling sprints, seat height directly affects the cyclist's metabolic and mechanical effectiveness in maintaining high power output. Numerous methods exist for calculating the most effective seat height. Nonetheless, all methods suggest that once the height and fore–aft position of the seat are correctly adjusted, the angle formed between the back of the femur and the leg (shank), with the foot and crank in the BDC position, should be between 150° and 155°. In other words, when the leg is in the BDC position of the pedaling cycle, the knee should be flexed between 25° and 30° (figure 11.11). Seat height affects not only the forward lean of the cyclist but also the length–tension and force–velocity properties and mechanical advantage of all the principal leg muscles used to propel the bike. Seat position also affects rider comfort, muscle activation patterns, and changes in variability of sitting postural control in submaximal cycling (Verma et al., 2016). No single optimal position, however, exists for all riders. Each rider must rely on continuous testing and trial-and-error experience to ensure maximal comfort and performance.

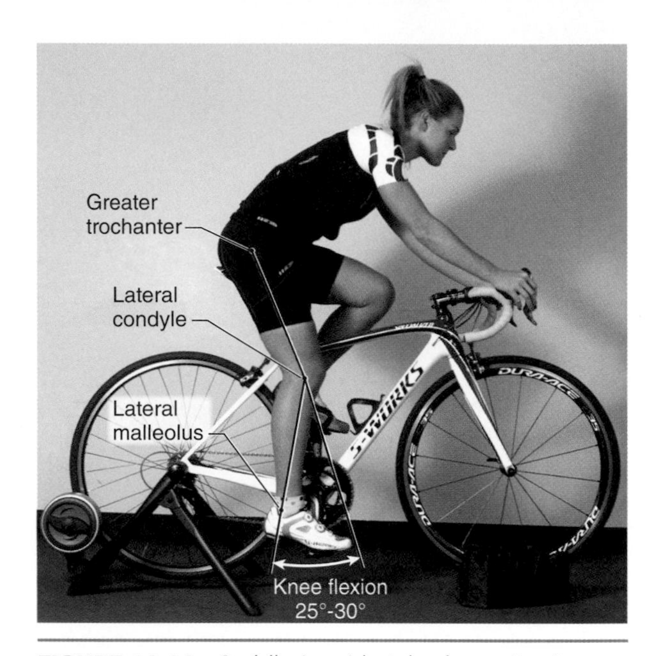

FIGURE 11.11 Saddle (seat) height determination.

Aerodynamic Drag

During training and racing, the greatest force resisting the forward movement of the cyclist is air or wind resistance. The cyclist must work against two types of aerodynamic drag: form drag and surface drag.

Form drag (also referred to as shape, profile, or pressure drag) is produced by the parting of air by the cyclist's body and the bike. The amount of form drag depends on the size, shape, and speed of the cyclist and the bike, as well as the orientation of the cyclist's body with respect to airflow. Using aerodynamic frame tubing, aerodynamic handlebars, and, most important, aerodynamic wheels can significantly reduce form drag produced by the bike. In addition, riders can significantly reduce their form drag by assuming the forward lean position. During time trials (when the cyclist races against the clock), the cyclist assumes a position with the back almost parallel to the ground and arms stretched out in front, with elbows and forearms as close as possible while still maintaining control of the bike. This streamlined position not only minimizes the frontal area of the cyclist's body exposed to the wind but also lengthens the principal hip extensors, thereby increasing their force contribution during the pedaling cycle.

Surface drag (also referred to as frictional resistance, skin resistance, skin-friction drag, or skin drag) is produced as air passes over the surface of the cyclist and the bike. For example, friction created between the air and the cyclist's skin and clothing slows the air and, therefore, resists forward motion. Although surface drag is less of a concern than form drag, particularly at high speeds, the use of skin suits, aerodynamic helmets, and aerodynamic shoe coverings is critical for maximizing time trial performance.

Because both form and surface drag increase as the square of the cyclist's forward velocity, small increases in speed, particularly at the high speeds achieved during racing, necessitate an enormous increase in power output by the cyclist.

Muscle Activation

Figure 11.12 presents the muscle activation patterns typical of well-trained cyclists, measured using electromyography. EMG records show that principal activity for the hip, knee, and ankle extensors occurs during the power phase, specifically in the first quadrant (first 90°) of the pedaling cycle. The high muscle activity recorded from the gluteus maximus, semitendinosus, vastus lateralis, gastrocnemius, and soleus during the initial portion of the power phase explains the timing of peak pedal forces and crank torque at about 100° of the pedaling cycle.

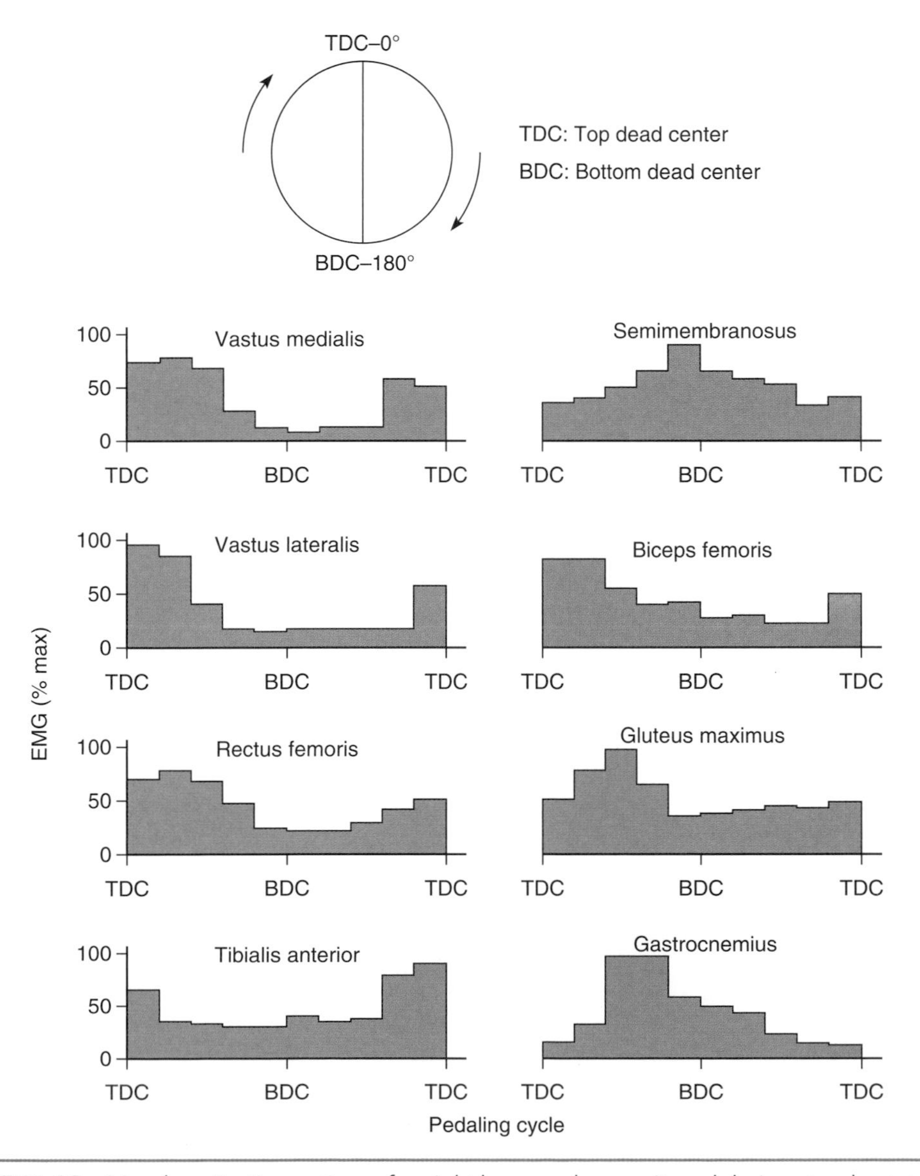

FIGURE 11.12 Muscle activation patterns for eight leg muscles monitored during steady-state cycling.

Adapted by permission from R.J. Gregor and S.G. Rugg, Effects of saddle height and pedaling cadence on power output and efficiency, in *Science of cycling*, edited by E.R. Burke, (Champaign, IL: Human Kinetics, 1986), 74.

The semitendinosus and gastrocnemius activity during the recovery phase primarily results from their role as knee flexors. Their activity helps sweep the leg through BDC and back up toward TDC. In contrast to the three uniarticular vasti muscles (vastus medialis, vastus lateralis, vastus intermedius), which work only as knee extensors, the rectus femoris tends to be more active during the recovery phase because of its added function as a hip flexor. This hip flexor activity helps bring the recovery leg up to TDC in preparation for the high hip, knee, and ankle extensor activity needed during the power phase of the next cycle. The high muscle activity in the soleus and gastrocnemius during the power phase serves two primary functions: (1) to resist the dorsiflexor torque exerted by the pedal on the ankle and (2) to transmit the large forces, and therefore torques, produced by the hip and knee extensors into the pedals. There appears to be limited coactivation of agonist–antagonist pairs at the hip, knee, and ankle joints, especially for the gastrocnemius and tibialis anterior at the ankle and the hamstrings and quadriceps at the knee (Jorge & Hull, 1986).

Research has shown that muscle coordination patterns play an important role with respect to cycling mechanics during tasks that involve maximal power output, mechanical efficiency, and fatigue. Increases in mechanical power during maximum-effort cycling are more due to muscle coordination than to increases in muscle EMG intensity (Wakeling et al., 2010). Increased mechanical efficiency is achieved by sequential peak muscle activation from knee to hip to ankle and by moderate intensities (i.e., 55%-60% of $\dot{V}O_2max$; Blake et al., 2012). Large fatigue-related power reduction during sprint cycling is associated with substantial reductions in EMG activity in biarticular muscles (gastrocnemius, rectus femoris) and coactivation of the gastrocnemius and the primary power-producing muscles (gluteus maximus, vastus lateralis, vastus medialis obliquus; O'Bryan et al., 2014).

American Football

The game of American (also Canadian) football is a form of gridiron football in which two teams of 11 players compete on a rectangular field (120 × 53.3 yd [109.7 × 48.7 m], including 10 × 53.3 yd [9.1 × 48.7 m] end zones at each end of the field) with goal posts at the back of each end zone. The football is a leather (or composite), oval-shaped (or more technically a prolate [elongated] spheroid) ball about 11 in. (27.9 cm) in length. The team in possession of the football (i.e., on offense) attempts to advance the ball, either by running or passing, down the field in a series of plays. The ultimate goal is to advance the ball past the opponents' goal line and score a touchdown (worth 6 points). After a touchdown, the scoring team attempts a conversion (also point-after-touchdown, or PAT) by kicking the ball over and through the goal post for 1 point or running a play to advance the ball from the 3 yd line to score 2 points. The defensive team tries to prevent the offensive team from advancing the ball. Further technical aspects of the game are beyond the scope of this text.

The first American football game was played between Princeton University and Rutgers University on November 6, 1869. In the nearly 150 years since its inception, American football has grown into the most popular sport in the United States. It is played at the youth, high school, college, and professional levels and followed by many millions of fans. American football also is popular in Canada and has a growing presence in Mexico and Japan, along with some countries in Europe and South America (in particular, Brazil).

General Concepts

Football involves a wide variety of movement forms, including running, throwing (passing), kicking (punting and placekicking for a kickoff or field goal), jumping, catching, tackling (in which a defensive player attempts to knock down or wrestle the ball carrier to the ground), and blocking (in which an offensive player tries to prevent a defensive player from tackling the ball carrier).

Applying the Concept

Football

The term *football* refers to a family of team sports, each of which involves some degree of kicking a ball with the foot. Generic use of the term typically refers to the specific sport most popular in a given country or region. Among the sports included in the football family are *association football* (also *soccer* or *fútbol*), *gridiron football* (also *American football* or *Canadian football*), *Australian rules football*, and *rugby football* (also *rugby*). Two of these sports, American football and soccer (association football), are discussed in detail in this chapter.

Applications

Despite the popularity of American football, scant research has been conducted on muscle activity during specific movement tasks. As a result, with the exception of passing (i.e., quarterback throwing the ball), all of the following movements are assessed using movement analysis (not confirmed by EMG evaluation).

All football players run—some (e.g., offensive linemen) to a limited degree, others to a greater extent (e.g., running backs, receivers, defensive backs), but rarely more than 40 yd (37 m) on a single play. The muscular and mechanical elements of running discussed in chapter 8 apply, with the added factor that the running patterns of football players often are dictated by the movements and intent of both teammates and opponents.

Kicking in football takes one of two forms: punting or placekicking. Punting involves the punter receiving (catching) the ball snapped by the offensive center, followed by a kick in which the punter extends his arms, drops the ball, and then kicks the ball either as far as possible or to a specific location (figure 11.13). The primary leg muscle action is provided by the powerful and sequential activity of the hip flexors and knee extensors.

Placekicking involves kicking the ball from a stationary tee (e.g., on the kickoff that begins a game or second half and following a score) or from a fixed position with the ball held by a placeholder (e.g., attempting a field goal or PAT; see figure 11.14). Most current placekickers use a side-foot soccer-style kick.

Jumping in football typically involves an offensive receiver who becomes airborne in attempting to catch the ball or a defensive player (e.g., linebacker or defensive back) trying to intercept (catch) the ball or knock it to the ground.

Offensive receivers and running backs and defensive linebackers and defensive backs require catching skills when the quarterback (in most cases) passes the ball. Offensive players catch the ball to advance it toward the goal line. Defensive players catch the ball

FIGURE 11.13 Football punting.

FIGURE 11.14 Football placekicking: *(a)* from a tee and *(b)* held by a placeholder.

(interception) to stop the opponent's offense and take possession of the ball for their own team. Catching a football can be particularly challenging since an opponent is usually trying to prevent the catch. If the receiver catches the ball, he has to avoid opponents or else be tackled. Since most football games are played outdoors and in the fall and winter seasons, inclement weather can add to the difficulty of catching a football.

Tackling is a movement task in which a defensive player attempts to stop an opponent carrying the ball, either by wrestling or knocking the ball carrier to the ground (figure 11.15). Given the size and speed of football players, especially at the collegiate and professional levels, great muscular strength and mobility are required to successfully tackle a ball carrier.

In an attempt to prevent a defensive player from tackling a ball carrier, offensive players, primarily offensive linemen, tight ends, and running backs, use blocking techniques to block or impede the movement of a potential tackler (figure 11.16). Blocking requires substantial leg, torso, and arm strength and effective balance and coordination.

As noted earlier, the only football skill with any appreciable research with regard to muscle activity is passing (figure 11.17). In the most comprehensive study of shoulder muscle activity in football throwing (passing) to date, Kelly and colleagues (2002) examined the EMG of nine shoulder girdle muscles. Fine-wire (indwelling) electrodes were inserted into three of the rotator cuff muscles (supraspinatus, infraspinatus, subscapularis), with surface electrodes placed on the pectoralis major, latissimus dorsi, biceps brachii, and anterior, middle, and posterior deltoid of 14 experienced male subjects with a range of skill levels

FIGURE 11.15 Football tackling.

(i.e., recreational to collegiate). The authors divided the passing motion into four phases: early cocking, late cocking, acceleration, and follow-through. Muscles were classified as either *stabilizers* (high-level stabilizers: supraspinatus, infraspinatus; moderate-level stabilizers: deltoids; low-level stabilizer: biceps brachii) or *accelerators* (subscapularis, latissimus dorsi, pectoralis major). The accelerator muscles exhibited consistently high activity during the acceleration and follow-through phases. Muscle activity was highest for all muscles (except for the anterior deltoid and pectoralis major) during the follow-through phase. Based on the results of their study, the authors identified five clinically relevant conclusions:

FIGURE 11.16 Football blocking.

1. Rapid increase in accelerator muscle activity in the acceleration and follow-through phases may be involved in observed incidence of pectoralis major ruptures in this population.
2. Two distinct patterns of muscle activity (i.e., accelerators and stabilizers) may have implications for modifications in training and rehabilitation programs.
3. Low-level activation of the biceps brachii questions its role in the football passing motion.
4. High muscle activation during the follow-through phase may implicate certain mechanisms of injury.
5. Differences in injury patterns seen in football passers as compared to baseball pitchers may be related to differences in muscle activation patterns and mechanics.

FIGURE 11.17 Football passing.

Research in Mechanics

Football Passing Versus Baseball Pitching

Especially at the youth and high school levels, the best throwers often play both quarterback in football and pitcher in baseball. What are the similarities and differences in the throwing motions of athletes doing both? Fleisig and colleagues (1996) conducted a study to examine the kinematics and kinetics of baseball pitching and football passing. The researchers compared the throwing motion of 26 high school and collegiate pitchers and 26 high school and collegiate quarterbacks using three-dimensional high-speed motion analysis.

The authors reported that while quarterbacks reached maximum shoulder external rotation earlier, baseball pitchers achieved maximum angular velocity of pelvic rotation, upper torso rotation, elbow extension, and shoulder internal rotation earlier and with greater magnitude than quarterbacks did. Quarterbacks took shorter strides and stood more erect (at ball release) than pitchers. During the cocking phase, quarterbacks exhibited greater elbow flexion and shoulder horizontal adduction. Pitchers produced greater compressive force at the elbow and greater compressive force and adduction torque at the shoulder than quarterbacks did. These mechanical differences may play a role in throwing performance and injury risk.

See the references for the full citation:

Fleisig, Escamilla, Andrews, Matsuo, Satterwhite, & Barrentine, 1996.

Golf

Golf is a striking sport in which players use a variety of clubs to hit a ball into a series of holes (typically 9 or 18) in an effort to complete the course in as few strokes as possible. Playing a hole in golf begins with a tee shot, a full-swing drive from a designated area. On short-length holes, the tee shot may reach the putting green (or *green*), a flat or gently rolling area of short-cut grass with a hole (or cup) 4.25 in. (10.8 cm) in diameter. On longer holes, a second or even a third drive (*fairway drive*) may be needed to near the green. For approach shots near the green, the golfer may make a chip, or layup, shot to reach the green. Once the ball is on or very near the green, the golfer putts the ball until it goes into the hole.

The modern game of golf originated in Scotland in the 15th century. The current standard of an 18-hole golf course was set at the Old Course in St. Andrews, Scotland, in 1764. Currently, there are more than 34,000 golf courses worldwide, with nearly half of those in the United States.

General Concepts

In most striking sports (e.g., baseball, softball, tennis, volleyball), the player attempts to hit a moving ball. Golf is unique in that players hit a stationary, or motionless, ball using a variety of strokes (figure 11.18). Mechanically, the golf swing is a multijoint task involving a proximal-to-distal transfer of momentum and energy from the player's legs and trunk, to the arms, and finally to the club, which then strikes the ball.

Most golf studies divide the full swing into five phases (Marta et al., 2012; see figure 11.19): (1) backswing phase, or takeaway (from the starting position, or address, to the top of the swing), (2) forward swing phase (from top of swing to horizontal positioning of the club in the early part of the downswing), (3) acceleration phase (from horizontal club position to ball impact), (4) early follow-through (from impact to horizontal club positioning), and (5) late follow-through (from horizontal club position to completion of the swing).

Applying the Concept

You're Never Too Old

Age unquestionably takes a toll on human movement potential. Older people, for example, show declines in muscle strength, balance, flexibility, and visual acuity as part of the aging process. Despite these losses, however, there are remarkable examples of capable older movers. Such is the case of PGA golf professional Gus Andreone, a resident of Sarasota, Florida, who celebrated his 106th birthday in 2017. Born in 1911, Andreone gave his first golf lesson in 1934. Over his lifetime, he has made eight holes-in-one, most recently in 2014 at the age of 103. In so doing, Andreone is the oldest known golfer to make a hole-in-one.

FIGURE 11.18 Golf strokes: *(a)* tee shot (drive), *(b)* fairway drive, *(c)* chip shot, *(d)* putt.

FIGURE 11.19 Phases of a golf swing.
David Cannon/Getty Images

Applications

Among the hundreds of research articles on the sport of golf, most studies on muscle activity and mechanics have focused on a full-swing drive rather than on chipping or putting. A comprehensive review of EMG during the golf swing has been reported by Marta and others (2012). We focus here on several studies, summarized by anatomical region. Some studies used surface EMG electrodes, while others used indwelling electrodes (see chapter 6). The following descriptions are for a right-handed golfer, whose left side (lead side) is forward and right side (trail side) is to the rear with respect to the ball.

Shoulder and Scapular Region

Of all the body regions, the shoulder and scapular areas have received the most research attention with respect to muscle activity in golfers. Much of the early pioneering EMG research was done by researchers at the Biomechanics Laboratory at Centinela Hospital Medical Center and the Kerlan-Jobe Orthopaedic Clinic in Inglewood, California (e.g., Jobe et al., 1986; Jobe et al., 1989; Kao et al., 1995; Pink et al., 1990).

In reviewing the literature on shoulder muscle activity in the golf swing, Escamilla and Andrews (2009) noted that while shoulder muscle activity was low during the backswing, moderate activity was found in the levator scapulae and lower and middle trapezius of the trail arm (to elevate and upwardly rotate the scapula) and in the serratus anterior of the lead arm (to protract and upwardly rotate the scapula). All regions of the trapezius (upper, middle, lower) were most active in the backswing compared to the other four phases. The infraspinatus and supraspinatus of the trail (right) shoulder also showed peak activity during the backswing, but with only ~25% of their maximum voluntary isometric contraction. This suggested low involvement of these two rotator cuff muscles in the trail arm. Rotator cuff activity may be higher in the lead shoulder.

During the forward swing phase, muscle activity was highest in the subscapularis, pectoralis major, latissimus dorsi, and serratus anterior of the trail arm (to adduct and internally rotate the trail arm and protract the scapula). In the lead arm, highest activity was seen in rhomboids and middle and lower trapezius (to help retract and stabilize the scapula). In the acceleration phase, muscle activity was higher, especially in the subscapularis, pectoralis major, latissimus dorsi, and serratus anterior as they power the trail arm forward. These muscles maintained high activity in the early follow-through phase, though now acting as decelerators through eccentric action. Overall, muscle activity decreased in the late follow-through phase.

Forearm

Farber and colleagues (2009) measured EMG activity in four forearm muscles of 10 amateur and 10 professional golfers. Fine-wire electrodes were inserted into the flexor carpi radialis, flexor carpi ulnaris, extensor carpi radialis brevis, and pronator teres of both forearms in each golfer.

Flexor carpi radialis (FCR) activity in the lead (left) arm in the amateur golfers peaked during the forward swing phase, while the FCR in the professional golfers peaked a bit later, during the acceleration phase. Peak FCR activity in the trail (right) arm was seen during the forward swing phase in both groups. Flexor carpi ulnaris (FCU) activity was similar in both groups and in both arms, peaking in all cases during the forward swing phase and diminishing thereafter.

Extensor carpi radialis brevis (ECRB) activity in the amateurs' lead arm peaked during the acceleration phase. In the professionals, ECRB activity was highest a bit earlier during the forward swing phase. In the trail arm, amateurs' ECRB activity again peaked during the acceleration phase. In the professional golfers, ECRB activity in the trail arm decreased during forward swing, increased during acceleration, and diminished thereafter.

Finally, pronator teres (PT) activity in the lead arm of both groups peaked during the acceleration phase but was significantly greater in the professional golfers. In contrast, PT activity in the trail arm of the amateur group was greater than in the professional group. The authors identified this difference in PT activity as the most important finding in their study. They concluded that "these data may help better explain the discrepancy in elbow injuries between amateur and professional golfers . . . [and] help clinicians to better understand and rehabilitate golf-related injuries and to develop appropriate exercise protocols to help prevent injury" (p. 401).

Trunk

Researchers emphasize the importance of strong trunk musculature in golfers for best performance and injury prevention. Pink and colleagues (1993) measured EMG activity of the bilateral abdominal (external) oblique and erector spinae muscles of 23 golfers. They found low activity in all muscles during the backswing and high and constant trunk muscle activity during the rest of the swing.

Watkins and others (1996) examined the bilateral activity of the abdominal (external) oblique, gluteus maximus, erector spinae, and upper and lower rectus abdominis during the golf swing of 13 professional male golfers. Among their conclusions were the following: (1) muscle activity is relatively low during the backswing (takeaway) phase, (2) during the forward swing phase, the gluteus maximus muscles, especially on the trailing right side, showed high activity in stabilizing the hips, (3) high activity is seen in all trunk muscles during the acceleration phase in generating power for the swing, (4) in general, trunk muscle activity diminished in the early follow-through phase, but remained relatively active in the abdominal (external) obliques, and (5) in the late follow-through, the abdominal (external) obliques remain active in decelerating trunk rotation at the end of the swing.

Lower Limb

Few studies have examined the activation patterns of lower-extremity musculature during the golf swing. One study (Bechler et al., 1995) identified the role of the hip extensors and abductors of the trail (right) leg and of the adductor magnus of the lead (left) leg in initiating pelvic rotation during the forward swing phase, as well as the importance of the lead-leg (left) hamstring muscles in maintaining a flexed knee and stable base for pelvic rotation. In addition, the authors reported that peak EMG activity in the hip and knee musculature preceded peak activity of the trunk and shoulder muscles. This confirms the sequential nature of muscular recruitment in the golf swing.

A study by Marta and others (2016) compared lower-extremity muscle activity in low- and high-handicap golfers. The researchers used surface EMG to measure bilateral activity of the biceps femoris, semitendinosus, gluteus maximus, vastus medialis and lateralis, rectus femoris, tibialis anterior, peroneus longus, and gastrocnemius medialis and lateralis. Overall, muscle activity was low to medium during the backswing phase. In the forward swing phase, medium-to-high peak activity was seen in the gastrocnemius medialis and lateralis in both legs, right biceps femoris, right semitendinosus, and right gluteus maximus muscle. During the acceleration phase, activity peaked in the left semitendinosus, left gluteus maximus, right vastus medialis, and right vastus lateralis. In the early and late follow-through phases, muscle activity decreased for all muscles.

In comparing the low-handicap (i.e., more skilled) and high-handicap (i.e., less skilled) golfers, the authors reported that low-handicap (LHc) group had shorter phase durations than the high-handicap (HHc) group in all phases except for the late follow-through phase, in which the LHc group had a longer duration. In terms of muscle activity, the authors found differences, especially on the lead (left) side. LHc golfers reached maximum activation of the left quadriceps during the forward swing phase earlier than the HHc group, which reached peak left quadriceps activity during the acceleration phase. LHc golfers also had stronger right gastrocnemius activity during the forward swing phase and lower muscle activity in the thigh muscles and right gastrocnemius during the early and late follow-through phases, suggesting that better golfers relax these muscles after ball impact.

Soccer

Soccer (also *association football*, or *fútbol*) is the world's most popular sport, both in terms of participants and fans. A 2006 survey by the Fédération Internationale de Football Association (FIFA) estimated that there are 265 million soccer players worldwide. If the trend of increasing participation has continued since then, there now may be 290 million people playing soccer, with millions more involved as officials and referees. The fastest growth in recent years, on a percentage basis, has been in women's participation. In terms of fans, an estimated 3.5 billion people follow soccer, primarily in Europe, Africa, Asia, and the Americas. Soccer's World Cup, contested every four years, is the most watched sporting event in the world.

General Concepts

In terms of movement, proficiency in soccer requires skillful execution of running, jumping, intentional and reactive change of direction (also cutting), specific defensive movements (e.g., slide tackling), heading, and, to a limited extent, throwing (by the goalie and on inbound passes). And, of course, kicking.

Among team sports involving a ball, soccer players run the greatest distance. On average, a soccer player logs 7 mi (11 km) of running during the course of a game. Some positions (e.g., midfielders) may run more than 9 mi (14 km) miles per game. Goalkeepers (also goaltenders, or goalies), who remain stationed near their team's goal, run much less.

The heading motion, unique to soccer, involves players intentionally using their heads to impact and propel the ball. The effects of heading remain controversial.

Applications

As a member of the football family, soccer's primary movement form is kicking. Soccer players execute a variety of kicks, some for long distance to advance the ball down the field, others of short distance in a confined area crowded with teammates and opponents, and some, perhaps most importantly, toward the opponent's goal in an attempt to score.

For analysis purposes, the soccer kick commonly is divided into phases. Phase 1 (preparation) starts with heel strike of the kicking leg and ends with toe-off of the kicking leg. Phase 2 (backswing) spans from toe-off of the kicking leg until maximum hip extension.

Research in Mechanics

Sex Differences in Muscle Activation and Injury

Research has convincingly demonstrated that women are at higher risk of anterior cruciate ligament (ACL) injury compared to men. One of the proposed factors for this disparity is differences in neuromuscular recruitment, or activation, patterns for specific movement tasks. Hanson and colleagues (2008) examined the muscle activation patterns in male and female soccer players during a side-step cutting maneuver. Among their findings were that during a change-of-direction cutting task, female collegiate soccer players exhibited more vastus lateralis EMG activity and had larger quadriceps-to-hamstrings coactivation ratios than their male counterparts, suggesting that women did not increase hamstrings activation to compensate for increased quadriceps activation. These differences in muscle recruitment may play a role in increased risk of ACL injury in women.

See the references for the full citation:

Hanson, Padua, Blackburn, Prentice, & Hirth, 2008.

Research in Mechanics

Heading in Soccer

Despite considerable research and clinical evidence, the advisability of heading in soccer and its relation to concussions, especially in children, remains equivocal and controversial. There is general consensus that female players are at greater risk of concussion than male players and that the most common mechanism for concussive injury is player-to-player contact, not heading (Gessel et al., 2007; Maher et al., 2014). Beyond that, however, opinions and recommendations differ, as seen in the conclusions of recent studies:

- "No single study has provided conclusive evidence for the relationship between heading and cognitive and other deficits. . . . Our analysis indicates no overall effect for heading a [soccer] ball on adverse outcomes" (Kontos et al., 2017).
- "Intentional (i.e., heading) and unintentional head impacts are each independently associated with moderate to very severe CNS symptoms" (Stewart et al., 2017).
- "Subconcussive head impacts routine in soccer heading are associated with immediate, measurable electrophysiological and cognitive impairments. Although these changes in brain function were transient, these effects may signal direct consequences of routine soccer heading on (long-term) brain health which requires further study" (DiVirgilio et al., 2016).
- "There is evidence of association between heading and abnormal brain structures, but the data are still preliminary" (Rodrigues et al., 2016).
- "Although banning heading from youth soccer would likely prevent some concussions, reducing athlete–athlete contact across all phases of play would likely be a more effective way to prevent concussions as well as other injuries" (Comstock et al., 2015).
- "There is no evidence that heading in youth soccer causes any permanent brain injury and there is limited evidence that heading in youth soccer can cause concussion. . . . While concussion from heading the ball without other contact to the head appears rare in adult players, some data suggest children are more susceptible to concussion from heading with biomechanical factors, less developed technique, and the immature brain's susceptibility to injury all potentially contributing" (O'Kane, 2016).
- "Results of this study. . . suggest an association between exposure to RHI [repetitive head impacts] while heading the ball in soccer and lack of improvement in cognitive performance. This indicates that, although cognitive performance improves in youth athletes over time, this benefit may be suppressed by the cumulative effects of RHI, especially those with high velocity" (Koerte et al., 2017).
- "The long-term consequences of repetitive heading in soccer are not well understood" (Maher et al., 2014).

See the references for the full citations:

Comstock, Currie, Pierpoint, Grubenhoff, & Fields, 2015.

DiVirgilio, Hunter, Wilson, Stewart, Goodall, Howatson, Donaldson, & Ietswaart, 2016.

Gessel, Fields, Collines, Dick, & Comstock, 2007.

Koerte, Nichols, Tripodis, Schultz, Lehner, Igbinoba, et al., 2017.

Kontos, Braithwaite, Chrisman, McAllister-Deitrick, Symington, Reeves, & Collins, 2017.

Maher, Hutchison, Cusimano, Comper, & Schweizer, 2014.

O'Kane, 2016.

Rodrigues, Lasmar, & Caramelli, 2016.

Stewart, Kim, Ifrah, Lipton, Bachrach, Zimmerman, Kim, & Lipton, 2017.

Phase 3 (leg cocking) goes from maximum hip extension to maximum knee flexion, followed by phase 4 (acceleration), which begins at maximum knee flexion and ends with ball contact. Phase 5 (follow-through) goes from ball contact until toe speed inflection.

FIGURE 11.20 *(a)* Instep kick; *(b)* side-foot kick.

As described in chapter 9, during the kicking movement, there is a proximal-to-distal sequencing of joint actions at the hip, knee, and foot. In such sequencing, the proximal segment (thigh) initially moves forward (flexes), with the more distal segment (shank) lagging behind. The proximal segment then decelerates. In so doing, it assists in accelerating the distal segment in its extension (Putnam, 1991).

The two most common soccer kicks are the instep kick, in which the player strikes the ball with the dorsum (i.e., upper surface) of the foot (figure 11.20*a*), and the side-foot kick, in which ball contact is made on the medial side of the midfoot (figure 11.20*b*). Most of the research on soccer kicking has focused on the instep kick.

Most of the EMG research related to soccer kicking understandably has focused on lower-extremity muscles, particularly those acting at the hip and knee joints. These include uniarticular hip flexors (e.g., iliopsoas), hip extensors (e.g., gluteus maximus), knee extensors (e.g., vastus medialis, vastus lateralis, vastus intermedius), and biarticular muscles, which act at two adjacent joints. The rectus femoris acts as a hip flexor and knee extensor, the gastrocnemius acts as a knee flexor and ankle plantar flexor, and the semitendinosus, semitendinosus, and long head of the biceps femoris act as hip extensors and knee flexors.

In general, during phase 1 (preparation), all of the lower-extremity muscles are active in supporting the body's weight while the kicking limb is in contact with the ground. In phase 2 (backswing), the thigh is extended by action of the hip extensors (gluteus maximus and hamstrings). Toward the end of phase 2, the hip flexors become active to eccentrically slow thigh movement as the segment reaches maximum hip extension. At the beginning of phase 3 (leg cocking), the hamstrings and gastrocnemius flex the knee. Toward the end of phase 3, the knee extensors act eccentrically to slow knee flexion. Then, making use of a stretch–shortening cycle, the knee extensors (especially the vastus medialis and vastus lateralis) act concentrically to forcefully extend the knee during phase 4 (acceleration). Phase 4 ends with foot–ball impact. During phase 5 (follow-through), hip extensors and knee flexors act to slow down joint motion of the hip and knee, respectively.

While muscle activity patterns are similar between the instep and side-foot kicks, a study by Brophy and colleagues (2007) identified a few differences. For the kicking leg, greater hamstring activity was noted for the side-foot kick during phase 5, and the tibialis anterior showed greater activity during the side-foot kick in phases 2, 3, and 4. Overall, there was greater activation in the instep kick for the iliacus, vastus medialis, gastrocnemius, and hip adductors. There was no main effect difference in the gluteus medius, gluteus maximus, and vastus lateralis.

A note of caution: There is considerable variability in the reported soccer research in terms of muscle activation and joint mechanics during kicking. In reviewing past studies, Scurr and others (2011) noted that reported data "demonstrates the variability in previous research, which limits conclusions regarding the importance of each muscle during soccer kicking" (p. 247).

Some of the variability can be accounted for by the conditions being studied. For example, differences in muscle activation have been reported in comparing maximum kicking with kicking for accuracy (Kellis & Kattis, 2007; Scurr et al., 2011) and between skill levels (Cerrah et al., 2011).

Kellis and Kattis (2007) reported greater overall muscle activity in quadriceps muscles (vastus medialis, vastus lateralis, rectus femoris) when kickers aimed for the top right of the goal and greater quadriceps activity when kicking the ball to the right side of the goal than to the left side (for right-footed kickers).

Cerrah and colleagues (2011) compared muscle activation patterns in soccer players at different skill levels and found that in comparison with amateurs, professional players had earlier biceps femoris activation during the swing phase, lower rectus femoris activation, and earlier and greater muscle activity in the vastus medialis and vastus lateralis. During knee extension, the professional players showed much greater vastus medialis and lateralis activity, compared to the amateur players. During the follow-through, professionals exhibited lower gastrocnemius activity. The authors concluded that superior kicking performance in professional players may not be due to muscle strength factors alone, but rather to differences in technique. Once a suitable level of strength is achieved, professionals excel not by getting stronger, but by practicing effective techniques.

Numerous studies have examined various aspects of the biomechanics of soccer kicking. Among the many findings are that knee extension velocity is the most significant predictor of ball velocity (Sinclair, Fewtrell, Taylor, Bottoms, et al., 2014), segmental coordination during maximal effort kicks is crucial for completing a successful kick (DeWitt & Hinrichs, 2012), effective upper body movements are a key factor in creating greater muscle action during kicking (Shan & Westerhoff, 2005), final speed, path, and spin of the ball depend on the quality of foot–ball contact (Kellis & Katis, 2007), and centrifugal effects due to the kicking hip's flexion angular velocity contribute substantially in generating rapid knee extension (Naito et al., 2010).

One final line of research has addressed the issue of differences in kicking with the preferred (also dominant) leg versus the nonpreferred (or nondominant) leg. Not surprisingly, higher ball velocities were obtained when using the preferred leg due to higher foot linear velocity and knee extension velocity (Sinclair, Fewtrell, Taylor, Atkins, et al., 2014) and better intersegmental movement pattern and a transfer of velocity from the foot to the ball (Dörge et al., 2002; Sinclair & Hobbs, 2016). In comparing bilateral differences in knee and ankle loading in the support limb, significantly higher knee extensor and abduction moments (torques) and higher patellofemoral contact forces were present in the support limb when kicking with the nonpreferred leg (Sinclair & Hobbs, 2016).

Swimming

Swimming is a form of locomotion that involves propelling the body through water by natural means (i.e., using arms and legs). Swimming is an excellent exercise for muscular and cardiorespiratory conditioning. It is not effective, however, for increasing bone density because the buoyancy provided by the water negates the impact forces needed for osteogenesis (i.e., bone formation). To increase bone density, weight-bearing activities and resistance training are far more effective.

The water's buoyancy, however, can benefit obese individuals concerned about the risk of musculoskeletal injury associated with land-based activities. Running in water, for example, produces leg and arm motions similar to those of land-based running but without the corresponding joint stress. Studies have also shown that swimming does not produce significant fat or weight loss when compared with land-based weight-bearing exercises. Although the reasons for the negligible fat and weight loss associated with aquatic exercise are unclear, it may be that the body retains the higher levels of fat for temperature regulation and buoyancy.

Competitive swimming has four movement patterns, or *strokes*: freestyle (also crawl), breaststroke, backstroke, and butterfly (figure 11.21).

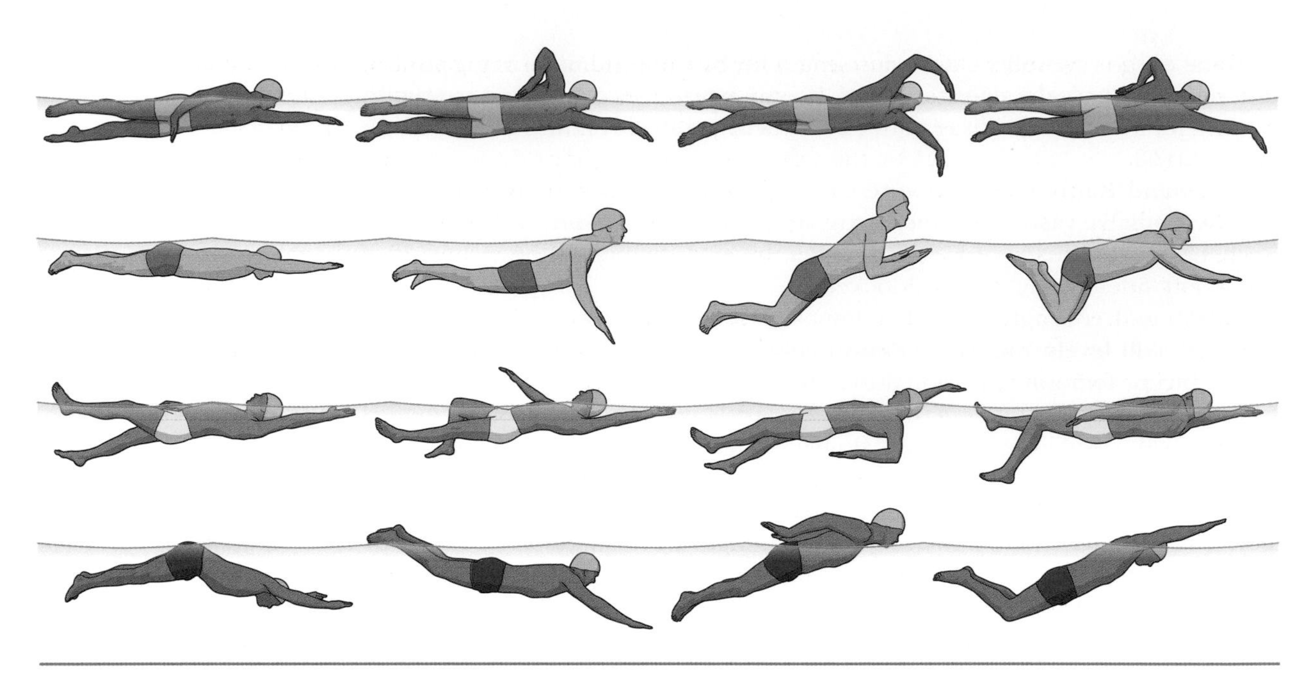

FIGURE 11.21 Swim strokes (from top to bottom): freestyle, breaststroke, backstroke, and butterfly.

General Concepts

Swimmers progress through the water by way of propulsive forces provided by the upper and lower extremities. The relative contribution and mechanical action of the extremities depends on the type of swim stroke being used and the swimmer's intent (i.e., recreation, exercise, rehabilitation, racing [sprint or endurance]).

Swimming Velocity and Momentum

Swimming velocity is largely determined by the product of stroke length and stroke rate. Depending on the duration and speed of the swimming event, a correct combination of stroke length and rate must be used to maximize performance. The current trend for skilled swimmers is to increase their velocity by first increasing their stroke length, followed by an increase in stroke rate. Longer strokes help swimmers use the forward momentum generated from their power stroke, thus enabling them to glide more during the stroke. If the stroke rate is too slow, however, the forward momentum will start to dissipate and the swimmer's forward progression will be marked by accelerations and decelerations. In addition to optimizing stroke length and stroke rate, swimmers must maintain a streamlined body position to obtain the maximum benefit from their forward momentum.

Resistive (Drag) Forces

Similar to cyclists, swimmers must also contend with drag forces, but in contrast to cyclists, they must overcome the greater viscosity of water. The total drag that resists a swimmer's forward motion is the sum of form drag, wave drag, and surface drag.

Form drag (also referred to as shape, profile, or pressure drag) is produced by the swimmer's body parting the water. The amount of form drag depends on the swimmer's size, shape, and speed and the body's orientation with respect to the flow of water. For both freestyle and backstroke, proper stroking mechanics require the body to rotate around its long axis. Rotating the body not only increases stroke effectiveness but also reduces form drag by decreasing the surface area of the body slicing through the water. To reduce form drag in the butterfly and breaststroke, swimmers need to minimize the vertical projection of their head and shoulders to prevent the hips and legs from sinking deeper in the water.

Wave drag is caused by the swimmer's body moving near or along the surface of the water. A portion of the water displaced by the swimmer moves up from a zone of high pressure to a zone of low pressure, creating a wave. Because the swimmer's kinetic energy provides the energy needed to form the waves, the waves present an opposing force to the swimmer. As the speed of the swimmer increases, the opposing force of the waves also increases.

When entering the water after a start or pushing off the wall from a turn, swimmers remain completely submerged in a streamlined position to reduce wave drag. The streamlined position referred to here is the same position used by divers as they enter the water: straight legs with ankles plantar flexed and arms held stretched over the head with hands together. The streamlined position actually serves a dual purpose by decreasing form drag and allowing swimmers to use their forward momentum to propel them through the water. In breaststroke and butterfly, the brief submerging of the body under the water's surface also reduces wave drag.

Surface drag (also referred to as frictional resistance, skin resistance, skin-friction drag, or skin drag) is produced as the water passes over the surface of the swimmer. Friction created between the water and the swimmer's skin and swimsuit slows the water and therefore resists forward motion. Shaving body hair and using new high-tech swimsuits (e.g., "sharkskin suits") enhance performance by reducing surface drag. Although surface drag is the smallest resistive force encountered by swimmers at high velocities, even a small improvement in performance can make a big difference in top competitions where the margin of victory often is measured in hundredths of a second.

Applications

Swimming and swimming-related sports (e.g., water polo) are performed in an aquatic environment and therefore are uniquely subject to hydrodynamic principles and effects. The mechanics of swimming involves the use of lift and drag forces in a fluid environment to propel the body forward through mechanical actions of the body's torso and extremities.

Stroke Mechanics

Similar to cycling, each swimming stroke can be divided into a power and recovery phase. The power phase occurs when the arm pulls through the water and propels the body forward. The recovery phase completes the stroke and repositions the arm for the next power phase.

To maximize forward propulsion during the power phase of all the strokes, swimmers are instructed not to use a straight pull (referred to as **paddling**) with the hand oriented 90° to the surface of the water. Instead, the angle of the hand (angle of attack, or pitch) should vary throughout the power phase, and the hands should follow a curvilinear path through the water. This stroke technique is known as **sculling**. In addition, during freestyle and backstroke, the whole body (i.e., shoulders, trunk, hips, and legs) rotates as one unit through a range of 70° to 90°, or 35° to 45° to each side. Controlled by the alternating power and recovery phases of each arm, this rotation reduces form drag, facilitates proper sculling motion of the hand and arm through the power phase, and reduces the muscle effort associated with pulling the arm through the recovery phase.

Propulsive Forces Produced by the Arms and Hands

Although the whole arm pulls through the water during each stroke, it is the hand that produces the largest propulsive force in swimming. The force acting on the hand as it passes through the water is normally divided into two components: a **lift force** that acts perpendicular to the hand's line of pull and a **drag force** that acts in the opposite (i.e., parallel) direction to the motion of the hand. The magnitude of the lift and drag forces is determined by the pitch (angle of attack) of the hand (figure 11.22). How can these forces be used for propulsion? Remember, just because the lift force acts perpendicular to the movement of the hand through the water does not mean that the lift force always acts perpendicular to the surface of the water.

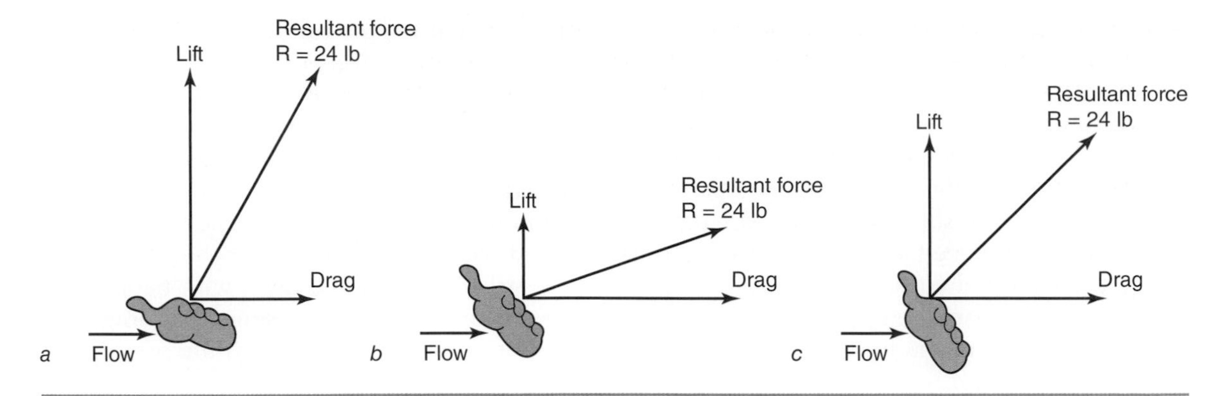

FIGURE 11.22 Hand angles and resultant force production in swimming: *(a)* pitch of 15°, *(b)* pitch of 30°, *(c)* pitch of 45°.

Adapted by permission from C. Colwin, *Breakthrough swimming* (Champaign IL: Human Kinetics, 2002), 38; Adapted by permission from R.E. Scheihauf, Swimming propulsion: A hydrodynamic analysis, in *American Swimming Coaches Association world clinic yearbook 1977* (Fort Lauderdale, FL: American Swimming Coaches Association, 1977), 53.

Research shows that the forward propulsive forces produced by lift and drag are greater when the swimmer uses transverse and vertical sculling motions of the hand throughout the power phase of the stroke (figure 11.23). Because the limb moves backward through the water, the drag force is directed forward. Because the hand is not held perpendicular to the water's surface, a portion of lift can be used for propulsion. Propulsion, therefore, is a continual interaction between the lift and drag forces as the hand follows a curvilinear path through the water. The shape and size of the hand, as well as the speed of the hand moving through the water, affect the magnitude of the propulsive lift and drag forces.

Propulsive Forces Produced by the Legs and Feet

Although the thigh, leg, and foot all contribute to the kick for all four strokes, the foot produces the largest propulsive force. When pushing off the wall after a turn or entering the water after a start, the dolphin kick (also butterfly kick) is popular because when combined with a streamlined body position, it allows the swimmer to move with great speed underwater

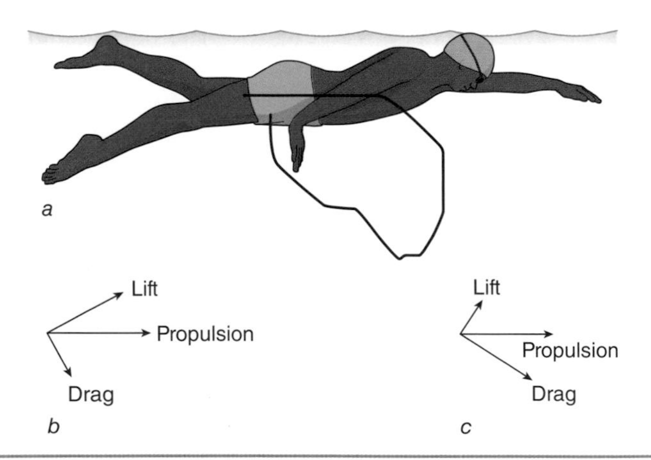

FIGURE 11.23 *(a)* Swimmer's fingertip trajectory pattern in the freestyle stroke, *(b)* lift-dominated propulsion, *(c)* drag-dominated propulsion.

Reprinted by permission from E. Maglischo, *Swimming fastest* (Champaign, IL: Human Kinetics, 2003), 22; Adapted by permission from American Swimming Coaches Association, *American Swimming Coaches Association world clinic yearbook 1977* (Ft. Lauderdale, FL: American Swimming Coaches Association, 1997), 53.

without using the arms (figure 11.24*a* and 11.24*b*). Both freestyle and backstroke use the flutter kick for propulsion (figure 11.24*c* and 11.24*d*). The flutter kick consists of asynchronous movement of the legs, which contrasts with the synchronous movement of both legs during the dolphin kick. Unlike the other three strokes, the breaststroke uses a propeller-like kick to drive the body forward. Regardless of the kicking technique used, each kick is effective because it produces a net forward thrust force.

Muscular Control of Stroke Mechanics

The power phase for each of the four primary swim strokes (freestyle, breaststroke, backstroke, butterfly) can be subdivided into a pull and a push phase. In freestyle, backstroke, and butterfly, the swimmer first pulls her body forward over the hands and then continues to push the body past the hands (Colwin, 2002). The pull phase of the breaststroke is similar to that of the other strokes, but the breaststroke differs during the push phase in that the body does not pass over the hands.

Even with the added challenge of immersing electrodes in water, muscle activity patterns during swimming have been recorded since the early 1960s. Although the four strokes are significantly different visually, each one uses glenohumeral adduction as a principal joint motion for moving the body forward through the water. The power phase for freestyle, butterfly, and breaststroke also consists of glenohumeral extension, whereas the power phase for backstroke includes glenohumeral flexion. These shoulder joint motions help explain why the primary muscles used across all four strokes are the pectoralis major, latissimus dorsi, and teres major. To assist with the power phase and perform a smooth and powerful recovery, the trapezius, deltoids, and elbow extensors and flexors also play an important role.

Although the longitudinal body rotation characteristic of both freestyle and backstroke is primarily controlled by the alternating stroke mechanics of each arm, the observed rectus abdominis and external oblique activity would appear to help stabilize the trunk and facilitate trunk rotation. Lower-limb muscle activity, including the gluteus maximus, semitendinosus, biceps femoris long head, rectus femoris, gastrocnemius, and tibialis anterior, has been recorded during swimming. As expected, when considering the variation across and within kicks, the specific muscle activation patterns can differ substantially.

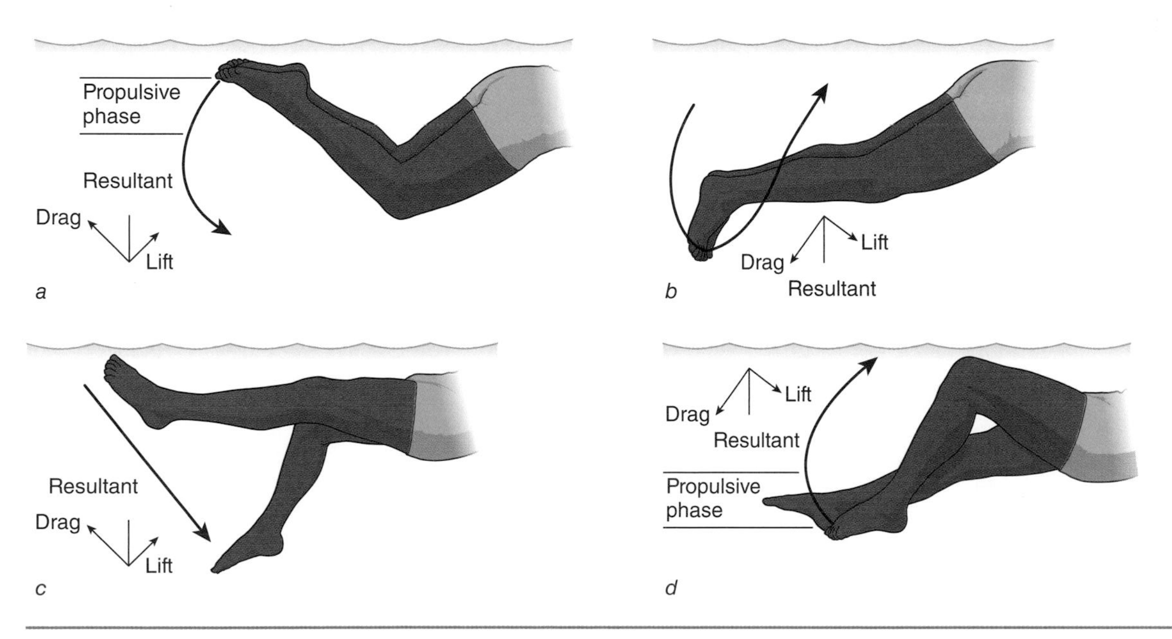

FIGURE 11.24 Dolphin kick *(a)* downbeat and *(b)* upbeat; flutter kick *(c)* downbeat and *(d)* upbeat.

Adapted by permission from E. Maglischo, *Swimming fastest* (Champaign, IL: Human Kinetics, 2003), 39.

Research in Mechanics

Swimmers and Shoulder Pain

Given the primary involvement of the shoulder in all swim strokes and the repetitive nature of the task, shoulder pain in swimmers should be neither surprising nor unexpected. Pain can cause alterations in muscle activation patterns and movement mechanics. A series of studies (Pink et al., 1993; Ruwe et al., 1994; Scovazzo et al., 1991) compared the EMG activity of 12 shoulder girdle muscles in competitive swimmers with shoulder pain to that of swimmers with normal (i.e., nonpainful) shoulders.

In the freestyle stroke, Scovazzo and colleagues (1991) reported that swimmers with shoulder pain showed (1) less muscle activity in the anterior and middle deltoids, rhomboids, and upper trapezius at hand entry, (2) less activity in the serratus anterior and more activity in the rhomboids during the pulling phase, (3) less activity in the anterior and middle deltoids and more activity in the supraspinatus at hand exit, and (4) less activity in the subscapularis at midrecovery. No differences in muscle firing pattern or amplitude were found in the posterior deltoid, supraspinatus, teres minor, pectoralis major, or latissimus dorsi when comparing swimmers with painful shoulders and normal shoulders.

In the butterfly stroke, Pink and others (1993) found that swimmers with shoulder pain had higher activity of the posterior deltoid and lower activity of the upper trapezius and serratus anterior at hand entry. This activity pattern allowed swimmers to position the humerus for a wider hand entry, which decreased the pain of supraspinatus impingement on the coracoacromial arch. In addition, the authors reported less activity in the teres minor and serratus anterior and more activity in the subscapularis and infraspinatus during the pulling phase for swimmers with painful shoulders. These muscle actions depressed the humeral head to avoid impingement. No differences in muscle activity were seen in the rhomboids, pectoralis major, latissimus dorsi, or the anterior and middle deltoids.

Ruwe and colleagues (1994) compared muscle activity during the breaststroke. They found increased internal rotator activity and decreased teres minor, supraspinatus, and upper trapezius activity in swimmers with painful shoulders. They concluded that these factors increase the risk of glenohumeral impingement.

All three studies noted that research on shoulder muscle activity provides information that can be used to develop accurate and effective preventive and rehabilitative exercise programs.

See the references for the full citations:

Pink, Jobe, Perry, Browne, Scovazzo, & Kerrigan, 1993.

Ruwe, Pink, Jobe, Perry, & Scovazzo, 1994.

Scovazzo, Browne, Pink, Jobe, & Kerrigan, 1991.

A comprehensive analysis of all four swimming strokes across different skill levels is beyond the scope of this text. For more detail, the reader is referred to Martens and colleagues (2015), who published a systematic review of the electromyography in the four competitive swimming strokes. Most, if not all, studies of EMG in swimmers note high variability in muscle activity patterns among individuals, warranting caution in interpreting and generalizing study results. Muscle activity patterns also vary with skill level (e.g., Leblanc et al., 2005; Vaz et al., 2016; Olstad, Zinner, et al., 2017), effort level (e.g., Olstad, Vaz, et al., 2017), and swimming speed (e.g., Chollet et al. 2008).

In brief, in the freestyle stroke, Pink and others (1991) found similar muscle activity at hand entry and exit, with the upper trapezius and rhomboids working synchronously (upper trapezius upwardly rotating the scapula and rhomboids retracting the scapula), all heads of the deltoid active in lifting and positioning the arm for hand exit, and the anterior and middle deltoids at hand entry. They found that the supraspinatus was active (with the anterior and middle deltoid) in abducting the humerus at hand entry and exit. Each of the rotator cuff muscles played a unique role. The pectoralis major and latissimus dorsi, together with serratus anterior and teres minor, were active during the propulsive phase. Subscapularis and serratus anterior activity occurred throughout the stroke cycle (therefore, these muscles were susceptible to fatigue). Rouard and Clarys (1995) reported lower latissimus dorsi activity and a different activation pattern of the anterior deltoid than Pink and colleagues (1991). These differences may be due to skill level of the swimmers in each study. Lauer and others (2013) reported activation patterns of the biceps and triceps brachii during the freestyle stroke, with the highest coactivation to stabilize the elbow during the early pull-through phase and when the hand enters the water.

In the breaststroke, Ruwe and colleagues (1994) reported, for the pull-through phase, (1) powerful pectoralis major activity to adduct and extend the upper arm, (2) eccentric teres minor activity in conjunction with concentric action of the pectoralis major to moderate internal rotation of the humerus, (3) latissimus dorsi activity to assist the pectoralis major with extension, adduction, and internal rotation, and (4) posterior deltoid activity to assist with extension and outward rotation of the humerus. During the recovery phase, the authors found (1) sequential action of the deltoid heads (anterior, middle, posterior) in synchrony with the supraspinatus to lift the arm and maintain glenohumeral congruency and avoid impingement, and (2) continuous serratus anterior activity, in concert with the upper trapezius, to stabilize, protract, and upwardly rotate the scapula.

In the backstroke, Pink and others (1992) found (1) similar patterns of muscle activation at hand entry and exit, with the deltoid, supraspinatus, and scapular muscles working to place the shoulder girdle in position for hand entry and exit, (2) latissimus dorsi, subscapularis, and teres minor activity primarily during the propulsive phase (i.e., in a pulling action), (3) unique rotator cuff activity, with the supraspinatus acting in conjunction with the deltoids, the infraspinatus with minimal activity, and the teres minor and subscapularis working together as a force couple in controlling humeral rotation, (4) constant teres minor and subscapularis activity throughout the stroke, which makes these muscles susceptible to fatigue, and (5) scapular muscle activity to minimize glenohumeral impingement and maximize congruency of the humeral head with the glenoid fossa.

In the butterfly stroke, Pink, Jobe, Perry, Kerrigan, and colleagues (1993) reported (1) deltoid and rotator cuff activity at hand entry with the humerus in an abducted, extended, and externally rotated position, along with rhomboid and upper trapezius active in retracting and upwardly rotating the scapula, (2) pectoralis major and latissimus dorsi activity to provide power during the propulsion phase, together with serratus anterior activity to help pull the body over the arm and posterior deltoid activity to begin lifting the arm out of the water

Applying the Concept

Remarkable Endurance

Between 1978 and 2013, endurance swimmer Diana Nyad made four unsuccessful attempts to swim from Cuba to the United States, a distance of more than 100 mi (161 km). On the morning of August 31, 2013, Nyad set out from Havana, Cuba, on her fifth attempt. After 53 hours of continuous swimming, braving jellyfish and cold water and battling incredible muscular and mental fatigue, Nyad reached the beach at Key West, Florida, on September 2, 2013. She swam a distance of about 110 mi (177 km). In so doing, she became the first person to successfully swim from Cuba to the United States without a shark cage. Age proved to be no barrier, since Nyad was 64 years old at the time. Nyad's mantra for her fifth, and finally successful, swim was "Find a way!"

at the end of propulsion, (3) middle and anterior deltoid, supraspinatus, and infraspinatus activity to rotate the arm during recovery, and (4) high serratus anterior and subscapularis activity throughout the stroke, which puts these muscles at risk of fatigue.

Tennis

Tennis is a racket sport in which players on opposite sides of a net hit a ball back and forth until one player is unable to return the ball. In singles tennis, there is one player on each side; doubles has two teams of two players on each side of the net. Each player uses a racket made up of a hand grip, neck, and elliptical frame. The frame holds tightly interwoven strings of synthetic material that form the hitting surface of the racket. For the first century of modern tennis, rackets were made of wood. In the late 1960s, rackets made of steel and aluminum were introduced, but were not widely used. By the 1980s, composite materials made of carbon graphite, ceramics, boron, and titanium became the norm. Because of their lighter weight, these materials allowed for larger frames and more powerful strokes.

Tennis is played on a flat, rectangular court measuring 78 ft (23.8 m) in length and 27 ft (8.2 m) in width for singles and 36 ft (11 m) in width for doubles. The court is divided into two halves by a net that is 3.5 ft (1.1 m) high at the posts (edge of the court) and 3 ft (0.9 m) high at the center of the court. Tennis is unique in its use of a variety of playing surface materials (e.g., grass, clay, and hard courts made of concrete or asphalt covered by an acrylic layer).

General Concepts

Tennis players employ several basic movement forms, including running, cutting (i.e., change of direction), striking, and occasionally jumping. A tennis rally begins with one player (the server) using an overhead striking motion to hit the ball over the net and into the opposite side of the court. The opponent (or receiver) attempts to strike the ball and return it to the server's side of the court. This back-and-forth action continues until a player is unable to return the ball or the ball lands out of bounds. During the rally, players run and cut (side to side and forward to back) to position themselves to strike the ball.

Applications

The striking movements in tennis consist of eight basic strokes: serve, forehand, backhand, volley, half volley, lob, overhead smash, and drop shot. (Note: Each stroke is described as follows for a right-handed player. For left-handed players, the body sides should be reversed.)

- Serve: The first shot of a rally, in which the server tosses the ball high in the air and forcefully drives it, using an overhead motion, toward the opponent's side of the court.
- Forehand: This stroke begins on the right side of the body, swings forward toward ball–racket contact, and continues across to the left side of the body. Virtually all players use a one-handed forehand. There have, however, been a few notable exceptions of professional players (e.g., Pancho Segura, Monica Seles) who used a two-handed forehand effectively.
- Backhand: This stroke begins on the left side of the body, swings forward toward ball–racket contact, and continues across to the right side of the body. Some players use a one-handed technique; others use two hands.
- Volley: A return shot (usually near the net) made by striking the ball in midair before it hits the ground. A volley typically involves a short, jabbing motion, rather than a full swing.
- Half volley: A return shot made low, near the ground, just after the ball has bounced, usually near the net.
- Lob: A high, arching shot hit deep into the opponent's court.
- Overhead smash: A high, hard shot hit with a servelike motion, often near the net.

Applying the Concept

Tennis Backhand: One Hand or Two?

A long-standing debate exists: Which is better, a one-handed or a two-handed backhand? Notable one-handers include John McEnroe, Pete Sampras, Roger Federer, and Justine Henin. Tennis legends Björn Borg, Andre Agassi, Rafael Nadal, Chris Evert, Jimmy Connors, and Serena Williams favor the two-handed backhand. Before the 1970s, virtually no one used a two-handed backhand. One-handers dominated tennis through the late '80s and early '90s. By the mid to late '90s, there was a fairly even distribution between the two. Since then, the two-handed backhand has become the preferred technique. Will the pendulum swing back toward one-handers in the future? Only time will tell.

From a performance perspective, there is general, though not unanimous, consensus that two-handed backhands allow for a more powerful stroke, greater stroke consistency, and a shorter stroke length (which allows for less time needed to prepare for a shot, but also may make it more difficult to hit open-angle, cross-court shots). One-handers may have an advantage in hitting low balls, performing better slices, and having a longer reach.

- Drop shot: A return shot hit to just clear and land near the net in the opponent's court. A drop shot usually is used as an unexpected shot when the opposing player is playing deep (i.e., far back from the net).

Research on muscle action during tennis strokes has focused primarily on four strokes: serve, forehand, backhand, and volley.

The serving motion is divided into four phases (or stages): windup, cocking, acceleration, and follow-through (figure 11.25). During the windup phase, there is low muscle activity in both shoulder and elbow muscles. Horizontal extension, abduction, and external rotation of the glenohumeral joint are reached passively through leg and trunk rotation and lateral

FIGURE 11.25 Tennis serve.
United States Tennis Association

trunk flexion (Morris et al., 1989; Ryu et al., 1988). As the serve progresses through the cocking phase, in which the shoulder abducts and externally rotates along with elbow flexion, increased activity is seen in most shoulder and elbow muscles, with notably high activity in the serratus anterior and supraspinatus and in the triceps, extensor carpi radialis brevis (ECRB), extensor digitorum communis (EDR), and extensor carpi radialis longus (ECRL).

During the acceleration phase, muscle activity is highest, both at the shoulder (subscapularis, pectoralis major, serratus anterior, latissimus dorsi) and elbow (ECRB, triceps, pronator teres, flexor carpi radialis). The subscapularis and pectoralis major are particularly active in producing powerful adduction and internal rotation. Trunk muscles (rectus abdominis, external oblique, internal oblique, erector spinae) are most active during the acceleration phase (Chow et al., 2009). In the follow-through phase, some muscles (e.g., subscapularis, pectoralis major, serratus anterior) show decreased activity, while other muscles (e.g., biceps brachii, middle deltoid, supraspinatus) are more active.

Kibler and colleagues (2007) evaluated muscle activation patterns in selected scapulohumeral muscles and found patterns of muscle activation in scapular stabilizers (upper trapezius, lower trapezius, serratus anterior), muscles that position the arm (anterior deltoid and posterior deltoid), and muscles of the rotator cuff group (supraspinatus, infraspinatus, teres minor). Among their reported patterns were that (1) the serratus anterior and upper trapezius were active in the early cocking phase, followed by lower trapezius activity in the late cocking phase, (2) the anterior deltoid was activated in early cocking, followed later by posterior deltoid activity, (3) the teres minor was activated early in the cocking phase, with the supraspinatus active in late cocking, and (4) the infraspinatus was activated in the follow-through phase. In summary, all muscles except the infraspinatus were active for more than 50% of the tennis serve motion.

The forehand and backhand tennis ground strokes each consists of three phases (or stages): racket preparation, acceleration, and follow-through. In the forehand stroke, muscle activity generally is low (Morris et al., 1989; Ryu et al., 1998) during the racket preparation phase. Of the shoulder and elbow muscles, only the ECRL shows moderate activity; all other muscles have low activity. During the acceleration phase, muscle activity is highest, with moderate to high activity in the biceps brachii, subscapularis, pectoralis major, serratus anterior, extensor carpi radialis (longus and brevis), brachialis, triceps, and flexor carpi radialis.

In the forehand follow-through phase, muscle activity decreases overall but remains high in the early follow-through in the biceps brachii and extensor carpi radialis brevis and moderate in the middle deltoid, supraspinatus, and infraspinatus (Morris et al., 1989; Ryu et al., 1988).

As with the forehand, the backhand exhibits lower muscle activity during the racket preparation phase. Muscle activity is predictably greatest during the acceleration phase, with highest activity seen in the middle deltoid, supraspinatus, and infraspinatus (Ryu et al., 1988) and the wrist extensors, brachialis, biceps brachii, and triceps brachii (Morris et al., 1989). During the early follow-through phase of the backhand stroke, activity in the extensor carpi radialis brevis remains high (Morris et al., 1989).

In making a volley stroke, the player strikes the ball in midair before it hits the ground, using a stabbing, or punching, motion. Research on the volley stroke has been limited. One of the most in-depth studies of upper-extremity muscle action in the volley was conducted by Chow and others (1999). The researchers measured EMG activity in 14 muscles, including the flexor carpi radialis (FCR), extensor carpi radialis (ECR), triceps brachii (TRB), anterior-middle and posterior-middle deltoids (AMD and PMD), and pectoralis major (PCM) of the dominant-arm-side, and left and right external oblique (EO), and erector spinae (ES) muscles of seven skilled males, all current or former intercollegiate tennis players.

Each volley was divided into six phases: (1) ready phase—200 ms period before ball release (from a tennis ball machine), (2) reaction phase—from ball release to initial racket movement, (3) backswing phase—from ball release to end of backswing, (4) forward swing phase—from end of backswing to ball impact, (5) stroke phase—from initial racket movement to ball impact, and (6) pushing phase—from initial racket movement to contralateral foot push-off.

Research in Mechanics

Factors Affecting the Mechanics of Tennis Strokes

Given the variety and complexity of tennis strokes, muscle activation patterns and stroke mechanics unsurprisingly are affected by many variables, including experience, equipment, fatigue, and injury.

Playing experience affects mechanics and muscle activity patterns in tennis players. Wei and colleagues (2016), for example, compared shock transmission and forearm EMG in experienced and recreational tennis players' backhand strokes. They concluded that recreational (i.e., less experienced) players transmit more shock impact from the racket to the elbow joint and have greater wrist flexor and extensor muscle activity (EMG) in the follow-through phase of the backhand stroke.

Equipment-related measures also affect tennis performance. For example, grip size, grip tightness, string tension, and ball impact location all have been associated with differences in muscle activity and stroke mechanics (e.g., Adelsberg, 1986; King et al., 2012).

As with most movement tasks, fatigue has been implicated in altered muscle activation and stroke mechanics in tennis. Several studies have examined the effects of fatigue in tennis. One study (Rota et al., 2014) reported decreased EMG amplitude in the pectoralis major and flexor carpi radialis for serves and forehand strokes in fatigued players, along with decreases in serve accuracy and velocity and forehand accuracy and consistency. Another study (Martin et al., 2016) found decreases in serve ball speed, ball impact height, and maximal joint angular velocities, as well as an increase in rating of perceived exertion, when comparing performance at the end of a match compared to its beginning. In addition, tennis players with low back pain show lower activation of trunk extensor muscles, compromised coactivation patterns, and less abdominal endurance when compared to asymptomatic players (Correia et al., 2016).

The injury most associated with tennis is so-called tennis elbow, more appropriately termed lateral epicondylitis (LE). Many studies have implicated the backhand stroke, and more specifically action of the wrist extensors (particularly the extensor carpi radialis brevis), with the onset of lateral epicondylitis in tennis players. For example, Kelley and colleagues (1994) reported that injured players (with LE) had significantly greater wrist extensor and pronator teres muscle activity during ball impact and early follow-through of the backhand stroke. Knudson and Blackwell (1994) found a significant difference in wrist angular velocity between tennis professionals without LE and intermediate players with a confirmed history of TE, and concluded that "the greater muscle stress in repeated eccentric muscle actions of the wrist extensors may be an important factor in developing TE [tennis elbow]" (p. 81).

Alizadehkhaiyat and others (2007) found "global upper limb weakness" in players with LE and an imbalance among forearm muscles in those with tennis elbow. While many have identified wrist extensor muscle action as a causal factor in lateral epicondylitis, others are more equivocal. In their review of the literature on EMG assessment of forearm muscle function in tennis players with and without LE, Alizadehkhaiyat and Frostick (2015) concluded the following:

> Despite indications of increased activity of wrist extensor muscles during all basic tennis strokes, insufficient evidence exists to support its aetiologic relationship with LE. While existing literature is suggestive of increased wrist extensor activity in less experienced single-handed backhand players due to suboptimal joint biomechanics, its association with the development of LE requires further evidence. (p. 884)

See the references for the full citations:

Adelsberg, 1986.

Alizadehkhaiyat, Fisher, Kemp, Vishwanathan, & Frostick, 2007.

(continued)

Research in Mechanics *(continued)*

Alizadehkhaiyat & Frostick, 2015.
Correia, Oliveira, Vaz, Silva, & Pezarat-Correia, 2016.
King, Kentel, & Mitchell, 2012.
Knudson & Blackwell, 1997.
Martin, Bideau, Delamarche, & Kulpa, 2016.
Rota, Morel, Saboul, Rogowski, & Hautier, 2014.
Wei, Chiang, Shiang, & Chang, 2006.

Among their primary findings, Chow and colleagues (1999) concluded the following:

- The ECR was more active than the FCR during the volley, suggesting wrist extension and abduction (i.e., cocking the wrist). Co-contraction of the ECR and FCR helped stabilize the wrist.
- TRB activity was greatest during the forward swing phase (from the end of backswing to ball impact) and is a distinct component of the volley's punching motion.
- The AMD and PMD were active in most stroke phases.
- The PCM acts during the forward swing phase in assisting the TRB with the punching action.
- EO activity during the active phases of a volley suggests that ipsilateral lateral flexion plays a role in initiating later movements.
- The decrease in bilateral EO activity with decreasing ball height suggests that trunk posture and movement during the backswing phase differ in high and low volleys.

In a follow-up study, Chow and others (2007) examined pre- and postimpact muscle activation in five upper-extremity muscles in the tennis volley across different ball speeds, ball types, and body side (i.e., forehand volley versus backhand volley). Their primary conclusion was that preimpact muscle activation differed from postimpact activation and that activation was affected by the interaction of ball speed, ball type, and body side.

Volleyball

Some sports evolve over time and do not have a definite point of origin. Other sports, such as basketball and volleyball, have a known beginning. Volleyball was invented by William G. Morgan on February 9, 1895, at a YMCA in Holyoke, Massachusetts. Interestingly, basketball was invented just four years earlier, by James A. Naismith, at a YMCA in Springfield, Massachusetts, a mere 7 mi (12 km) south of Holyoke. Morgan called his new game Mintonette, but the game was renamed within a year to its current name of volleyball. The original rules were somewhat different than those currently in effect, but the nature of the game was the same: to hit a ball over a net in an effort to ground the ball on the opponent's side of the net.

Volleyball is a team sport played on a court 9 by 18 m (29.5 by 59.1 ft) separated by a net, the top of which is 2.43 m (~8 ft) above the court for men and 2.24 m (~7 ft 4 in) for women. There are two teams of six players each. In brief, volleyball consists of the best two out of three or three out of five set (formerly called games) matches. A point is scored for winning a rally, which begins with a serve. The team receiving the serve typically bumps or passes the ball to its team's setter, who then sets the ball to an attacker who spikes the ball over the net. The serving team tries to block the spike. Each team is limited to three touches after the ball crosses the net. Play continues back and forth until the ball lands inside the boundaries of the court or an error (e.g., performing four touches, hitting the ball so it lands outside the court boundaries) is made. For a more detailed discussion of the rules, techniques, and strategies in volleyball, the reader is referred to the many full texts available on the sport.

A newer form of volleyball, beach (also sand) volleyball, was first played in 1915 on Waikiki Beach in Hawaii. While employing most of the same techniques as indoor volleyball, beach volleyball has some differences. It is played on a slightly smaller sand court with only two players per team (with no substitutions), and it has different permissible types of ball contact. Beach volleyball became a Summer Olympic sport in 1996.

General Concepts

Volleyball players use a combination of running, jumping, and striking skills to execute the game's basic movements, which include serving, passing (also bumping or digging), setting, spiking, and blocking (figure 11.26). Players run to position themselves on both offense and defense, jump to spike, serve, set and block, and strike the ball forcefully (e.g., spike or serve) or softly (e.g., bump, pass, set). Of all the volleyball skills, spiking and serving have received the most research attention (e.g., Coleman et al., 1993; Reeser et al., 2010; Rokito et al., 1998; Serrien et al., 2016).

Applications

Spiking is volleyball's most powerful movement. Several types of spikes exist: a straight-ahead spike, in which the hitting arm's follow-through remains on the hitter-hand side of the body; the cross-body spike, in which the spiker hits the ball at an angle by adducting the hitting arm across the body's midline; and the roll shot, in which the spiker puts top spin on the ball using a slower, more controlled arm movement. In the straight-ahead and cross-body spikes, the spiker tries to hit the ball with maximum velocity. Peak ball velocity has been significantly correlated with maximal isokinetic torque by the internal rotators of the dominant shoulder, height at which the player contacted the ball, and the muscular training hours performed weekly (Forthomme et al., 2005), as well as with angular velocity of the humerus in the hitting arm (Coleman et al., 1993).

For analysis purposes, the spike can be divided into phases. Coleman and colleagues (1993) used a seven-phase system, consisting of an approach, plant, takeoff, flight, hitting action, landing, and recovery. Reeser and others (2010) modified Coleman's system and described a four-phase system using approach, arm cocking, arm acceleration, and follow-through phases. The approach begins about 3 m (10 ft) from the net. The spiker takes two or three steps, lowers her center of gravity by flexing the hips and knee, and plants her trailing leg next

FIGURE 11.26 Volleyball movements: *(a)* passing, *(b)* setting, *(c)* spiking, and blocking.

Applying the Concept

Record Spike Speeds

In volleyball, the faster the ball is spiked, the more difficult it is to block or dig. The highest reported spike speed for women is 64 mph (103 kph) by Yanelis Santos (Cuba). For men, the record is held by Matey Kaziyski (Bulgaria) at 82 mph (132 kph). To reach such speeds requires remarkable muscle power and joint coordination.

to the lead leg (either with a long, low skip or a small jump). During this time, the spiker swings her arms back in extension in preparation for glenohumeral flexion (i.e., swinging the arms up and forward). The spike progresses to takeoff, in which the spiker explosively extends her hips and knees and plantar flexes her ankles. The flight phase begins when both feet leave the ground. For a right-handed spiker, the hitting action begins with abduction and external rotation of the hitting arm and extension (lowering) of the nonhitting arm. The spiker cocks her arm (maximum external rotation) and then reverses (accelerates) the arm (with elbow fully extended) with glenohumeral internal rotation, radioulnar pronation, and wrist flexion to impact the ball at or near its apex. Following ball contact, the spiker follows through and lands on the court (using either one or two legs), absorbing the impact with simultaneous flexion of the hip, knee, and ankle, controlled by eccentric action of each joint's extensor muscles.

In terms of specific muscle action, Rokito and colleagues (1998) measured EMG for eight muscles (anterior deltoid, supraspinatus, infraspinatus, teres minor, subscapularis, teres major, latissimus dorsi, pectoralis major), using indwelling intramuscular electrodes, during the spiking movements of 15 volleyball players (10 male, 5 female) who played at either the professional or collegiate level. They used a five-phase system specific to the shoulder (table 11.1). Among their results and conclusions, the authors noted that (1) the anterior deltoid and supraspinatus worked together throughout all phases of the volleyball spike in elevating and positioning the upper arm (humerus), (2) the infraspinatus and teres minor functioned independently during the acceleration phase with the arm in an overhead position, and (3) EMG activity was greater for most muscles during each phase of the spike (compared to a slow serve) because the range of motion, while comparable, took place in less time, which may increase risk of injury for the spiker's shoulder.

A variety of volleyball serving styles are in current use: the underhand serve (used primarily at lower skills levels), standing topspin serve (rarely used above the high school level), standing float serve (ball hit with no spin to create an erratic trajectory, similar to a knuckleball in baseball), jump serve (hit with forceful topspin; this is the most popular serve at the collegiate, professional, and international levels [figure 11.27]), and jump float serve (which combines elements of the jump serve with a softer, floating action on the ball).

In addition to summarizing muscle action at the shoulder during the spike, Rokito and

TABLE 11.1 Phases of a Volleyball Serve and Spike

Phase	Beginning	End
Windup	Shoulder abduction and extension	Initiation of external shoulder rotation
Cocking	Initiation of external shoulder rotation	Maximal external shoulder rotation
Acceleration	Forceful internal shoulder rotation	Ball impact
Deceleration	Ball impact	Upper arm perpendicular to trunk
Follow-through	Upper arm perpendicular to trunk	All arm motion is complete

From Rokito et al. (1998).

FIGURE 11.27 Volleyball jump serve.

colleagues (1998) also reported EMG for the volleyball serve. At the time of that study, the predominant serving style was the standing float style. The authors wrote, "In volleyball, the server does not strive to impart maximal velocity to the ball. Instead, the serving motion is slow and the objective is to place the ball over the net in a floating trajectory" (p. 261). That is no longer the case. In the past two decades, that style has largely been abandoned, especially at higher skill levels, and replaced with the jump serve. Thus, the EMG serve-related data reported in Rokito's study is largely obsolete, since their study predated the popularization of the jump serve. However, since the kinematics of the jump serve are similar to those for the straight-ahead and cross-body spikes (Reeser et al., 2010), we might expect the muscle involvement in the jump serve to be consistent with that of the spike, as just described.

Dance Applications

Dancing combines athleticism, grace, and artistic composition. Although dancers can choose to specialize in ballet, modern, aerobic, or jazz dancing, they also have the freedom to combine different dance forms to express their own artistic imagination or that of a choreographer. This great diversity of forms, however, helps explain in part why dancing has not been the subject of as much research as other movement forms (e.g., walking, running, cycling, swimming).

Most of the dance literature and research focuses on technique, biomechanical flaws in technique that may lead to injury, injury rehabilitation, and film or video analysis of movement. Fortunately, more studies are now using EMG analysis to record and study the way dancers use their muscles to perform basic and complex movements. By documenting the major muscle groups performing fundamental movements, dancers and instructors will be able to diagnose muscle weaknesses, design more effective training programs, and focus on fundamental dance movements to enhance performance and reduce injury. Detailed discussion of muscle activation during dance movements can be found in several sources (e.g., Clippinger, 2016; Krasnow et al., 2011).

One of the fundamental dance movements in classical ballet, and perhaps one of the most studied, is the **plié**. The French word *plié* literally means bent or bending. There are two principal pliés: the **grand plié**, or full bending of the knees (the knees should be bent until the thighs are nearly horizontal; see figure 11.28*a*), and the **demi plié**, or half bending of the knees (figure 11.28*b*). In addition, both forms of plié should be performed isokinetically during both the lowering, or eccentric, phase and the rising, or concentric, phase.

In classical ballet, there are five basic foot positions. Every step or movement starts and ends in one of these positions. For example, in first position, the hips are laterally rotated, with the heels touching so the feet are parallel with the frontal (coronal) plane. If the grand plié is started with the feet in first position, then proper form requires that the heels lift from the floor during the lowering phase, but not until the flexing of the knees forces the dancer to allow her heels to leave the floor. Likewise, the heels must be lowered to the floor during the rising phase. Every grand plié thus passes through the demi plié position, with the knees half extended and the heels on the floor. In dance training, numerous repetitions of the plié improve strength, flexibility, balance, timing, trunk alignment and stability, and coordination of the joint movements. In addition, the plié is often the first and last element of other movements such as **relevé** (heels off of the ground, figure 11.29*a*), **pirouette** (a complete rotation of the body on one foot, figure 11.20*b*), and jumps, such as the **grand jeté** (figure 11.29*c*).

Eccentric and Concentric Phases of the Grande Plié

The grand plié consists of two phases: the lowering, or eccentric, phase and the rising, or concentric, phase. The lowering phase can be further divided into two parts: (1) start to heel-

FIGURE 11.28 *(a)* Grand plié and *(b)* demi plié.

Sean Nel/Hemera/Getty Images

FIGURE 11.29 *(a)* Relevé, *(b)* pirouette, *(c)* grand jeté.
(a) © Milkos/Dreamstime; *(b)* David Sacks/Stone/Getty Images; *(c)* Hiroyuki Ito/Hulton Archive/Getty Images

off, during which time both the forefoot and heel remain in contact with the floor, and (2) heel-off to midcycle, during which time the heel rises with progressive metatarsophalangeal (MP) joint dorsiflexion (hyperextension). The rising phase is also divided into two parts: (1) midcycle to heel-on, during which time the heel is lowered to the floor, and (2) heel-on to end (i.e., the starting position), during which time both the forefoot and heel are in contact with the floor. Midcycle is the lowest body position during the grand plié, occurring when the dancer reaches maximum hip flexion, hip abduction, knee flexion, and MP dorsiflexion (hyperextension). Maximum ankle dorsiflexion occurs at heel-off during the lowering phase and at heel-on during the rising phase.

Muscle Activation in a Grande Plié

The grand plié closely resembles a resistance training movement similar to a modified squat. Using the same approach we used for resistance training, let's determine the principal muscle groups used during the grand plié. The concentric phase consists of hip extension coupled with hip adduction, knee extension, and ankle plantar flexion. The principal muscles being trained, therefore, are the hip extensors and hip adductors, knee extensors, and ankle plantar flexors. Using tables 4.6 and 4.7, we see that the following muscles are being trained during a grand plié:

▸ Hip extensors

Gluteus maximus

Semimembranosus

Semitendinosus

Biceps femoris, long head

Adductor magnus, posterior fibers

▸ Hip adductors

Adductor magnus

Adductor longus

Adductor brevis

Pectineus

Gracilis

Gluteus maximus, inferior fibers

▸ Knee extensors

Vastus lateralis

Vastus intermedius

Vastus medialis

Rectus femoris

▸ Ankle plantar flexors

Gastrocnemius

Soleus

Peroneus longus

Peroneus brevis

Tibialis posterior

Flexor hallucis longus

Flexor digitorum longus

Plantaris

Research by Trepman, Gellman, Micheli, and De Luca (1998) reports activity from muscles associated with each of these groups during the grand plié, including the gluteus maximus, hamstrings, adductors, vastus lateralis and medialis, and gastrocnemius. Muscle activity was recorded from each muscle or muscle group during both the lowering (eccentric) phase and the rising (concentric) phase. Their findings include the following:

- Knee flexion and thigh abduction during the descent are controlled by eccentric action of the quadriceps and adductors, respectively.
- The hamstrings stabilize the hips and knees at midcycle.
- Ascent is produced by action of the quadriceps and adductors, with high levels of muscle activity early in the rising phase.
- Isometric action of the tibialis anterior stabilizes the dorsiflexed ankle during portions of the lowering and rising phases.

Muscle activation patterns during all dance movements are affected by numerous factors, including skill level, muscle strength, flexibility, technique (e.g., arm positions, upper body posture, degree of turnout at the hip, and range of motion at each joint), speed of motion, and footwear. Variations in EMG patterns therefore are to be expected, as they would be in any complex human movement form.

Concluding Comments

The dynamics of the common forms of exercise, sport, and dance discussed in this chapter serve as examples of how any physical conditioning, exercise, or sport task can be analyzed. In some sports, such as cycling and swimming, it is important to study not only the athletes but also the equipment, clothing, and physical environment that directly affect athlete performance. Although many dance movements require great athleticism, the complex artistic component often makes research difficult.

Having completed this and earlier chapters, you now have the tools to perform fundamental movement analyses on your own. Comprehensive movement analysis encompasses many disciplines, including functional anatomy, biomechanics, physiology, and psychology. By understanding movement from all these perspectives, you have the foundation necessary to pursue advanced work in any of the numerous areas that involve the science of human movement.

Go to the web study guide to access critical thinking questions for this chapter.

Suggested Readings

Bartlett, R. (2014). *Introduction to sports biomechanics: Analyzing human movement patterns* (3rd ed.). London: Routledge.

Blazevich, A.J. (2017). *Sports biomechanics: the basics: Optimizing human performance* (3rd ed.). London: Bloomsbury Sport.

Bompa, T., & Buzzichelli, C. (2015). *Periodization training for sports* (3rd ed.). Champaign, IL: Human Kinetics.

Bompa, T.O., Di Pasquale, M., & Cornacchia, L.J. (2003). *Serious strength training* (2nd ed.). Champaign, IL: Human Kinetics.

Burke, E.R. (2002). *Serious cycling* (2nd ed.). Champaign, IL: Human Kinetics.

Burke, E.R. (Ed.). (2003). *High-tech cycling* (2nd ed.). Champaign, IL: Human Kinetics.

Cheung, S.S., & Zabala, M. (Eds.). (2017). *Cycling science.* Champaign, IL: Human Kinetics.

Clippinger, K. (2016). *Dance anatomy and kinesiology* (2nd ed.). Champaign, IL: Human Kinetics.

Colwin, C.M. (2002). *Breakthrough swimming*. Champaign, IL: Human Kinetics.

Davids, K., Hristovski, R., Araújo, D., Serre, N.B., Button, C., & Passos, P. (Eds.). (2014). *Complex systems in sport*. London: Routledge.

Delavier, F. (2010). *Strength training anatomy* (3rd ed.). Champaign, IL: Human Kinetics.

Fleck, S.J., & Kraemer, W.J. (2014). *Designing resistance training programs* (4th ed.). Champaign, IL: Human Kinetics.

Haff, G.G., & Triplett, N.T. (Eds.). (2015). *Essentials of strength training and conditioning* (4th ed.). Champaign, IL: Human Kinetics.

Hay, J.G. (1993). *The biomechanics of sports techniques* (4th ed.). Englewood Cliffs, NJ: Prentice Hall.

Laws, K. (2008). *Physics and the art of dance* (2nd ed.). New York: Oxford University Press.

McGinnis, P. (2013). *Biomechanics of sport and exercise* (3rd ed.). Champaign, IL: Human Kinetics.

Troup, J.P., Hollander, A.P., Strasse, D., Trappe, S.W., Cappaert, J.M., & Trappe, T.A. (Eds.). (2011). *Biomechanics and medicine in swimming VII*. New York: Routledge.

Vanlandewijck, Y.C., & Thompson, W.R. (Eds.). (2016). *Training and coaching the Paralympic athlete*. Hoboken, NJ: Wiley-Blackwell.

Vorontsov, A.R., & Rumyantsev, V.A. (2000). Propulsive forces in swimming. In V. Zatsiorsky (Ed.), *Biomechanics in sport* (pp. 205-231). Malden, MA: Blackwell Science.

Vorontsov, A.R., & Rumyantsev, V.A. (2000). Resistive forces in swimming. In V. Zatsiorsky (Ed.), *Biomechanics in sport* (pp. 184-204). Malden, MA: Blackwell Science.

12

Clinical Applications

Objectives

After studying this chapter, you will be able to do the following:

- Describe concepts of musculoskeletal injury
- Explain techniques for injury prevention
- Discuss the role of movement in the rehabilitation of injuries
- Describe how human movement relates to specific cases of injury
- Explain how prosthetic and orthotic devices can improve movement

Clinicians, including physicians, physical therapists, nurses, athletic trainers, occupational therapists, and others, deal with patients whose injuries or diseases compromise their everyday lives. Most movement-related disorders are a consequence of musculoskeletal injuries or conditions. Our focus in this chapter is limited to how human movement is involved in the prevention, production, and recovery from injuries. Full coverage of the clinical diagnosis, treatment, and rehabilitation of injuries is beyond the scope of our discussion and is well covered by other sources. The emphasis in this chapter, as throughout this book, is on the role of human movement.

Concepts of Musculoskeletal Injury

Injury pervades everyday life. Although people sustain injuries of varying severity, and some more frequently than others, virtually no one is spared the pain, distraction, and incapacity caused by injury at some time in their life. Many musculoskeletal injuries are movement related. Excessive and violent movements are involved in a vast array of injuries, from twisted ankles and knees to dislocated shoulders and fingers.

Our working definition of **injury** is damage sustained by tissues of the body in response to physical trauma. This definition is less encompassing than generally accepted notions of injury, but is useful in the context of the mechanics of musculoskeletal injury. The term *injury* usually is associated with negative consequences. In some situations, however, injury may be involved in events with positive consequences. In the bone remodeling process (see Wolff's law, chapter 2), for example, bone must first be injured (resorbed) to prepare it for subsequent positive adaptive changes.

Many musculoskeletal injuries involve movement of specific joints or the entire body. We describe the cause of an injury by its **mechanism**, the fundamental physical process responsible for a given action, reaction, or result. Exemplar injury mechanisms include hyperextension, hyperflexion, valgus rotation, compression, distraction (i.e., being pulled apart), torsion or twisting, and direct impact.

Exploration of the biomechanics of injury is an interdisciplinary endeavor. Among the disciplines involved are anatomy, physiology, mechanics, kinesiology, medicine, engineering, and psychology. The problem of musculoskeletal injuries and conditions cannot be addressed effectively by any single discipline examining injury in isolation. To ensure optimal progress and outcomes in addressing the clinical issues related to musculoskeletal injury, an interdisciplinary approach is essential (Whiting & Zernicke, 2008).

Injury Prevention

Despite the inevitability of injury, prevention plays an important role in the overall clinical approach to injury. After all, the best injury is the one that never happens. Injury prevention programs can be in the form of education (i.e., providing safety information to workers, supervisors, parents, coaches, athletes), policy implementation, worksite or school modification, and training to adopt safer movement patterns and techniques.

In terms of training, one of the first steps toward injury prevention is a thoughtful and informed program of preparation for any movement task. Such a program includes proper technique instruction, gradual progression of implementation (e.g., intensity, load, volume), developmental appropriateness, and personalization in terms of an individual's level of skill and experience, conditioning, and personal medical history.

One approach to prevention, used generally as a preventive technique but more specifically in advance of orthopedic surgery, is *prehabilitation* (also *pre hab*).

> By improving an individual's functional capacity through increased physical activity before an anticipated orthopaedic procedure, it seems reasonable to assume that the individual will maintain a higher level of functional ability and rebound more

Applying the Concept

Clinical Gait Analysis

Clinical gait analysis (CGA) is an approach to assessing gait to help clinicians understand gait abnormalities and make informed recommendations as part of the clinical decision-making process. CGA can be qualitative (observational) or quantitative (measured). Observational gait analysis (OGA) involves trained visual assessment of in-person or video-recorded gait. A variety of systems can be used to formalize the OGA process, including the Rancho Los Amigos system, Physician Rating Scale, and Edinburgh Visual Gait Score system. Quantitative CGA incorporates data measurement systems to measure muscle activity (EMG), kinematics, and kinetics. The goal of clinical gait analysis is to provide clinicians with information to be considered in conjunction with patient status and clinician experience in making informed clinical judgments and decisions.

> rapidly in the rehabilitation process. Prehabilitation is the process of enhancing functional capacity of the individual to enable him or her to withstand the stressor of inactivity associated with an orthopaedic procedure. A generic prehabilitation program incorporates the components of warm-up, a cardiovascular component, resistance training, flexibility training, and practicing functional tasks. (Ditmyer et al., 2002, p. 43)

Another approach to reducing injury involves the use of preparticipation screening tests to identify risk factors. (On a cautionary note, care should be taken in interpreting results of such tests as being predictive versus diagnostic.) One of the most widely used screening instruments is the Functional Movement Screen (FMS), which tests seven basic movement patterns:

1. Deep squat
2. Hurdle step
3. In-line lunge
4. Shoulder mobility
5. Active straight-leg raise
6. Trunk stability pushup
7. Rotary stability

The FMS was designed as a preparticipation screen and return-to-sport test to determine whether or not an athlete has the necessary movement capacity to participate in sport activities at a level of minimum competency (Cook et al., 2006; Cook et al., 2014a; Cook et al., 2014b). The FMS system has been widely researched, and results are equivocal. Among several articles reviewing the efficacy of the FMS, authors have disparately concluded that FMS has limited ability to predict athletic performance; however, authors have concluded that

- there is moderate evidence to support use of the FMS total score to predict injury risk in team sports (Kraus et al., 2014),
- the strength of association between FMS scores and subsequent injury does not support its use as an injury prediction tool (Moran et al., 2017),
- FMS is not an adequate field test for athletic performance (Parchmann & McBride, 2011),
- the FMS can be used to assess the movement patterns of athletes and make decisions related to interventions for performance enhancement (Minick et al. 2010), and

- development and criterion validation of consistent grading procedures must precede research regarding the relation between FMS performance and injury rates and that the FMS should be used with caution with respect to directing strength and conditioning programs (Whiteside et al., 2016).

One of the first steps in designing an injury prevention program is to identify potential risk factors. Hewett and colleagues (2005), for example, conducted a prospective study to identify risk factors for anterior cruciate ligament (ACL) injuries in female athletes. They followed a group of 205 female athletes in the high-risk sports of basketball, soccer, and volleyball. The study participants were evaluated for 3-D kinematics (joint angles) and joint moments while performing a jump-landing task. Of the 205 participants, 9 subsequently suffered ACL injuries. The 9 injured athletes showed higher dynamic knee valgus and higher joint moments in the initial testing when compared to the 196 noninjured athletes. The authors concluded that monitoring neuromuscular control at the knee may provide information helpful in designing targeted interventions for at-risk athletes.

Numerous studies over the past two decades have demonstrated the efficacy of injury prevention programs involving mostly young female athletes. Many of these programs focus on ACL injuries in women, since they are three to six times more likely to sustain ACL injuries compared to men. One of the most successful programs was that reported by Mandelbaum and others (2005). More than 1,000 female soccer players ages 14 to 18 were given a sport-specific training intervention consisting of education, stretching, plyometrics, and sport-specific agility drills that replaced their traditional warm-up. Nearly 2,000 female athletes served as age- and skill-matched controls. During the first soccer season, there was an 88% lower incidence of ACL injuries in the intervention group. In the second season, the intervention group had a 74% lower rate of ACL injury compared to controls. Clearly, the training intervention played a role in lowering the risk of ACL injuries.

Rehabilitation

Rehabilitation is an area of medicine that aims to enhance and restore functional ability to those with injuries, physical impairments, or disabilities. In a human movement context, rehabilitation seeks to restore a person's ability to move in a functional and efficient manner, and emphasizes restoration of mechanical function, proprioception, and neuromuscular control.

Applying the Concept

Movement Following Spinal Cord Injury

Until recently, paralysis due to spinal cord injury was believed to be irreversible. Ongoing research, however, provides hope. Collaborating research labs in the United States and Russia, for example, have developed novel strategies to allow people with complete paralysis to recover some movement ability. The procedure involves a noninvasive stimulation strategy of painless transcutaneous electrical stimulation combined with a pharmacological (drug) intervention (Gerasimenko et al., 2015). Most recently, the research group has used the following method:

> Non-invasive stimulation technology, painless cutaneous enabling motor control (pcEmc), to determine the feasibility of re-establishing functional brain spinal cord connectivity that enables a subject with complete motor paralysis to move upon volitional intent and perform work that can assist a robotic exoskeletal device in generating over-ground stepping. (Gad et al., 2017, p. 2)

Many rehabilitation models and programs have been proposed and implemented. We present one exemplar model that highlights important elements in the rehabilitation process. Voight and colleagues (2014) describe a three-phase rehabilitation model. Phase 1 seeks to restore static stability through proprioception and kinesthesia with a goal of restoring proprioception. In this phase, exercises are used that challenge the patient and are neither too easy nor too difficult. Therapeutic exercises are prescribed using a four-by-four method, with four positions (non–weight bearing [supine or prone], quadruped, kneeling, and standing) and four types of resistance (unloaded with core activation, unloaded without core activation, loaded with core activation, and loaded without core activation).

Once proprioception and kinesthesia are adequately restored, phase 2 seeks to restore dynamic stability, with a goal of encouraging preparatory agonist–antagonist coactivation. Patients are progressively challenged with exercises that transition from bilateral to unilateral stance, eyes open to eyes closed, and stable to unstable surfaces.

Phase 3 seeks to restore and develop reactive neuromuscular control, with a goal of initiating reflex muscular stabilization. In this phase, the level of difficulty is increased by transitioning from slow-speed to fast-speed activities, from low-force to high-force activities, and from controlled to uncontrolled activities (Voight et al., 2014).

Common Injuries

The following sections focus on movement-related aspects of injury mechanisms and neuromusculoskeletal conditions. Consideration of the diagnosis and treatment of injuries, while important, is beyond the scope of our discussion.

Anterior Cruciate Ligament

Few injuries have received more clinical and research attention than those of the anterior cruciate ligament (ACL). This is not surprising given the important role the ACL plays in knee joint function. The ACL is a complex ligament that connects the femur and tibia. Proximally, the ACL attaches to the medial surface of the lateral femoral condyle. The ACL attaches distally to the anterior surface of the mid-tibial plateau (figure 12.1). The ACL consists of two bands, or bundles: anteromedial (AM) and posterolateral (PL). Each band plays a unique role in stabilizing the tibiofemoral articulation. In knee flexion, the AM band is taut, whereas the PL band is relatively lax. With knee extension, the PL band becomes taut and the AM band remains taut, bus less so than the PL.

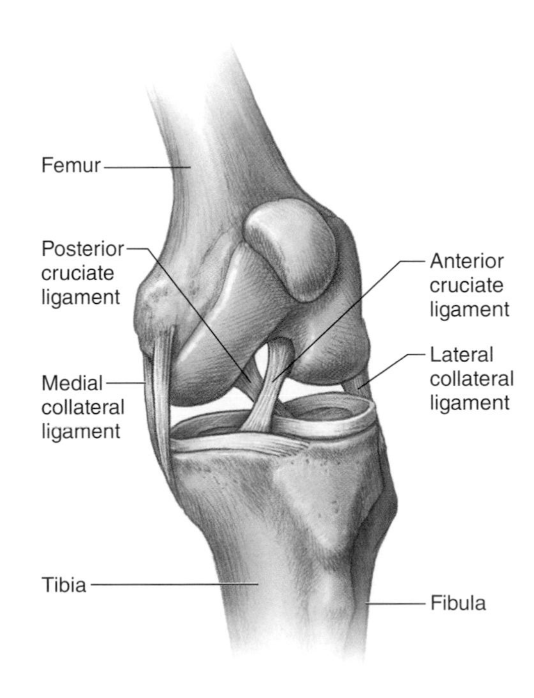

FIGURE 12.1 Knee joint showing the anterior cruciate ligament.

The ACL acts as the primary restraint to anterior tibial translation (i.e., forward movement of the tibia relative to the femur). It also acts as a secondary restraint to internal tibial rotation. The ACL's role as a secondary restraint to varus–valgus angulation and external rotation is less clear, though generally accepted. It also works in concert with the posterior cruciate ligament (PCL) to limit knee hyperextension and hyperflexion.

ACL injury happens most often in response to valgus loading in combination with external tibial rotation or to hyperextension with internal tibial rotation. The first mechanism (*valgus rotation*) typically happens in what is termed a *noncontact injury*, in which the foot is planted on the ground, the tibia is externally rotated, the knee is near full extension, and the knee collapses into a valgus (i.e., knock-kneed) position. The situation is made worse if a force is applied to the knee while the foot is in contact with the ground (*contact injury*). This is common in contact sports such as American football, rugby, and soccer when another player impacts the lateral aspect of the knee, accentuating the valgus loading and rotation.

The second mechanism involves knee hyperextension with internal tibial rotation. Although this is a less common mechanism overall, hyperextension may be the predominant mechanism in certain populations such as basketball players or gymnasts, whose injuries often occur as they violently hyperextend their knee on landing from a jump.

Shoulder Pathologies

Given the complexity of the shoulder complex's anatomy (figure 12.2) and the shoulder's inherent instability, shoulder injuries are inevitable. Shoulder symptoms are third (after back and knee problems) in terms of number of physician visits. Many of these shoulder problems include glenohumeral impingement, or rotator cuff lesions. The shoulder's rotator cuff muscle group includes the subscapularis, supraspinatus, infraspinatus, and teres minor (see figure 4.12 and table 4.4). Rotator cuff (RC) problems are a common source of pain and dysfunction in people who use overhead movements in their work or play.

An impingement syndrome occurs when increased pressure within a confined anatomical space adversely affects the enclosed structures. Glenohumeral impingement is a broad term covering two primary types: subacromial impingement and internal impingement.

Subacromial impingement refers to shoulder abduction that results in suprahumeral structures (most notably the distal supraspinatus tendon, subacromial bursa, and proximal tendon of the long head of the biceps brachii) being forcibly pressed against the anterior surface of the acromion and the coracoacromial ligament (collectively the *coracoacromial arch*). Subacromial contact pressures are elevated in patients with impingement syndrome; maximal contact pressure develops with the arm in a hyperabducted position or with the arm adducted across the chest with the arm internally rotated.

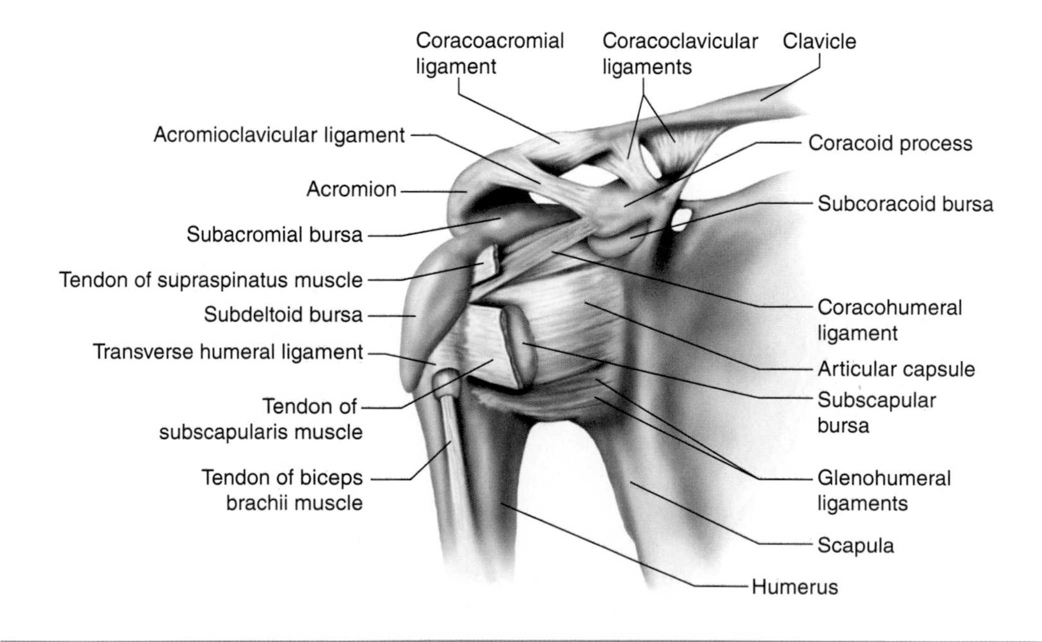

FIGURE 12.2 Glenohumeral joint structures.

Internal impingement happens when the supraspinatus tendon contacts the posterosuperior rim of the glenoid fossa. This mechanism may be significant in the development of rotator cuff pathologies. Internal impingement often happens in throwing, when the shoulder is abducted and externally rotated (e.g., cocking phase in an overhead throw).

Impingement pathologies fall into two broad age-based categories. Impingement in those younger than 35 years usually happens to participants in sports (e.g., swimming, water polo, baseball, or American football) or occupations (e.g., carpentry, painting) involving extensive overhead movements. Older individuals are more likely to suffer from the effects of degenerative processes that lead to bone spur formation, joint capsular thinning, decreased tissue perfusion, and muscular atrophy.

Repeated abduction places large stresses on the musculotendinous and capsuloligamentous structures and eventually leads to tissue microtrauma. Continued mechanical loading further weakens the tissues and hastens their failure. Tissue failure, in turn, contributes to glenohumeral instability and greater joint movement. This increases the chance of humeral subluxation (i.e., dislocation) that further aggravates the impingement condition. As a result, the person is trapped in an unfortunate loop of joint deterioration and compromised function.

Rupture of musculotendinous structures in the rotator cuff typically results from a chain of events that begins with minor inflammation that progresses with continued overuse to advanced inflammation, microtearing of tissue, and partial or complete rupture. Compromised tissue integrity and fatigue contribute to altered movement mechanics, and these modified movements further stress the involved tissues and hasten their eventual failure. The supraspinatus is the mostly commonly injured muscle in the rotator cuff group. Less frequently, other cuff muscles suffer damage. Supraspinatus injury, in particular, is associated with repeated, and often violent, overhead movement patterns (e.g., throwing, striking, hammering).

Hip Fractures

Proximal femoral fractures, or hip fractures, are classified according to their location. Most hip fractures happen in the femoral neck or in the intertrochanteric region between the greater and lesser trochanters (figure 12.3). Hip fractures in the young usually result from high-energy impacts, most commonly a result of motor vehicle crashes. These injuries often are accompanied by hip dislocation.

The most common mechanism for femoral neck fracture is direct trauma to the hip (e.g., impact from a fall). Hip fractures in older people are associated with falls, often caused by tripping or unsteady gait. This association raises an intriguing question: Does hip fracture cause the fall, or does the impact of landing from a fall cause the bone to break? In most cases, the force of impact precipitates the fracture (Parkkari et al., 1999), with only rare instances of a spontaneous fracture causing a fall. These rare cases usually are associated with severe osteoporosis.

The energy created by a fall is much greater than that necessary to fracture a bone. Because hip fractures occur in less than 5% of falls, other tissues obviously absorb considerable energy. This observation is substantiated by the fact that risk of hip fracture is lower in people with a higher body mass index, or BMI (weight/height^2). In addition, other factors, such as breaking the fall with outstretched arms or eccentric action of the quadriceps, likely are involved in attenuating the forces.

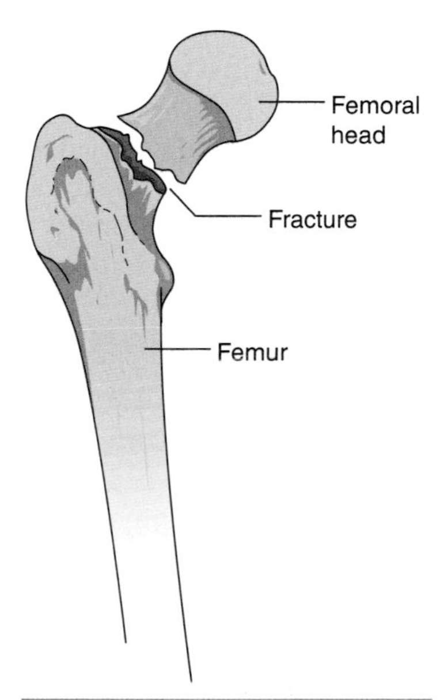

FIGURE 12.3 Hip fracture.

Applying the Concept

Chronic Traumatic Encephalopathy

In 1928, Harrison Martland authored an article in the *Journal of the American Medical Association* with the simple title "Punch Drunk." In 1949, British neurologist Macdonald Critchley published a paper titled "Punch-Drunk Syndromes: The Chronic Traumatic Encephalopathy of Boxers." In a 1957 follow-up paper, Critchley presented a detailed discussion of medical aspects of boxing, with emphasis on neurological events. The term used by Critchley, **chronic traumatic encephalopathy** (CTE), has become the most common descriptor of neurological damage caused by repetitive head trauma. At that time, neurological damage from repeated blows to the head was thought to be exclusive to boxing. Over the last 60 years, and particularly in the last decade, CTE has been linked to a wide variety of situations, particularly contact sports, including American football, boxing, wrestling, rugby, ice hockey, lacrosse, soccer, and skiing.

Clinical manifestations of CTE include memory disturbance, Parkinson-like symptoms, behavioral and personality alterations, and gait and speech abnormalities. From a neuropathological perspective, CTE is characterized by atrophy of nervous system structures, including the cerebral hemispheres, medial temporal lobe, thalamus, and brain stem (McKee et al., 2009). The exact mechanisms by which multiple acute traumatic brain injuries may develop, over time, into CTE and its associated tau protein malformation and development of neurofibrillary tangles (NFTs) have not been fully defined (Ling et al., 2015). Despite sharing symptoms with other neurodegenerative conditions, "chronic traumatic encephalopathy is a neuropathologically distinct, slowly progressive tauopathy with a clear environmental etiology" (McKee et al., 2009, p. 709).

Head Injuries

Many head injuries are directly or indirectly related to movement characteristics associated with head acceleration and deceleration. Head injuries occur in response to the sudden application of forces to the head or its connected structures. Numerous interrelated factors combine to determine the exact mechanisms of injury. These factors include the type of force and its magnitude, location, direction, duration, and rate.

The forces causing head injuries are characterized as direct or indirect. Direct (contact) loading results from impact, as seen in a boxer's punch. Indirect loading occurs when forces are transmitted to the head through adjacent structures such as the neck (e.g., whiplash mechanism). Whether direct or indirect, an applied force either accelerates or decelerates the head. Force applied to a stationary head will tend to accelerate its mass, whereas forces acting in opposition to the head's motion will decelerate it. A forceful blow to the head typifies an acceleration mechanism. The deceleration mechanism is involved when the head's motion is abruptly stopped by an unyielding surface. These acceleration and deceleration mechanisms often are implicated in brain injury caused by head trauma.

The effects of forces also are categorized by the type of head motion that occurs in response to loading. Forces directed through the head's center of mass cause linear translation of the head, while forces acting off center (e.g., on the chin) result in neck and head rotation.

Traumatic brain injury (TBI) is a term that covers numerous conditions arising from direct impact or acceleration and deceleration of the head. Specific TBI injuries include focal lesions, contusions, laceration, intracranial hematoma (i.e., bleeding), brain damage from increased intracranial pressure (ICP), and a variety of diffuse brain injuries. TBI is a major public health issue that because of the absence of visible injury has been characterized by the U.S. Centers for Disease Control and Prevention (CDCP) as a silent epidemic.

Prosthetics and Orthotics

Human movement can be compromised by many factors, including damage to or loss of a limb as a result of trauma, disease, or congenital abnormalities. In such cases, recovery can be greatly facilitated by use of an orthotic or prosthetic device. **Orthotics** involves the use of devices (e.g., splints, braces) that assist in supporting anatomical structures. Prosthetics provides artificial devices that replace missing body parts (e.g., an amputated leg). Professionals in these areas, orthotists and prosthetists, work collaboratively with other health care providers such as physicians, physical therapists, and occupational therapists to prescribe, design, produce, and manage devices to assist patients in regaining movement function.

Orthotic devices can restrict motion, assist motion, reduce joint loading, provide protection and support during movement tasks and injury rehabilitation, and control, limit, or immobilize an extremity or joint. Orthotic design is specific to the body region in need of support (figure 12.4). For example, ankle–foot orthoses (AFOs) support the foot and ankle region (e.g., following ankle surgery), thoracolumbosacral orthoses (TLSOs) brace the trunk (e.g., to prevent progression of scoliosis), and foot orthoses inserted in the shoe support the arch of a person with *pes planus* (i.e., flat feet). Foot orthoses can be purchased over the counter from a fixed set of shapes and sizes or custom fit specifically for an individual's foot structure.

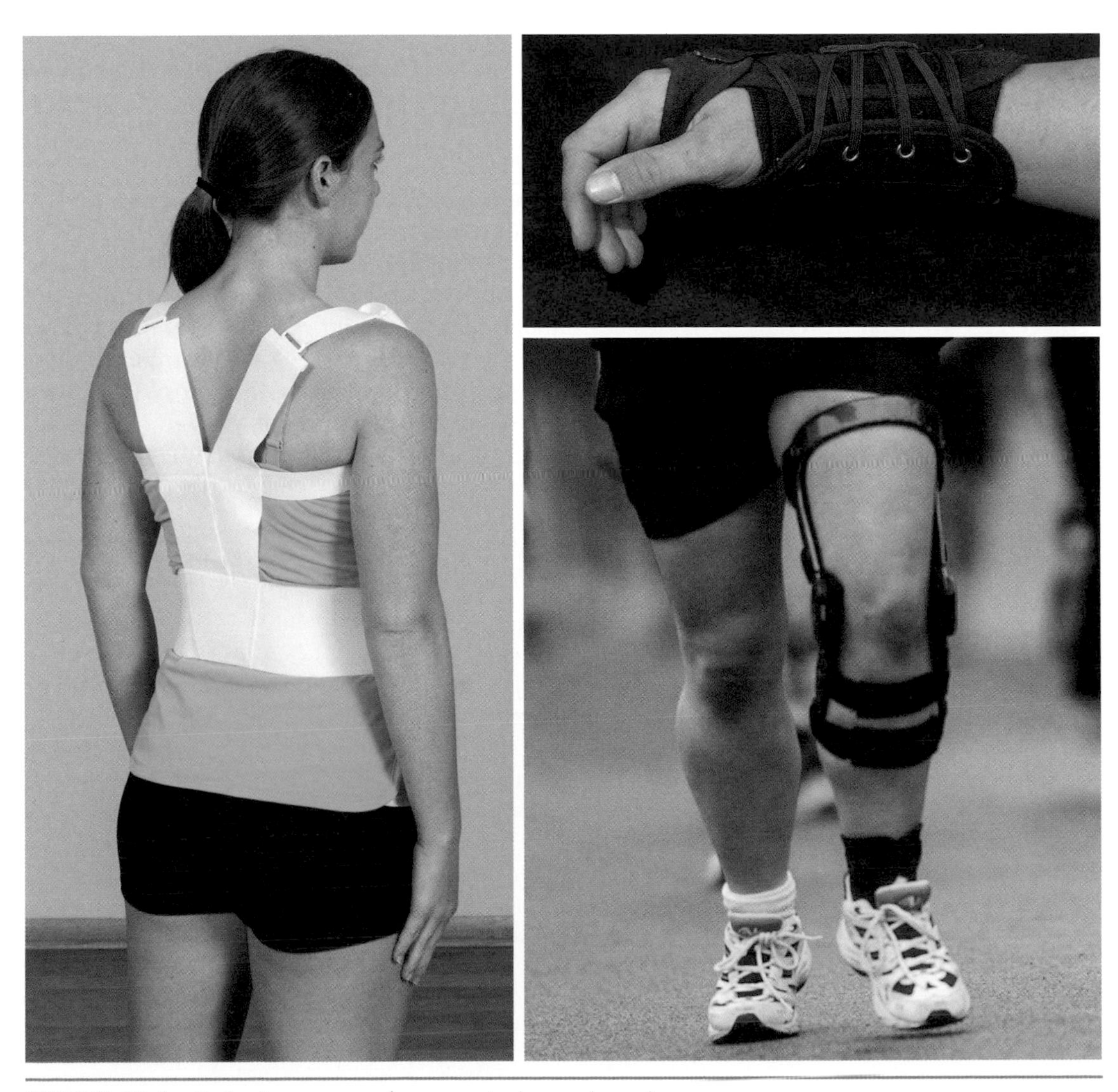

FIGURE 12.4 Orthotics: posture brace, wrist brace, knee brace.

Prosthetic devices serve as replacements for missing or damaged body structures (figure 12.5). Long ago, the prosthesis for a person with an amputated leg was simply a piece of wood (i.e., "peg leg"). Fortunately, advances in technology and materials science (e.g., new plastics and carbon fiber) have allowed for development of sophisticated prostheses. Current limb prostheses may by controlled by miniature motors, myoelectric signals, or even the mind (see sidebar). Prosthetic limbs allow people with amputations to lead movement-filled lives.

FIGURE 12.5 Prosthetic leg.

Limb prostheses are used when a person is missing all or part of a limb, usually as a result of trauma, disease, or a congenital condition. The primary goal of prosthetic design is to restore as much movement function as possible. The type of prosthesis depends on the level of amputation. Lower extremity prostheses include devices for hip, knee, and ankle disarticulations, and trans-tibial (i.e., below-the-knee, or BK) and trans-femoral (i.e., above-the-knee, or AK) amputations. Upper extremity prostheses are needed for shoulder, elbow, and wrist disarticulations, and trans-radial (i.e., below-the-elbow, or BE) and trans-humeral (i.e., above-the-elbow, or AE) amputations.

The debilitating pain of advanced osteoarthritis of the hip and knee severely limits a person's mobility. Joint replacement surgery (arthroplasty), in which the damaged structures are replaced by artificial materials, provides substantial pain relief and restores movement function in most cases. Because of the load-bearing responsibilities of the lower extremities, it is not surprising that the hip and knee are the leading arthroplasty sites. In light of continued advances in biomaterials and surgical techniques, the advent of computer-assisted design and manufacturing, and an aging population, the number of arthroplasties will continue to increase.

Research in Mechanics

Mind-Controlled Prostheses

Technological advancements can make things real that were once thought to be science fiction. Such is the case of a mind-controlled prosthetic arm with moving fingers. Researchers at Johns Hopkins University have developed a brain–machine interface prosthetic arm driven by electrocorticography (ECoG) to use the existing functional anatomy of sensorimotor cortical populations to control individual finger movements in real time (Hotson et al., 2016). This development has the potential to allow people who have lost arms due to injury or disease to regain their manipulative ability by using a device to control a dexterous modular prosthetic arm. This device is powered by online neural decoding of individual finger movements using a high-density ECoG array.

See the references for the full citation:

Hotson, McMullen, Fifer, Johannes, Katyal, Para, Armiger, Anderson, Thakor, Wester, & Crone, 2016.

Total hip replacement (THR) involves excision of the femoral head and part of the neck and enlargement of the acetabulum. A metallic femoral prosthesis is inserted into the medullary canal of the femur. The prosthesis may be cemented into the canal using methyl methacrylate. An alternative cementless technique uses a prosthesis with porous structure that encourages bony ingrowth.

Although traditional THR is reasonably successful in older, relatively inactive patients, in younger, active patients, it offers poor long-term outcomes. An alternative procedure known as hip resurfacing may be more promising for younger people. Hip resurfacing involves a metal-on-metal prosthesis that minimizes bone loss, alleviates hip pain, and allows patients to return to work and leisure activities and participate in sports. In resurfacing, the neck and head of the femur are not removed. The prosthesis fits like a cap on the head of the femur and a matching cup fits precisely into the recess of the acetabulum.

Concluding Comments

Many clinical options are available to identify, diagnose, treat, and rehabilitate musculoskeletal injuries and conditions that affect our ability to move. The best first step is to take preventive action to reduce the risk of injury. Following injury, a well-designed rehabilitation program often can improve movement function and allow the mover to return to action, whether in normal activities of daily living (ADLs) or more challenging movement tasks.

Go to the web study guide to access critical thinking questions for this chapter.

Suggested Readings

Ellenbecker, T.S., & Wilk, K.E. (2017). *Sport therapy for the shoulder.* Champaign, IL, Human Kinetics.

Hoogenboom, B.J., Voight, M.L., & Prentice, W.E. (2014). *Musculoskeletal interventions: Techniques for therapeutic exercises* (3rd ed.). New York: McGraw-Hill.

Kirtley, C. (2006). *Clinical gait analysis: Theory and practice.* London: Churchill Livingstone.

Los Amigos Research and Education Center. (2001). *Observational gait analysis.* Downey, CA: Author.

Loudon, J.K., Manske, R.C., & Reiman, M.P. (2013). *Clinical mechanics and kinesiology.* Champaign, IL: Human Kinetics.

Lusardi, M.M., Jorge, M., & Nielsen, C.C. (2012). *Orthotics and prosthetics in rehabilitation* (3rd ed.). Philadelphia: Saunders.

May, B.J., & Lockard, M.A. (2011). *Prosthetics & orthotics in clinical practice: A case study approach.* Philadelphia: F.A. Davis.

McGill, S. (2016). *Low back disorders* (3rd ed.). Champaign, IL: Human Kinetics.

Moore, K.L., Dalley, A.F., & Agur, A.M.R. (2013). *Clinically oriented anatomy* (7th ed.). Philadelphia: Lippincott Williams & Wilkins.

Price, J. (2019). *The BioMechanics method for corrective exercise.* Champaign, IL: Human Kinetics.

Taktak, A., Ganney, P., Long, D., & White, P. (2014). *Clinical engineering: A handbook for clinical and biomedical engineers.* Oxford: Academic Press.

Whiting, W.C., & Zernicke, R.F. (2008). *Biomechanics of musculoskeletal injury* (2nd ed.). Champaign, IL: Human Kinetics.

13

Ergonomics Applications

Objectives

After studying this chapter, you will be able to do the following:

- Define and describe the goals of ergonomics
- Explain ergonomic methods of analysis
- Describe the concept of human–machine interface
- Explain ergonomic aspects of lifting tasks
- Describe lifting safety guidelines and assessment
- List and explain ergonomically related overuse conditions

> Ergonomics (or human factors) is the scientific discipline concerned with the understanding of interactions among humans and other elements of a system, and the profession that applies theory, principles, data and methods to design in order to optimize human well-being and overall system performance. (IEA, 2017)

In simpler terms, **ergonomics** is the study of how humans interact with their immediate environment. Since a primary mode of human interaction is through movement, a basic knowledge of ergonomics is essential to a broad understanding of dynamic human anatomy.

Ergonomic Concepts

As an interdiscipline, ergonomics spans a variety of disciplines, including human anatomy, biomechanics, physiology, psychology, sociology, medicine, and engineering. Ergonomics is composed of three primary domains: organizational, cognitive, and physical. The organizational domain deals with organizational design, policies, and processes as they relate to workplace communication, work design and systems, networking, and teamwork. The cognitive domain involves mental processes, including perception, memory, reasoning, and motor response.

The physical domain is most relevant to our study of dynamic human anatomy. This domain integrates anthropometric, biomechanical, and physiological concepts as they relate to human movement, primarily in occupational settings. The principles of physical ergonomics have been used extensively in the design of consumer and industrial products, assessment of manual materials handling tasks, development of workplace safety guidelines, workplace design, and diagnosis of work-related medical conditions.

Goals

The primary goals of ergonomics are to improve productivity, improve efficiency, enhance safety, reduce injury risk, and reduce cost. Ergonomic interventions can improve both the quantity and quality of worker output and increase efficiency by facilitating production in a time-effective manner.

Safety enhancement and injury risk reduction are at the core of most ergonomic programs. One of the most common worker risks is a class of conditions collectively known as **musculoskeletal disorders** (MSDs). MSDs are the risk factor most closely associated with human movement tasks.

Applying the Concept

Musculoskeletal Disorders

What constitutes a musculoskeletal disorder (MSD)? The U.S. Bureau of Labor Statistics defines the term as follows:

> Musculoskeletal disorders (MSDs) include cases where the nature of the injury or illness is pinched nerve; herniated disc; meniscus tear; sprains, strains, tears; hernia (traumatic and nontraumatic); pain, swelling, and numbness; carpal or tarsal tunnel syndrome; Raynaud's syndrome or phenomenon; musculoskeletal system and connective tissue diseases and disorders, when the event or exposure leading to the injury is overexertion and bodily reaction, unspecified; overexertion involving outside sources; repetitive motion involving microtasks; other and multiple exertions or bodily reactions; and rubbed, abraded, or jarred by vibration. (2016b)

Research consistently has shown that ergonomic analysis and intervention can result in significant cost savings by reducing health care costs, lost work time, workers' compensation claims, and human error. Many ergonomic interventions are relatively inexpensive and are therefore cost effective for businesses, agencies, and workers alike.

Many governmental agencies and professional organizations have issued safety guidelines addressing specific ergonomic issues and recommendations. These guidelines cover a multitude of industrial and service areas, including agriculture, apparel and footwear, baggage handling, computer workstations, construction, health care, product manufacturing, metalwork foundries, meatpacking, mining, poultry processing, printing, sewing, shipyards, and telecommunications (U.S. Department of Labor, n.d.).

Methods of Analysis

Ergonomic analyses can be reactive or proactive. A reactive analysis addresses an existing problem or situation. A proactive analysis seeks to anticipate potential problems and make changes that prevent these problems.

An ergonomic analysis typically involves several steps, the first of which is identification of risk factors. General risk factors associated with movement-related ergonomic problems include awkward postures, repetitive motions, forceful exertions, pressure points, and sustained static postures (NIOSH, 2007).

In conducting an ergonomic assessment, the first step is to identify risk factors specific to the situation being assessed. Risk factors may be systematic (i.e., evident in the overall work environment) or specific to an individual. In assessing the work environment of a computer data-entry operator, for example, potential risk factors might include keyboard height and inclination, monitor height (relative to the operator's line of sight), distance, brightness, lack of arm and wrist support, chair design and support, and operator posture.

Once the ergonomic risk factors have been identified, the ergonomist must identify possible changes (e.g., new or adjusted keyboard, monitor, chair) to improve comfort and safety. After the changes have been implemented, the worker should be re-evaluated to ensure that the modifications have achieved the ergonomic goals.

Risk factors that are common to a group of workers can be addressed through either engineering or administrative controls. Engineering controls involve improving worker conditions by modifying tasks, adjusting movement patterns, redesigning workstations or tools, and providing protective equipment, as needed. Administrative controls include development and implementation of procedures and processes that can reduce risk such as job rotation (i.e., varying work tasks) and appropriate work breaks (e.g., rest or stretching breaks).

Numerous analysis approaches have been used to identify ergonomic problems and find solutions. Among those approaches are surveys and questionnaires, iterative prototyping, meta-analysis, work sampling, and a wide range of computer-based models applicable to specific tasks or systems.

Human–Machine Interface

Many occupations involve human interaction with a machine or device, in what is termed a **human–machine interface**. Examples include computer or keyboard operators (figure 13.1), assembly-line and construction workers, medical technicians and clinicians, and automobile mechanics.

Lifting

As discussed in chapter 9, lifting is a basic movement skill in sport, recreational, household, and occupational settings. It is used to move an object from one position to another. Various techniques are used to perform a lift depending on the characteristics of the object, lifter anthropometrics, environment, and task (see chapter 9).

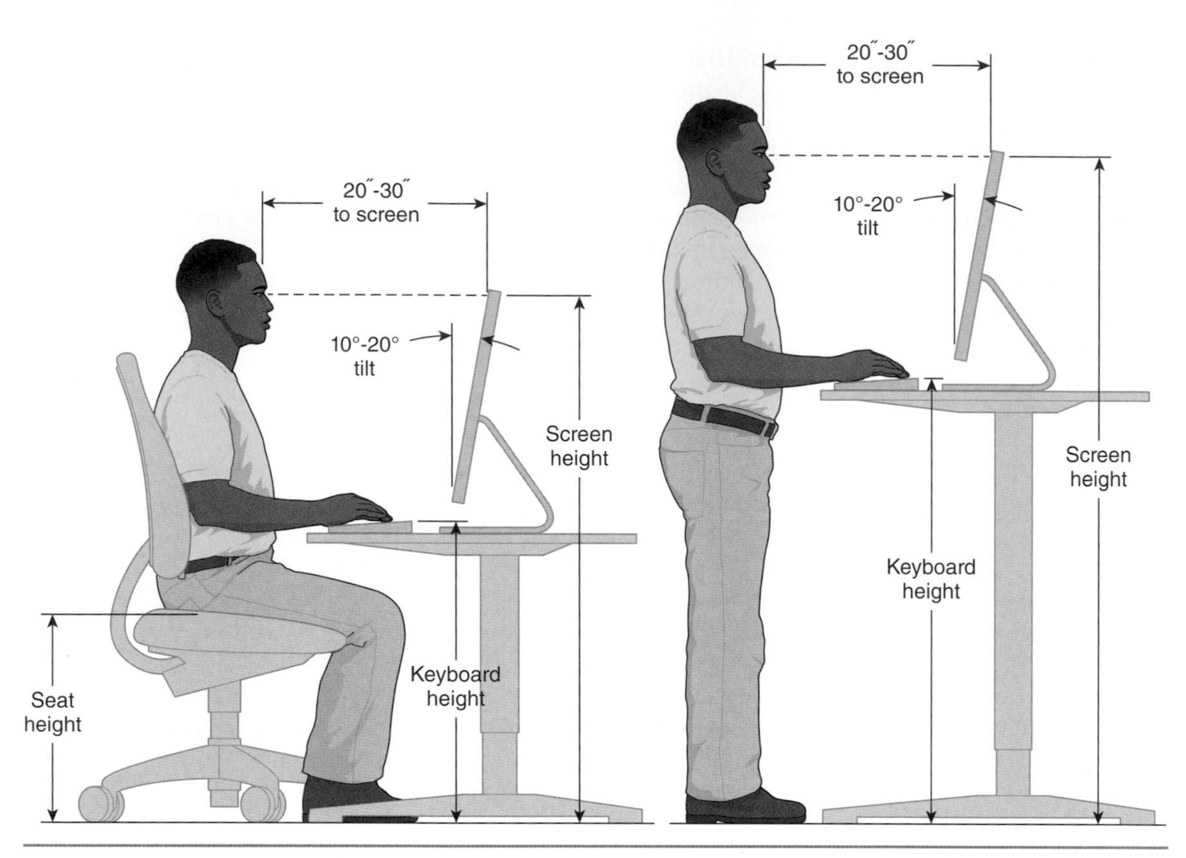

FIGURE 13.1 Computer workstation anthropometrics for seated (*left*) and standing (*right*) operators.

Lifting Technique

While no lifting task is risk free, guidelines for safer lifting are commonly available. For occupational tasks handling materials, these guidelines often include lifting techniques such as elimination of unnecessary lifts, work organization that allows for physical demands and work pace to increase gradually, minimizing the distances that loads are lifted and lowered, and positioning pallet loads of materials at a height that allows workers to lift and lower within their individual power zone.

Additional suggestions include avoiding manually lifting or lowering loads to and from the floor, adopting special strategies for handling unstable or heavy loads (e.g., use of mechanical assistance devices, repacking containers to reduce weight and balance contents, team lifting), reducing lifting frequency and the amount of time workers perform lifting tasks, clearing space to improve access, and wearing appropriate clothing and safety equipment (DHHS, 2007).

Lifting Injuries and Safety Recommendations

According to the U.S. Bureau of Labor Statistics (2016), the occupations with the highest musculoskeletal disorder (MSD) rates are ones that involve significant lifting. These occupations include emergency medical technicians and paramedics, nursing assistants, firefighters, light truck or delivery services drivers, non–farm animal caretakers, laborers, and freight, stock, and materials movers.

The relationship between lifting mechanics and injury risk is complex because of the many mechanical, physiological, and psychophysical factors involved. The U.S. National Institute for Occupational Safety and Health (NIOSH), recognizing the prevalence and costliness of lifting injuries, issued lifting guidelines in 1981, followed by revised guidelines in 1993. These guidelines provide an equation (Waters, Putz-Anderson, Garg, & Fine, 1993) for calculating lifting limits based on several factors, including object weight, distance of the object from the body, object height, movement distance, angular displacement from the midsagittal plane,

Research in Mechanics

Biomechanics of Lifting

Research on the biomechanics of lifting emerged during the late 1960s and early 1970s. One of the pioneering works, "A Biomechanical Model for Analysis of Symmetric Sagittal Plane Lifting," was published by Chaffin and Baker in 1970. Since then, biomechanical models have become more sophisticated by considering, for example, EMG, optimization, and complex modelling to predict trunk muscle forces and lumbar spine loading (Gagnon, Larivière, & Loisel, 2001).

Recent studies have focused on differences in lifting mechanics between groups in moving boxes from ground level. Plamondon and colleagues, for example, published a series of studies comparing lifting characteristics in experts versus novices and women versus men (Plamondon, Delisle, et al., 2014; Plamondon et al., 2017; Plamondon, Larivière, et al., 2014). These researchers reported that expert lifters exhibited less spinal flexion and more knee bending during lifting, but they noted small effect on external back-loading variables (e.g., peak resultant moment and peak asymmetrical moment at L5/S1). In comparing gender effects, Plamondon and colleagues (2017) found that lifting techniques used by female and male subjects were similar with respect to task duration and cumulative loading, but different in terms of interjoint coordination patterns. They concluded, "Considering the female coordination pattern likely stretched posterior passive tissues when lifting boxes from the ground, potentially leading to higher risk of injury, the reason for this sex effect must be identified so that preventive interventions can be proposed" (p. 93).

See the references for the full citations:

Chaffin & Baker, 1970.

Gagnon, Larivière, & Loisel, 2001.

Plamondon, Delisle, Bellefeuille, Denis, Gagnon, & Larivière, 2014.

Plamondon, Larivière, Denis, Mecheri, & Nastasia, 2017.

Plamondon, Larivière, Denis, St-Vincent, & Delisle, 2014.

lifting frequency, lifting duration, and energy expenditure. The NIOSH lifting equation calculator is available online; an app is available through iTunes and Google Play.

Although the NIOSH lifting equation provides one useful component in evaluating lifting dynamics, it applies only to certain two-hand lifting tasks. The equation does not apply to any tasks that involve lifting with one hand or unstable objects, last more than 8 hours, and use high-speed motions, or that are performed while seated or kneeling, while working in a restricted work space, or on slippery floors. More complex models and equations are required for assessing these special case situations.

Based on extensive research, McGill (2015) presents detailed guidelines for injury prevention, many of which address lifting-related issues:

- Design work tasks that facilitate variety (i.e., don't do too much of any single thing).
- Avoid a fully flexed or bent spine and rotated trunk when lifting.
- Select a posture to minimize the reaction torque on the low back by keeping the external load near the body.
- Minimize the weight being lifted.
- Do not immediately perform strenuous exertions after periods of prolonged flexion.
- Avoid lifting or spine bending after rising from bed.
- Prestress and stabilize the spine during light lifting tasks.
- Avoid twisting while generating high twisting torques.

- Use momentum when exerting force to lower spinal loads.
- Avoid prolonged sitting.
- Adopt appropriate rest strategies.
- Provide protective clothing to foster joint-conserving postures.
- Practice joint-conserving kinematic movement patterns.
- Maintain a reasonable level of physical fitness.

Overuse Conditions

Overuse injuries exemplify a broad class of conditions caused by repeated application of force with insufficient time for recovery. Such conditions are referred to by a variety of names, including cumulative trauma disorders, repetitive stress syndromes, chronic microtrauma, overuse syndromes, repetitive motion disorders, and repetitive strain injuries. Overuse conditions related to workplace events are an important focus of ergonomic analysis.

Several overuse conditions are described briefly here, with emphasis on the movement-related factors responsible for each condition. Details of prevention and treatment of overuse injuries, while important, are beyond the scope of our discussion.

- **Tendinitis** results from excessive friction between a tendon and its sheath and in an inflammatory response. The reaction may be acute (in response to a limited session or event) but more likely is chronic (i.e., a result of repeated overuse). In addition to the tendon itself, related structures that facilitate tendon sliding (e.g., peritenon, tendon sheath, associated bursa) may also become inflamed and subsequently injured.
- **Osteoarthritis** (OA), also called **degenerative joint disease** (DJD), is a noninflammatory condition that affects synovial joints, especially those involved in weight bearing (e.g., hip, knee, ankle). OA is characterized by deterioration of the articular cartilage on the joint surfaces and bone outgrowth on joint surfaces and margins. OA of the hip, in particular, is a major cause of disability, especially in older people. Predictably, one of the primary risk factors for OA is activity-related joint loading. Repetitive high loads, as seen in heavy physical activity, certain sports, and occupations involving heavy lifting, may predispose an individual to OA. Other risk factors include gender, genetic predisposition, developmental deformities, nutrition, obesity, smoking, and traumatic injury.
- **Carpal tunnel syndrome** is a common overuse condition involving the hand and wrist caused by repetitive movements, primarily flexion and extension of the fingers and wrist (see sidebar).
- **Glenohumeral (GH) impingement** involves increased pressure, within the confined space of the glenohumeral joint, that deleteriously affects the joint's structures (see chapter 12).
- **Epicondylitis** results from overuse of structures on either the medial or lateral aspect of the elbow. Elbow epicondylitis is somewhat of a misnomer, since the term's suffix (*-itis*) implies an inflammatory condition, yet little evidence has been found of inflammatory markers at the involved epicondyles. More correctly, epicondylitis is a degenerative, rather than an inflammatory, process. Nonetheless, medial and lateral epicondylitis are overuse conditions involving intracytoplasmic calcification, collagen fiber splitting and kinking, and abnormal fiber cross-links (Kannus & Jozsa, 1991).

Lateral epicondylitis (sometimes referred to as *tennis elbow*) has been associated with faulty stroke mechanics (especially the backhand), off-center ball contact, grip tightness, and racket vibration in tennis players, but it also occurs in other striking sports (e.g., racquetball, golf) and in occupations involving repetitive motions of the wrist and elbow (e.g., carpentry, surgery). Medial epicondylitis, while much less common, afflicts tennis players in the forehand and serve strokes, throwers whose movement patterns involve high-velocity valgus extension at the elbow, and occupations requiring repetitive motions.

- **Plantar fasciitis** is an inflammatory condition of the plantar fascia in the midfoot or at its insertion on the medial tuberosity of the calcaneus that involves microtears or partial rupture of fascial fibers. In most cases, plantar fasciitis develops in response to repeated loading (e.g., running) in which compressive forces flatten the longitudinal arch of the foot. Forces in the plantar fascia during running have been estimated to be 1.3 to 2.9 times body weight (Scott & Winter, 1990). The flattening of the arches stretches the plantar fascia and absorbs the compressive load. Plantar fasciitis may result from repetitive compression of the foot arches in workers carrying heavy loads.

Applying the Concept

Carpal Tunnel Syndrome

One of the most debilitating overuse disorders is carpal tunnel syndrome (CTS), a condition first reported by Paget in 1854 (Lo et al., 2002). It is characterized by swelling within the carpal tunnel that creates a compressive neuropathy affecting the median nerve (figure 13.2). Like other entrapment syndromes, CTS involves increased pressure within a confined space. The inextensible borders formed by the carpal bones and the flexor retinaculum (also transverse carpal ligament) preclude an increase in tunnel size. Inflammation and edema in response to repeated loading compress neurovascular tissues and compromise their function. Of greatest consequence is compression of the median nerve, which results in sensory symptoms of numbness, tingling, burning, and pain in the wrist and the radial 3 1/2 fingers.

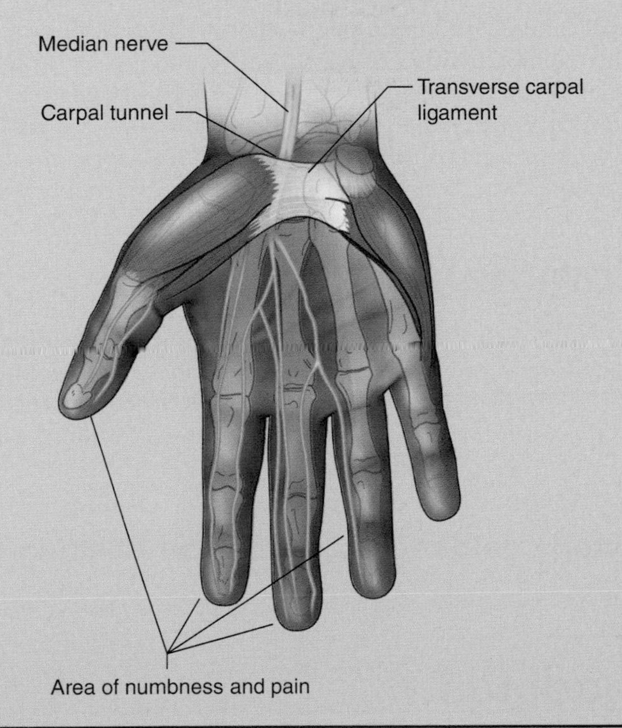

FIGURE 13.2 Carpal tunnel.

Symptoms of CTS are associated with specific movement patterns (e.g., performing assembly work, typing, playing a musical instrument, polishing, sanding, scrubbing, and hammering). Carpal tunnel syndrome has been documented in workers across a diverse range of jobs, including keyboard operators, sheet metal workers, supermarket checkers, sheep shearers, fish-processing workers, and sign language interpreters.

Plantar fasciitis is hastened or worsened by lack of ankle strength and flexibility. Tightness of the calcaneal tendon, for example, limits ankle dorsiflexion and results in greater plantar fascial stress. In addition to strength and flexibility, other factors associated with plantar fasciitis include overtraining, leg-length discrepancies, fatigue, fascial inextensibility, and poor movement mechanics. Excessive pronation during running provides an instructive example of how a pathological movement pattern contributes to plantar fasciitis. During pronation of the foot and ankle, the subtalar joint everts, causing plantar fascial elongation and increased tissue stress. Repetition of this pathological loading leads to microdamage and attendant inflammation.

- **Calcaneal tendinitis** involves the calcaneal (Achilles) tendon, the largest and strongest tendon in the body, which is formed by merging of the distal tendons of the triceps surae group (i.e., gastrocnemius, soleus) about 6 cm proximal to its insertion on the posterior surface of the calcaneus. Frequent and repeated loading of the calcaneal tendon, in activities such as running and jumping, predisposes it to overuse pathologies, most commonly peritenonitis (inflammation of the peritenon), insertional disturbances (e.g., bursitis or insertion tendinitis), myotendinous junction injury, or tendonopathies.

The calcaneal tendon experiences high forces. As evidence, Fukashiro and colleagues (1995), using an implanted tendon force transducer, reported peak calcaneal force of 2,233 N (502 lb; 228 kg) in a squat jump, 1,895 N (426 lb; 193 kg) in a countermovement jump, and 3,786 N (851 lb; 386 kg) in hopping. Lifting tasks involving heavy weights also can generate high calcaneal tendon forces. Despite the high magnitude loads across a spectrum of movement tasks, Komi and colleagues (1992) suggested that the loading rate may be more clinically relevant than the loading magnitude.

- Cervical myofascial pain (CMP) arises from irritation to neck muscles (e.g., trapezius, rhomboids, supraspinatus, infraspinatus, levator scapulae) and their surrounding fascia. CMP can be caused by injuries or overuse and is characterized by trigger points located in the muscle. Trigger points are hypersensitive areas in a palpable tight band of muscle fibers. A related condition, tension headache, is caused by muscle contractions in the head and neck regions, caused by a variety of stressors, including prolonged staring at a computer screen.

Concluding Comments

Ergonomics, an interdiscipline encompassing human anatomy, physiology, biomechanics, engineering, and medicine, has proven potential to enhance the health, well-being, and productivity of individuals and lessen the financial and human costs to businesses and governmental agencies. In the future, ergonomics will play an essential role in the success of society's institutions.

Go to the web study guide to access critical thinking questions for this chapter.

Suggested Readings

Guastello, S.J. (2013). *Human factors engineering and ergonomics: A systems approach* (2nd ed.). Boca Raton, FL: CRC Press.

Haight, J.M. (Ed.). (2013). *Ergonomics and human factors engineering*. Park Ridge, IL: American Society of Safety Engineers.

Hedge, A. (Ed.). (2016). *Ergonomics workplace design for health, wellness, and productivity.* Boca Raton, FL: CRC Press.

Kroemer, K.H.E., & Kroemer, H.B. (2000). Ergonomics: How to design for ease and efficiency (2nd ed.). London: Pearson.

Kumar, S. (Ed.). (2007). *Biomechanics in ergonomics* (2nd ed.). Boca Raton, FL: CRC Press.

McCauley-Bush, P. (2013). *Ergonomics: Foundational principles, applications, and technologies.* Boca Raton, FL: CRC Press.

McGill, S. (2015). *Low back disorders: Evidence-based prevention and rehabilitation* (3rd ed.). Champaign, IL: Human Kinetics.

Reilly, T. (2009). *Ergonomics in sport and physical activity: Enhancing performance and improving safety.* Champaign, IL: Human Kinetics.

Sanders, M.S., & McCormick, E.J. (1993). *Human factors in engineering and design* (7th ed.). New York: McGraw-Hill.

Stack, T., Ostrom, L.T., & Wilhelmsen, C.A. (2016). *Occupational ergonomics: A practical approach.* Hoboken, NJ: Wiley.

Stave, G.M., & Wald, P.H. (Eds.). (2017). *Physical and biological hazards of the workplace* (3rd ed.). Hoboken, NJ: Wiley.

Tillman, B., Tillman, P., Rose, R.R., & Woodson, W.E. (2016). *Human factors and ergonomics design handbook* (3rd ed.). New York: McGraw Hill.

Wickens, C.D., Lee, J., Liu, Y.D., & Gordon-Becker, S. (2004). *An introduction to human factors engineering* (2nd ed.). Upper Saddle River, NJ: Prentice Hall.

Glossary

abduction—Joint motion in the frontal plane (relative to anatomical position) that takes a segment away from the body's midline.

acetylcholine—Chemical neurotransmitter used by neuron terminal endings. The neurotransmitter used by all lower motor neurons that innervate skeletal muscles.

actin—Contractile protein that forms the backbone of the thin filaments within a sarcomere.

action—Internal state in which a muscle actively exerts a force, regardless of whether it shortens or lengthens. Also *contraction*.

action potential—Electrical signal that passes along the membrane of a neuron or muscle fiber. May also be called a *nerve impulse* with respect to a neuron or a *muscle action potential* as it relates to a muscle fiber.

active support—Contribution of muscle action to joint stability.

adaptive postural control—Strategy to maintain balance by modifying movement in response to situational changes.

adduction—Joint motion in the frontal plane (relative to anatomical position) that moves a segment toward the body's midline.

adenosine triphosphate—High-energy molecule used to supply energy for muscle contraction and other bodily functions.

adipocytes—Fat cells.

aerodynamic force—Force related to the motion of air and objects moving through the air.

agonist—Muscle actively involved in producing or controlling a movement.

amenorrhea—Absence of menstrual cycles.

amphiarthrosis—Functional classification of a joint with limited movement.

anatomical position—Erect posture with head facing forward and arms hanging straight down with palms facing forward.

anatomy—Study of the structure of organisms.

angular displacement—Angular measure from the starting position to the finishing position of an angular movement.

angular kinetic energy—(1/2) × (moment of inertia) × (angular velocity)^2.

angular momentum—Product of a body's mass moment of inertia and its angular velocity.

angular motion—Motion in which a body rotates about an axis. Also *rotational motion*.

annulus fibrosus—Layered fibrocartilage network surrounding the nucleus pulposus in an intervertebral disc.

antagonist—Muscle acting in opposition to a movement.

anticipatory postural control—Strategy to maintain balance using anticipatory actions.

aponeuroses—Sheets of tendonlike material that cover a muscle's surface or connect a muscle to another muscle, or muscle to bone.

appendicular skeleton—Subsystem of the skeletal system containing bones of the pelvic and pectoral girdles and the limbs.

area moment of inertia—Measure of a body's resistance to bending.

arthritis—Joint inflammation.

arthroplasty—Joint replacement through surgical procedures.

articular cartilage—Smooth, shiny layer of hyaline cartilage covering the joint surfaces of articulating bones.

atrophy—To decrease in size.

axial skeleton—Subsystem of the skeletal system containing the skull, spinal column, and thoracic cage.

axis of rotation—Line (imaginary) about which joint rotation occurs.

balance—Maintenance of postural stability or equilibrium. Also *postural control*.

ballistics—Study of projectile motion.

base of support—Area within the outline of all contact points with the ground.

biarticular—Having action at two joints.

biomechanics—Application of mechanical principles to the study of biological organisms and systems.

body—Any collection of matter.

buoyant force—Equal and opposite force exerted by a liquid against the weight of a body, allowing the body to float.

bursa—Fluid-filled sac that helps cushion or reduce friction.

cadence—Gait tempo measured in steps per minute.

calcaneal tendinitis—Calcaneal pathology caused by frequent and repeated loading of the calcaneal (Achilles) tendon.

cancellous bone—Bone with high porosity (low density). Also *spongy* or *trabecular bone*.

cardiac muscle—Muscle tissue in the heart that is responsible for generating the forces that pump blood.

carpal tunnel syndrome—Overuse condition involving the hand and wrist, caused by repetitive movements, primarily flexion and extension of the fingers and wrist.

cartilage—Stiff connective tissue whose ground substance is nearly solid (of three types: hyaline, fibrocartilage, elastic).

cartilaginous joint—Structural classification of a joint bound by cartilage.

center of gravity—Point at which the weight of a body may be considered a concentrated mass without altering the motion of the body.

center of mass—Point about which the mass of a body is equally distributed.

cerebral palsy—Nonprogressive condition of muscle dysfunction and paralysis caused by brain injury at or near the time of birth.

chronic traumatic encephalopathy (CTE)—Neurological brain damage caused by repetitive head trauma.

circumduction—Special form of angular motion in which the distal end of a limb or segment moves in a circular pattern about a relatively fixed proximal end, tracing out a cone-shaped pattern.

coactivation—Simultaneous action of agonists and antagonists at a given joint.

co-contraction—See *coactivation*.

compact bone—Bone with high density (low porosity). Also *cortical bone*.

compliance—Measure of the relationship between strain and stress. Inverse of stiffness.

compression—Action tending to push together.

concentric—Shortening muscle contraction. The torque produced by the muscle is greater than the external torque; therefore, the muscle is able to shorten while overcoming the external load.

connective tissue—Class of tissue that provides support and protection and binds tissues together (e.g., bone, tendons, ligaments, cartilage, adipose, blood).

conservation (momentum, energy)—No net gain or loss (of momentum or energy).

contractile component—Structure within a muscle that can produce force. For example, the fundamental contractile component of a skeletal muscle is the sarcomere.

contractility—A muscle's ability to generate a pulling, or tension, force.

contraction—Internal state in which a muscle actively exerts a force, regardless of whether it shortens or lengthens. Also *action*.

coordination—Muscles working together with correct timing and intensity to produce or control a movement.

cortex—Hard outer covering of a bone. Also *cortical shell*.

cortical bone—Bone with high density (low porosity). Also *compact bone*.

cortical shell—Protective outer bony surface. Also *cortex*.

countertorque—A torque acting in the opposite direction to an applied torque.

crawling—Form of infant locomotion with progression on hands and stomach.

creeping—Form of infant locomotion with movement on hands and knees.

curvilinear motion—Motion along a curved line.

deformational energy—Energy stored when a body is deformed. Also *strain energy*.

degenerative joint disease—See *osteoarthritis*.

delayed onset muscle soreness (DOMS)—The muscle soreness that may occur 48 to 72 hours after an exercise session.

demi plié—Dance movement involving half bending of the knees.

density (bone)—Measure of the hard tissue in bone (hydroxyapatite crystal per unit volume); inverse of porosity.

depression—Movement of a structure in an inferior, or downward, direction.

diaphysis—Shaft of a long bone.

diarthrosis—Functional classification of a freely movable joint.

digitization—Process used to quantitatively identify the location (coordinates) of anatomical landmarks using computerized analysis systems.

distance—Scalar measure of how far a body has moved.

dorsiflexion—Joint motion at the ankle where the foot moves toward the lower leg.

double support—Period when the body's weight is supported by both legs.

downward rotation—Scapular rotation in which the inferior border of the scapula moves downward and inward.

drag force—Component of force that acts parallel to the direction of an object's motion in a fluid.

dynamic equilibrium—State in which the accelerations are balanced according to Newton's second law of motion (i.e., force = mass × acceleration).

dynamic posture—Posture of motion (as seen in walking and running).

eccentric—Lengthening muscle contraction. The torque produced by the muscle is less than the external torque, but the torque produced by the muscle causes the joint movement to occur more slowly than the external torque would tend to make the limb move.

economy—Amount of metabolic energy required to perform a given amount of work.

edema—An abnormal accumulation of fluid within a structure, organ, or tissue.

efficiency—Amount of mechanical output produced for a given amount of metabolic input (i.e., how much work can be done using a given amount of energy).

elastic cartilage—Flexible cartilage found in areas where extensibility is needed (e.g., external ear and respiratory system).

elasticity—A tissue's ability to return to its original length and shape after an applied force is removed.

electromyography—Study of the electrical activity of muscles.

elevation—Movement of a structure in a superior, or upward, direction.

endomysium—Fibrous connective tissue sheath that surrounds each muscle fiber located within a fascicle in a skeletal muscle.

energy (mechanical)—Ability or capacity to do mechanical work.

epicondylitis—Tissue damage from overuse of structures on either the medial or lateral aspect of the elbow.

epimysium—Fibrous connective tissue sheath that surrounds a whole skeletal muscle.

epiphyseal growth plate—Region of hyaline cartilage found between the diaphysis and epiphysis in developing bone.

epiphysis—An end of a long bone (plural: epiphyses).

equilibrium—State of balance between opposing forces or actions.

equinus—Foot position in which the forefoot is lower than the heel. In gait, at initial contact, the foot is plantar flexed.

ergonomics—Study of how humans interact with their immediate environment.

eversion—Joint motion at the intertarsal joints that results in the sole of the foot being moved away from the body's midline.

excitability—Describes the ability of muscle to respond to a stimulus. Also *irritability*.

excitation–contraction coupling—Sequence of events involved in producing a muscle contraction, from exocytosis of the lower motor neuron through the interaction between actin and myosin filaments.

exocytosis—Process of fusion of a vesicle to the presynaptic membrane and the subsequent release of its neurotransmitter.

extensibility—Describes the ability of muscle to lengthen (or stretch) and, as a consequence, to generate force over a range of lengths.

extension—Joint motion in the sagittal plane (relative to anatomical position) in which the angle between articulating segments increases.

external rotation—Joint motion in the transverse plane (relative to anatomical position) that rotates a segment's anterior surface away from the body's midline. Also *lateral rotation*.

extracellular matrix—Noncellular component of a tissue.

fascicle—Bundle or collection of fibers, usually of muscle or nerve fibers.

fibrocartilage—Strong and flexible cartilage that reinforces stress points and serves as filler material in and around joints.

fibrous joint—Structural classification of a joint bound by connective tissues composed primarily of collagen fibers.

fibrous joint capsule—Fibrous encasement surrounding a synovial joint.

fine motor skills—Skills involving intricate movements of small joints.

first-class lever—Lever system with the axis between the resistance force and the effort force.

flexion—Joint motion in the sagittal plane (relative to anatomical position) in which the angle between articulating segments decreases.

flight phase—Period of the running gait cycle when both feet are off the ground (i.e., period of nonsupport).

float phase—See *flight phase*.

fluid mechanics—Branch of mechanics dealing with the properties and behavior of gases and liquids.

force—Mechanical action or effect applied to a body that tends to produce acceleration. A push or pull.

force–velocity relationship—Property of skeletal muscle that shows its force production capability is dependent on its contraction velocity.

form drag—Resistive force produced when an object parts the medium through which it is passing.

friction—Resistance developed at the interface of two surfaces, acting opposite the direction of motion or impending motion.

frontal plane—Plane dividing the body into anterior and posterior sections.

gait—Form of locomotion, usually referring to walking or running.

gait cycle—Sequential occurrence of a stance and swing phase for a single limb.

general motion—Combined linear and angular motion.

glenohumeral impingement—Pinching of tissues in the glenohumeral joint.

glycolytic—The ability to metabolize glucose for energy.

gouty arthritis (gout)—Joint inflammation caused by uric acid crystals embedded in joint structures, leading to irritation and inflammation.

grand jeté—Ballet movement in which the dancer jumps high in the air, extends one leg, and lands on that leg.

grand plié—Dance movement involving full bending of the knees.

gravitational potential energy—Energy possessed by a body as a result of its position.

gross motor skills—Skills involving movement and control of the limbs.

ground reaction force—Equal and opposite force exerted by the ground against an object contacting it.

ground substance—Nonfibrous component of the extracellular matrix.

growth plate—See *epiphyseal growth plate*.

hamstrings (muscle group)—Group of three muscles on the posterior thigh (semitendinosus, semimembranosus, biceps femoris).

hematopoiesis—Process of blood cell production.

histology—Study of tissue structure.

human–machine interface—Interaction between a human and a machine or device.

hyaline cartilage—Cartilage found on joint surfaces, on anterior surfaces of the ribs, and in areas of the respiratory system. Also serves as the precursor to bone in the developing fetus.

hyperextension—Joint motion in the sagittal plane (relative to anatomical position) in which the angle between articulating segments increases beyond anatomical position.

hypertrophy—To increase in size.

idealized force vector—Single vector used to represent many vectors.

impingement syndrome—Pathological condition in which pressure increases within a confined anatomical space and the enclosed tissues are detrimentally affected.

inertia—Resistance to a change in a body's state of linear motion.

initial contact—First foot contact with the ground to end the swing phase and begin the stance phase.

injury—Damage sustained by body tissues in response to physical trauma.

innervation—Connection between a nerve fiber (axon including the terminal endings) and another structure such as a muscle fiber.

insertion—Site of musculotendinous attachment on the more movable end of a bone.

intercellular substance—Noncellular component of a tissue.

internal impingement—Form of glenohumeral impingement in which the supraspinatus tendon contacts the posterosuperior rim of the glenoid fossa.

internal rotation—Joint motion in the transverse plane (relative to anatomical position) that rotates a segment's anterior surface inward toward the body's midline. Also *medial rotation*.

interosseous membrane—Collagenous tissue binding two bones together (e.g., tibiofibular joint).

intervertebral disc—Structure between two adjacent vertebrae composed of an inner gelatinous mass (nucleus pulposus) and a surrounding layered fibrocartilage network (annulus fibrosus).

inversion—Joint motion at the intertarsal joints that results in the sole of the foot being moved inward toward the body's midline.

irritability—Describes the ability of muscle to respond to a stimulus. Also *excitability*.

isoinertial—Represents constant resistance. A more accurate term than *isotonic* for human muscle contractions.

isokinetic—Describes a contraction performed with a constant angular velocity.

isometric—Refers to constant length of the musculotendinous unit and hence no limb movement. Torque produced by the muscle is equal and opposite to the external torque.

isotonic—Literally means constant tension. This condition does not occur in intact human subjects (i.e., *in vivo*) because the level of muscle force varies continuously and rarely, if ever, is constant throughout a movement.

kinematics—Description of motion with respect to space and time, without regard to the forces involved.

kinesiology—Study of the art and science of human movement.

kinetic energy—Energy possessed by a body by virtue of its motion.

kinetic friction—Frictional resistance created while an object is moving (i.e., sliding, rolling) along a surface.

kinetics—Assessment of motion with regard to forces and force-related measures.

knee extensor mechanism—Functional unit, made up of the quadriceps muscles, patella, and ligament attachment to the tibia, that works to extend the knee.

kyphosis—Sagittal plane spinal deformity characterized by excessive flexion, usually in the thoracic region.

labrum—U-shaped ring of fibrocartilage around the rim of a joint.

lacuna—Small pocket or space.

laminar flow—Flow characterized by a smooth, parallel pattern of fluid motion.

lateral flexion—Sideways bending of the vertebral column in the frontal plane.

lateral rotation—See *external rotation*.

length-tension relationship—Property of skeletal muscle that shows its force production capability is dependent on the length of the muscle's contractile and noncontractile structures.

lever—Rigid structure, fixed at a single point, to which two forces are applied at two different points.

lever arm—See *moment arm*.

lift force—Component of force that acts perpendicular to the direction of an object's motion in a fluid.

ligament—Connective tissue that connects bone to bone.

linear displacement—Straight-line vector measure from the starting point to the finishing point of a movement.

linear kinetic energy—$(1/2) \times (\text{mass}) \times (\text{linear velocity})^2$.

linear momentum—Product of a body's mass and its velocity.

linear motion—Motion along a straight or curved line.

line of force action—Line along which a force acts, extending infinitely in both directions along the line of a finite force vector.

load—Force applied externally to a body.

locomotion—Process of moving from one place to another.

lordosis—Sagittal plane spinal deformity characterized by excessive extension, usually in the lumbar region.

lower motor neuron—Nerve cell that originates in the central nervous system and innervates a skeletal muscle.

magnus force—Force created by the spin of an object that creates a deviation in its normal trajectory.

marrow—Loose connective tissue found in the cavities and spaces in bone.

mass—Amount of matter, or substance, constituting a body.

mass moment of inertia—Measure of a body's resistance to rotation about an axis.

material mechanics—Branch of mechanics dealing with the internal response of materials (tissues) to external loads.

maximum voluntary contraction (MVC)—Maximum amount of force a muscle can generate in a voluntary action.

mechanism—Fundamental physical process responsible for a given action, reaction, or result.

medial epicondylitis—Inflammation at the medial epicondyle of the humerus, often due to excessive throwing.

medial rotation—See *internal rotation*.

median plane—Sagittal plane that divides the body in half. Also *midsagittal plane*.

meniscus—Fibrocartilage pad interposed between bones to provide shock absorption and improve bony fit (plural: menisci).

mesenchymal cells—Unspecialized cells that later differentiate into specific cell types.

midsagittal plane—Sagittal plane that divides the body in half. Also *median plane*.

mobility (joint)—Ability of a joint to move through a range of motion.

mobility (movement)—Ability to move readily.

modeling—Formation of new bone.

moment—Effect of a force that tends to cause rotation about an axis, or bending of a deformable structure.

moment arm—Perpendicular distance from the axis of rotation to the line of force action. Also *torque arm* and *lever arm*.

moment of inertia—Resistance to a change in a body's state of angular motion.

momentum—Property of a moving body that is determined by the product of the body's mass and its velocity.

motion segment—Two adjacent vertebrae and the intervening intervertebral disc.

motor—(1) To produce or cause movement, (2) involving muscular movement, (3) related to nerve conduction from neurons to muscles.

motor behavior—Study of the behavioral aspects of movements, including development, learning, and control.

motor control—Study of the neural, physical, and behavioral aspects of movement.

motor development—Study of changes in movement behavior throughout the life span.

motor learning—Study of how motor skills are learned.

motor milestones—Landmark events in the progression of learning muscular control of movement.

motor skill—Voluntary movement used to complete a desired task action or achieve a specific goal.

motor unit—A single lower motor neuron plus all the muscle fibers it innervates.

multiarticular—Having action at more than three joints.

muscle control formula—Step-by-step procedure for determining the involved muscles and their action for any joint movement.

muscle (motor) abundancy—Reframing of the muscle (motor) redundancy problem to emphasize the potential evolutionary advantages of abundant number of available options (i.e., degrees to freedom).

muscle (motor) redundancy—Situation in which more muscles are available to perform an action than are minimally necessary.

muscle synergy—Cooperative action of several muscles working together as a single unit.

musculoskeletal disorders—Class of injuries and conditions to the body's anatomical structures, usually caused by acute trauma or overuse.

musculotendinous junction—Region where a muscle and tendon connect. Also *myotendinous junction*.

musculotendinous unit—Combined unit including a skeletal muscle and its tendons.

myomesin—Structural protein that forms the M line within a sarcomere and helps maintain the alignment of the myosin filaments.

myosin—Contractile protein that forms the thick filaments within a sarcomere.

myotendinous junction—Region where a muscle and tendon connect. Also *musculotendinous junction*.

net moment—Sum of all the moments (torques) acting on a body. Also *net torque*.

net torque—See *net moment*.

neutralization—Canceling, or neutralizing, action of two or more muscles.

neutral position—Reference position for joints in anatomical position.

nilotic stance—Posture in which a person stands on one leg with the opposite leg used to brace the standing knee (i.e., like a flamingo).

nomenclature—A generally agreed upon system or set of terms.

noncontractile component—Structure within a muscle that, by itself, cannot produce force (e.g., the connective tissue sheaths within a skeletal muscle).

normal force—Component of force acting on an object, perpendicular to the surface.

nucleus pulposus—Gelatinous mass on the inside of an intervertebral disc.

oligomenorrhea—Irregular menstrual cycles.

opposition—Ability of the thumb to work with the other four fingers to perform grasping movements.

organ—Structure composed of two or more tissues that have a definite form and function.

organ system—Group of organs that work together to perform specific functions.

origin—Site of musculotendinous attachment on the less movable end of a bone.

orthotic—A device used to support or protect anatomical structures.

osteoarthritis—Degradation of articular cartilage caused by mechanical action. The most common type of arthritis.

osteoblast—Mononuclear bone cells that produce new bone material.

osteoclast—Large, multinucleated cells that break down, or resorb, bone.

osteocyte—Mature bone cells that are smaller and less active than osteoblasts.

osteoid—Organic portion of the extracellular matrix in bone.

osteon—Fundamental structural unit of compact bone.

osteopenia—Mild to moderate bone loss.

osteoporosis—Severe bone loss that increases the risk of fracture.

osteoprogenitor cells—Undifferentiated mesenchymal cells with the ability to produce daughter cells that can differentiate to become osteoblasts.

osteotendinous junction—Region where a bone and tendon connect.

overload—The need for a person to exercise at increasing levels of difficulty in order to maintain progress.

oxidative—The ability to utilize oxygen for aerobic metabolism.

paddling—Stroke technique where the swimmer uses a straight pull with the hand oriented 90° to the surface of the water.

palpation—To examine by touch.

parasagittal plane—Any sagittal plane offset from the midline to one side or the other.

passive support—Contribution of noncontractile tissues to joint stability.

patellar tracking—Sliding movement of the patella in the intercondylar groove of the femur.

peak bone mass—Highest amount of bone mass during one's lifetime.

pectoral girdle—Bony complex made up of the clavicle and scapula.

pedaling cycle—One revolution of the pedal starting from and returning to the top dead center position of the crank arm.

pelvic girdle—Bony complex made up of the ilium, ischium, and pubis.

pennation—Angle formed between the muscle's line of pull and the orientation of the muscle fibers.

perception–action coupling—Response of an infant to environmental cues.

perimysium—Fibrous connective tissue sheath that surrounds each fascicle within a skeletal muscle.

periosteum—Fibrous connective tissue covering bone except at the joint surfaces.

perturbation—Disturbance of motion.

physiology—Study of the function of body parts and systems.

pirouette—Complete turn of the body on one foot.

plantar fasciitis—Inflammatory condition of the plantar fascia in the midfoot, or at its insertion on the medial tuberosity of the calcaneus, which involves microtears or partial rupture of fascial fibers.

plantar flexion—Joint motion at the ankle where the foot moves away from the lower leg.

plasticity—Ability to adapt to environmental stimuli and events.

plié—Literally translated, it means bent or bending. A bending of the knee or knees.

plyometrics—Form of training that consists of stretch–shortening cycles. In other words, cycles of eccentric and concentric contractions, such as box jumps.

point mass—Concentrated mass at a single point.

polar moment of inertia—Measure of a body's resistance to twisting, or torsion.

porosity (bone)—Measure of the soft tissue in bone; inverse of density.

postural control—Maintenance of postural stability, or equilibrium. Also *balance*.

postural sway—Small movement fluctuations used to maintain posture.

posture—Alignment or position of the body and its parts.

potential energy—Energy created by a body's position or deformation.

power—Rate of work production.

primary curvature (spine)—Spinal curvature first evident in the fetus and maintained after birth by the thoracic and sacral regions of the spine.

prime mover—Muscle most responsible for producing or controlling a given joint movement.

projectile—Object moving through space subject only to the effects of gravity and air resistance.

projection (center of gravity)—Line drawn from the center of gravity vertically downward to the base of support.

pronation (foot and ankle)—Combination of ankle dorsiflexion, subtalar eversion, and external rotation of the foot.

pronation (forearm)—Rotation of the radius over the ulna (from anatomical position).

prone—Lying posture with face pointed down.

prosthesis—A device used to replace a missing body part.

protraction—Movement of a structure anteriorly, or toward the front of the body.

quadriceps (muscle group)—Group of four muscles on the anterior thigh (vastus medialis, vastus lateralis, vastus intermedius, rectus femoris).

quadriceps avoidance—Gait strategy adopted by people with ACL deficiency to avoid activation of muscles in the quadriceps group.

radial deviation—Abduction of the wrist and hand such that the thumb moves closer to the radius.

range of motion—Measure of joint mobility.

reactive postural control—Strategy to maintain balance (e.g., to prevent a fall) using muscle action.

rectilinear motion—Motion along a straight line.

recumbent—Lying posture.

relevé—Rising on the toes by plantar flexing the ankle.

remodeling—Adaptation of existing bone through a process of resorption and replacement.

resistance training—Contracting muscles against resistance to enhance muscle strength, power, endurance, or size.

resorption—Breakdown or demineralization of bone.

retraction—Movement of a structure posteriorly, or toward the back of the body.

rheumatoid arthritis—Autoimmune condition leading to joint inflammation.

rotational motion—Motion in which a body rotates about an axis. Also *angular motion*.

rotator cuff (muscle group)—Group of four muscles surrounding the glenohumeral joint (subscapularis, supraspinatus, infraspinatus, teres minor).

running—Form of upright locomotion characterized by alternating periods of single support and a flight phase.

sagittal plane—Plane dividing the body into left and right sections.

SAID principle—Specific adaptation to imposed demands. Tissues and systems in the human body adapt, either structurally or physiologically, in a way that is specific to the mechanical or metabolic demands placed on them.

sarcolemma—Membrane covering a skeletal muscle cell (fiber).

sarcomere—Contractile unit within skeletal muscle that spans from one Z disc to the next.

sarcoplasmic reticulum—Network of tubes and sacs (known as *terminal cisternae*) that surrounds each myofibril within a muscle fiber. The terminal cisternae store the calcium needed for muscle contraction.

scapulohumeral rhythm—Coordinated action of the humerus and scapula in facilitating glenohumeral abduction.

scoliosis—Lateral (frontal plane) curvature of the spine.

screw-home mechanism—Tibiofemoral rotation during the final few degrees of knee extension.

sculling—Stroke technique where the swimmer uses a curvilinear pull while varying the hand angle (angle of attack).

secondary curvature (spine)—Spinal curvature developed primarily after birth in the cervical and lumbar regions of the spine.

second-class lever—Lever system with the resistance force between the axis and the effort force.

sesamoid bone—Bones embedded in a tendon. Sometimes referred to as "floating" bones (e.g., patella).

shear—Action that tends to produce horizontal sliding of one layer over another.

single support—Period when the body's weight is supported by a single leg.

size principle (Henneman's)—Recruitment of motor units in order from the smallest and most aerobic motor units to the largest and least aerobic motor units.

skeletal muscle—Muscle tissue responsible for maintaining posture and producing movement.

sliding filament model—Model used to describe how a sarcomere produces force. The final steps associated with excitation–contraction coupling that describe the interaction between the actin and myosin filaments needed to produce force.

smooth muscle—Muscle tissue that facilitates substance movement through tracts in the circulatory, respiratory, digestive, urinary, and reproductive systems.

spastic diplegia—Pathological gait characterized by abnormally flexed, adducted, and internally rotated hips; hyperflexed knees; and equinus of the foot and ankle.

spongy bone—Bone with high porosity (low density). Also *trabecular* or *cancellous bone.*

sprain—Injury to a ligament.

stability (joint)—Ability of a joint to resist dislocation.

stability (movement)—Ability to resist movement.

stabilization—Muscle action to maintain or stabilize a position.

stance phase—Period during which the foot is in contact with the ground during gait.

static equilibrium—State of balance in which there is no net acceleration. Usually refers to stationary, or nonmoving, bodies.

static friction—Frictional resistance created while an object is not moving.

static postural control—Strategy to maintain a static posture, with the body's center of gravity projection kept within the base of support.

static posture—Postures involving little or no movement.

steady-state posture—Postures involving slight movement or swaying.

stem cells—See *mesenchymal cells.*

step—Period from initial contact of one leg to initial contact of the opposite leg.

step time—Duration of a single step; inverse of cadence.

stiffness—Measure of the relationship between stress and strain (i.e., how much a body deforms in response to a given load).

strain (injury)—Musculotendinous injury typically produced when too much force is transmitted through the musculotendinous unit.

strain (mechanical)—Deformation or change in length and shape of a tissue.

strain energy—Energy stored when a body is deformed. Also *deformational energy.*

strain-rate dependent—Property of a tissue dictating that its mechanical response to loading depends on the rate at which the tissue is deformed.

stress—Internal resistance developed in response to an externally applied force (load).

stretch–shortening cycle—Eccentric contraction immediately followed by a concentric contraction.

striated muscle—See *skeletal muscle.*

stride—Period from initial contact of one leg to initial contact of the same leg (one stride equals two steps).

subacromial impingement—Form of glenohumeral impingement in which suprahumeral structures (notably the distal supraspinatus tendon, subacromial bursa, and proximal tendon of the long head of the biceps brachii) are forcibly pressed against the anterior surface of the coracoacromial arch.

subluxation—Partial joint dislocation.

supination (foot and ankle)—Combination of ankle plantar flexion, subtalar inversion, and internal rotation of the foot.

supination (forearm)—Rotation of the radius over the ulna (back to anatomical position).

supine—Lying posture with face pointed up.

surface drag—Resistive force produced directly over the surface of the object as it passes through a fluid medium.

suture joint—Fibrous joint connecting interlocking bones of the skull.

swing phase—Period in which the foot is not in contact with the ground during gait.

symphysis—Joint between two bones separated by a fibrocartilage pad.

synarthrosis—Functional classification of a joint with no movement.

synchondrosis—Joint bound by hyaline cartilage.

syndesmosis—Joint bound by ligaments.

synovial fluid—Viscous fluid found in synovial joints that provides lubrication and reduces friction.

synovial joint—Structural classification of a joint containing a fibrous joint capsule, synovial membrane, synovial cavity, synovial fluid, and articular cartilage.

synovial joint cavity—Space between the bones in a synovial joint.

synovial membrane—Thin membrane, on the inner surface of the fibrous joint capsule of a synovial joint, that produces synovial fluid.

takeoff—Last foot contact with the ground to end the stance phase and begin the swing phase. Also *toe-off.*

tangential force—Component of force acting parallel to a body's surface.

temporal analysis—Mechanical assessment based on time duration.

tendinitis—Inflammation of a tendon, usually due to overuse. Also *tendonitis.*

tendon—Cordlike connective tissue that connects muscle to bone.

tension—Action tending to pull apart.

third-class lever—Lever system with the effort force between the axis and the resistance force.

tissue—Group of cells with similar function and its surrounding noncellular material.

titin—Structural protein that extends from the Z disc to both the myosin filament and the M line.

toe-off—Last foot contact with the ground to end the stance phase and begin the swing phase. Also *takeoff.*

torque—Effect of a force that tends to cause twisting, or torsion, about an axis. Also used as a synonym for *moment* as the effect of a force that tends to cause rotation of a body about an axis of rotation.

torque arm—See *moment arm.*

total mechanical energy—Sum of a body's linear kinetic energy, angular kinetic energy, and positional potential energy.

trabecular bone—Bone with high porosity (low density). Also *spongy* or *cancellous bone.*

trajectory—Path along which a projectile travels.

transfer (momentum, energy)—Exchange of momentum or energy from one body to another.

translational motion—See *linear motion.*

transverse plane—Plane dividing the body into superior and inferior sections.

transverse tubules (T-tubules)—Invaginations of the sarcolemma that pass through the muscle fiber between the myofibrils.

traumatic brain injury (TBI)—General term used to describe numerous brain conditions arising from direct impact or acceleration-deceleration of the head.

Trendelenburg gait—Pathological gait characterized by pelvic drop during early to midstance due to weakness or paralysis of the hip abductors.

triarticular—Having action at three joints.

triceps surae (muscle group)—Group of two muscles on the posterior lower leg (gastrocnemius, soleus).

tropomyosin—Regulatory protein located on the actin filament. In the relaxed skeletal muscle, it covers the myosin-head binding sites on the actin molecules.

troponin—Regulatory protein that binds to both tropomyosin and actin. When combined with calcium, troponin influences tropomyosin to initiate muscle contraction.

turbulent flow—Flow characterized by a chaotic pattern of fluid motion.

ulnar deviation—Adduction of the wrist and hand such that the little (fifth) finger moves closer to the ulna.

uniarticular—Having action at one joint.

upward rotation—Scapular rotation in which the inferior border of the scapula moves upward and outward.

viscoelasticity—Describes a tissue's response to loading that is both strain-rate dependent and elastic.

viscosity—Resistance to flow.

walking—Form of upright locomotion in which at least one foot is always in contact with the ground.

wave drag—Resistive force produced when an object is moving near or along the surface of a fluid medium, usually water, causing waves to form.

weight—A measure of the effect of gravity on a mass.

Wolff's law—The capacity of bone to adapt its structure to imposed loads.

work—Mechanical measure of force multiplied by displacement.

zygapophysis—Articular process between adjacent vertebrae (plural: zygapophyses).

References

Adelsberg, S. (1986). The tennis stroke: An EMG analysis of selected muscles with rackets of increasing grip size. *American Journal of Sports Medicine, 14*(2), 139-142.

Ahamed, N.U., Sundaraj, K., Ahmad, B., Rahman, M., Ali, A., & Islam, A. (2014). Significance of the electromyographic analysis of the upper limb muscles of cricket bowlers: Recommendations from studies of overhead-throwing athletes. *Journal of Mechanics in Medicine and Biology, 14*(4), 1-32.

Alizadehkhaiyat, O., Fisher, A.C., Kemp, G.J., Vishwanathan, K., & Frostick S.P. (2007). Upper limb muscle imbalance in tennis elbow: A functional and electromyographic assessment. *Journal of Orthopaedic Research, 25*(12), 1651-1657.

Alizadehkhaiyat, O., & Frostick, S.P. (2015). Electromyographic assessment of forearm muscle function in tennis players with and without lateral epicondylitis. *Journal of Electromyography and Kinesiology, 25*(6), 876-886.

American College of Sports Medicine. (2017). *ACSM's guidelines for exercise testing and prescription* (10th ed.). Philadelphia: Lippincott Williams & Wilkins.

Balazs, G.C., Pavey, G.J., Brelin, A.M., Pickett, A., Keblish, D.J., & Rue, J.P. (2015). Risk of anterior cruciate ligament injury in athletes on synthetic playing surfaces: A systematic review. *American Journal of Sports Medicine, 43*(7), 1798-1804.

Barrentine, S.W., Fleisig, G.S., Whiteside, J.A., Escamilla, R.F., & Andrews, J.R. (1998). Biomechanics of windmill softball pitching with implications about injury mechanisms at the shoulder and elbow. *Journal of Orthopaedic and Sports Physical Therapy, 28*(6), 405-415.

Bartlett, R. (2000). Principles of throwing. In V.M. Zatsiorsky (Ed.), *Biomechanics in sport: Performance enhancement and injury prevention* (pp. 365-380). Oxford: Blackwell Science.

Bartlett, R. (2014). *Introduction to sports biomechanics: Analyzing human movement patterns* (3rd ed.). London: Routledge.

Bartlett, R.M., Stockill, N.P., Elliott, B.C., & Burnett, A.F. (1996). The biomechanics of fast bowling in men's cricket: A review. *Journal of Sports Sciences, 14*, 403-424.

Basmajian, J.V., & DeLuca, C. (1985). *Muscles alive* (5th ed.). Baltimore: Williams & Wilkins.

Bechler, J.R., Jobe, F.W., Pink, M., Perry, J., & Ruwe, P.A. (1995). Electromyographic analysis of the hip and knee during the golf swing. *Clinical Journal of Sport Medicine, 5*(3), 162-166.

Benjuya, N., Melzer, I., & Kaplanski, J. (2004). Aging-induced shifts from a reliance on sensory input to muscle cocontraction during balanced standing. *Journal of Gerontology, series A, Biological Sciences and Medical Sciences, 59*(2), 166-171.

Bergman, R.A., Thompson, S.A., Afifi, A.K., & Saadeh, F.A. (1988). *Compendium of human anatomic variation.* Baltimore: Urban & Schwarzenberg.

Bernstein, N.A. (1967). *The co-ordination and regulation of movements.* Oxford: Pergamon Press.

Blache, Y., & Monteil, K.M. (2013). Effect of arm swing on effective energy during vertical jumping: Experimental and simulation study. *Scandinavian Journal of Medicine and Science in Sports, 23*(2), e121-129.

Blake, O.M., Champoux, Y., & Wakeling, J.M. (2012). Muscle coordination patterns for efficient cycling. *Medicine & Science in Sports & Exercise, 44*(5), 926-938.

Blazevich, A.J. (2017). *Sports biomechanics: The basics: Optimizing human performance* (3rd ed.). London: Bloomsbury Sport.

Bohannon, R.W. (1997). Comfortable and maximum walking speed of adults aged 20-79 years: Reference values and determinants. *Age and Ageing, 26*(1), 15-19.

Bonacci, J., Saunders, P.U., Hicks, A., Rantalainen, T., Vicenzino, B.T., & Spratford, W. (2013). Running in a minimalist and lightweight shoe is not the same as running barefoot: A biomechanical study. *British Journal of Sports Medicine, 47*(6), 387-392.

Branch, T., Partin, C., Chamberland, P., Emeterio, E., & Sabetelle, M. (1992). Spontaneous fractures of the humerus during pitching: A series of 12 cases. *American Journal of Sports Medicine, 20*(4), 468-470.

Brophy, R.H., Backus, S.I., Pansy, B.S., Lyman, S., & Williams, R.J. (2007). Lower extremity muscle activation and alignment during the soccer instep and side-foot kicks. *Journal of Orthopaedic & Sports Physical Therapy, 37*(5), 260-268.

Burstein, A.H., & Wright, T.M. (1994). *Fundamentals of orthopaedic biomechanics*. Baltimore: Williams & Wilkins.

Cappozzo, A., Felici, F., Figura, F., & Gazzani, F. (1985). Lumbar spine loading during half-squat exercises. *Medicine & Science in Sports & Exercise, 17*(5), 613-620.

Cappozzo, A., & Marchetti, M. (1992). Borelli's heritage. In A. Cappozzo, M. Marchetti, & V. Tosi (Eds.), *Biolocomotion: A century of research using moving pictures* (pp. 33-47). Rome: Promograph.

Cech, D.J., & Martin, S. (2011). *Functional movement development across the life span* (3rd ed.). St. Louis: Elsevier Saunders.

Cerrah, A.O., Gungor, E.O., Soylu, A.R., Ertan, H., Lees, A., & Bayrak, C. (2011). Muscular activation patterns during the soccer in-step kick. *Isokinetics and Exercise Science, 19*(3), 181-190.

Chaffin, D.B., & Baker, W.H. (1970). A biomechanical model for analysis of symmetric sagittal plane lifting. *AIIE Transactions, 2*(1), 16-27.

Chiu, L.Z.F, Bryanton, M.A., & Moolyk, A.N. (2014). Proximal-to-distal sequencing in vertical jumping with and without arm swing. *Journal of Strength and Conditioning Research, 28*(5), 1195-1202.

Cholewicki, J., McGill, S.M., & Norman, R.W. (1991). Lumbar spine loads during the lifting of extremely heavy weights. *Medicine & Science in Sports & Exercise, 23*(10), 1179-1186.

Chollet, D., Seifert, L.M., & Carter, M. (2008). Arm coordination in elite backstroke swimmers. *Journal of Sports Sciences, 26*(7), 675-682.

Chow, J.W., Carlton, L.G., Lim, Y.T., Shim, J.H., Chae, W.S., & Kuenster, A.F. (1999). Muscle activation during the tennis volley. *Medicine & Science in Sports & Exercise, 31*(6), 846-854.

Chow, J.W., Knudson, D.V., Tillman, M.D., & Andrew, D.P.S. (2007). Pre- and post-impact muscle activation in the tennis volley: Effects of ball speed, ball size and side of the body. *British Journal of Sports Medicine, 41*, 754-759.

Chow, J.W., Park, S-A., & Tillman, M.D. (2009). Lower trunk kinematics and muscle activity during different types of tennis serves. *Sports Medicine, Arthroscopy, Rehabilitation, Therapy and Technology, 1*(1), 24.

Clippinger, K. (2016). *Dance anatomy and kinesiology* (2nd ed.). Champaign, IL: Human Kinetics.

Coleman, S.G.S., Benham, A.S., & Northcott, S.R. (1993). A three-dimensional cinematographical analysis of the volleyball spike. *Journal of Sports Sciences, 11*(4), 295-302.

Colwin, C.M. (2002). *Breakthrough swimming*. Champaign, IL: Human Kinetics.

Comstock, R.D., Currie, D.W., Pierpoint, L.A., Grubenhoff, J.A., & Fields, S.K. (2015). An evidence-based discussion of heading the ball and concussions in high school soccer. *JAMA Pediatrics, 169*(9), 830-837.

Conway, J.E., Jobe, F.W., Glousman, R.E., & Pink, M. (1992). Medial instability of the elbow in throwing athletes: Treatment by repair or reconstruction of the ulnar collateral ligament. *Journal of Bone and Joint Surgery, 74A*, 67-83.

Cook, G., Burton, L., & Hoogenboom, B. (2006). Preparticipation screening: The use of fundamental movements as an assessment of function—Part 1. *North American Journal of Sports Physical Therapy, 1*(2), 62-72.

Cook, G., Burton, L., Hoogenboom, B.J., & Voight, M. (2014a). Functional movement screening: the use of fundamental movements as an assessment of function—part 1. *International Journal of Sports Physical Therapy, 9*(3), 396-409.

Cook, G., Burton, L., Hoogenboom, B.J., & Voight, M. (2014b). Functional movement screening: the use of fundamental movements as an assessment of function—part 2. *International Journal of Sports Physical Therapy, 9*(4), 549-563.

Cooper, C., Campion, G., & Melton, L.J., III. (1992). Hip fractures in the elderly: A world-wide projection. *Osteoporosis International, 2*(6), 285-289.

Correia, J.P., Oliveira, R., Vaz, J.R., Silva, L., & Pezarat-Correia, P. (2016). Trunk muscle activation, fatigue and low back pain in tennis players. *Journal of Science and Medicine in Sport, 19*(4), 311-316.

Criswell, E. (2010). *Cram's introduction to surface electromyography* (2nd ed.). Burlington, MA: Jones & Bartlett Learning.

Critchley, M. (1949). Punch-drunk syndromes: The chronic traumatic encephalopathy of boxers. In *Neuro-chirurgie: Hommage à Clovis Vincent* (pp. 161-174). Paris: Maloine.

Critchley, M. (1957). Medical aspects of boxing, particularly from a neurological standpoint. *British Medical Journal, 1*, 357-362.

Cunningham, A. (Ed.) (2002). *Guinness World Records 2002*. London: Guinness World Records.

Czerniecki, J.M., Gitter, A.J., & Beck, M.C. (1996). Energy transfer mechanisms as a compensatory strategy in below knee amputee runners. *Journal of Biomechanics, 29*(6), 717-722.

Darwin, C. (1998). *The expression of the emotions in man and animals* (3rd ed.). New York: Oxford University Press.

de Araujo, G.G., Manchado-Gobatto, F.B., Papoti, M., Camargo, B.H.F., & Gobatto, C.A. (2013). Anaerobic and aerobic performances in elite basketball players. *Journal of Human Kinetics, 42*, 137-147.

Deutsch, K.M., & Newell, K.M. (2005). Noise, variability, and the development of children's perceptual-motor skills. *Developmental Review, 25*(2), 155-180.

DeWitt, J.K., & Hinrichs, R.N. (2012). Mechanical factors associated with the development of high ball velocity during an instep soccer kick. *Sports Biomechanics, 11*(3), 382-290.

Dimiano, D.L., Martellotta, T.L., Sullivan, D.J., Granata, K.P., & Abel, M.F. (2000). Muscle force production and functional performance in spastic cerebral palsy: Relationship of cocontraction. *Archives of Physical Medicine and Rehabilitation, 81*(7), 895-900.

Ditmyer, M.M., Topp, R., & Pifer, M. (2002). Prehabilitation in preparation for orthopaedic surgery. *Orthopaedic Nursing, 21*(5), 43-51.

DiVirgilio, T.G., Hunter, A., Wilson, L., Stewart, W., Goodall, S., Howatson, G., Donaldson, D.I., & Ietswaart, M. (2016). Evidence for acute electrophysiological and cognitive changes following routine soccer heading. *EBioMedicine, 13*, 66-71.

Dörge, H.C., Andersen, T.B., Sørensen, H., & Simonsen, E.B. (2002). Biomechanical differences in soccer kicking with the preferred and the non-preferred leg. *Journal of Sports Sciences, 20*(4), 293-299.

Dragoo, J.L., Braun, H.J., Durham, J.L., Chen, M.R., & Harris, A.H. (2012). Incidence and risk factors for injuries to the anterior cruciate ligament in National Collegiate Athletic Association football: Data from the 2004-2005 through 2008-2009 National Collegiate Athletic Association Injury Surveillance System. *American Journal of Sports Medicine, 40*(5), 990-995.

Enoka, R.M. (2002). *Neuromechanics of human movement* (3rd ed.). Champaign, IL: Human Kinetics.

Enoka, R.M. (2015). *Neuromechanics of human movement* (5th ed.). Champaign, IL: Human Kinetics.

Escamilla, R.F. (2001). Knee biomechanics of the dynamic squat exercise. *Medicine & Science in Sports & Exercise, 33*(1), 127-141.

Escamilla, R.F., & Andrews, J.R. (2009). Shoulder muscle recruitment patterns and related biomechanics during upper extremity sports. *Sports Medicine, 39*(7), 569-590.

Everett, T., & Kell, C. (2010). *Human movement: An introductory text* (6th ed.). Edinburgh: Churchill Livingstone.

Faigenbaum, A.D., Kraemer, W.J., Blimkie, C.J.R., Jeffreys, I., Micheli, L.J., Nitka, M., & Rowland, T.W. (2009). Youth resistance training: Updated position statement paper from the National Strength and Conditioning Association. *Journal of Strength and Conditioning Research, 23*(S5), S60-S79.

Faigenbaum, A.D., & Micheli, L.J. (2017). *Youth strength training*. Indianapolis: American College of Sports Medicine.

Falconer, K., & Winter, D.A. (1985). Quantitative assessment of co-contraction at the ankle joint in walking. *Electromyography and Clinical Neurophysiology, 25*(2-3), 135-149.

Farber, A.J., Smith, J.S., Kvitne, R.S., Mohr, K.J., & Shin, S.S. (2009). Electromyographic analysis of forearm muscles in professional and amateur golfers. *American Journal of Sports Medicine, 37*(2), 396-401.

Feltner, M.E., Bishop, E.J., & Perez, C.M. (2004). Segmental and kinetic contributions in vertical jumps performed with and without an arm swing. *Research Quarterly for Exercise and Sport, 75*(3), 216-230.

Fiatarone, M.A., Marks, E.C., Ryan, N.D., Meredith, C.N., Lipsitz, L.A., & Evans, W.J. (1990). High-intensity strength training in nonagenarians: Effects on skeletal muscle. *Journal of the American Medical Association, 263*(22), 3029-3034.

Fitts, P.M. (1964). Categories of human learning. In A.W. Melton (Ed.), *Perceptual-motor skills learning* (pp. 243-285). New York: Academic Press.

Fleisig, G.S., Escamilla, R.F., Andrews, J.R., Matsuo, T., Satterwhite, Y., & Barrentine, S.W. (1996). Kinematic and kinetic comparison between baseball pitching and football passing. *Journal of Applied Biomechanics, 12*, 207-214.

Forthomme, B., Croisier, J.-L., Ciccarone, G., Crielard, J.-M., & Cloes, M. (2005). Factors correlated with volleyball spike velocity. *American Journal of Sports Medicine, 33*(10), 1513-1519.

Fryer, J.C.J., Quon, J.A., & Vann, R.D. (2017). A proposed in vitro model for investigating the mechanisms of 'joint cracking': A short report of preliminary techniques and observations. *Journal of the Canadian Chiropractic Association, 61*(1), 32-39.

Fukashiro, S., Komi, P.V., Jarvinen, M., & Miyashita, M. (1995). In vivo Achilles tendon loading during jumping in humans. *European Journal of Applied Physiology and Occupational Physiology, 71*, 453-458.

Fuzhong, L., Harmer, P., Fitzgerald, K., Eckstrom, E., Stock, R., Galver, J., Maddalozzo, G., & Batya, S.S. (2012). Tai Chi and postural stability in patients with Parkinson's disease. *The New England Journal of Medicine, 366*(6), 511-519.

Gad, P., Gerasimenko, Y., Zdunowski, S., Turner, A., Sayenko, D., Lu, D.C., & Edgerton, V.R. (2017). Weight bearing over-ground stepping in an exoskeleton with non-invasive spinal cord neuromodulation after motor complete paraplegia. *Frontiers in Neuroscience, 11*, 1-8.

Gagnon, D., Larivière, C., & Loisel, P. (2001). Comparative ability of EMG, optimization, and hybrid modelling approaches to predict trunk muscle forces and lumbar spine loading during dynamic sagittal plane lifting. *Clinical Biomechanics, 16*(5), 359-372.

Gallant, M.P., Tartaglia, M., Hardman, S., & Burke, K. (2017). Using Tai Chi to reduce fall risk factors among older adults: An evaluation of a community-based implementation. doi:10.1177/0733464817703004

Gerasimenko, Y.P., Lu, D.C., Modaber, M., Zdunowski, S., Gad, P., Sayenko, D.G., . . . Edgerton, V.R. (2015). Noninvasive reactivation of motor descending control after paralysis. *Journal of Neurotrauma, 32*(24), 1968-1980.

Gessel, L.M., Fields, S.K., Collines, C.L., Dick, R.W., & Comstock, R.D. (2007). Concussions among United States high school and collegiate athletes. *Journal of Athletic Training, 42*(4), 495-503.

Gil-Agudo, A., Ama-Espinosa, A.D., & Crespo-Ruiz, B. (2010). Wheelchair basketball quantification. *Physical Medicine Rehabilitation Clinics of North America, 21*(1), 141-156.

Gowan, I.D., Jobe, F.W., Tibone, J.E., Perry, J., & Moynes, D.R. (1987). A comparative electromyographic analysis of the shoulder during pitching: Professional versus amateur pitchers. *American Journal of Sports Medicine, 15*(6), 586-590.

Gregor, R.J., Fowler, E.G., & Childers, W.L. (2011). Applied biomechanics of cycling. In D.J. Magee, R.C. Manske, J.E. Zachazewski, & W.S. Quillen (Eds.), *Athletic and Sport Issues in Musculoskeletal Rehabilitation* (pp. 187-216). St. Louis: Elsevier Saunders.

Grimm, D., Grosse, J., Wehland, M., Mann, V., Reseland, J.E., Sundaresan, A., & Corydon, T.J. (2016). The impact of microgravity on bone in humans. *Bone, 87*, 44-56.

Gullberg, B., Johnell, O., & Kanis, J.A. (1997). Worldwide projections for hip fracture. *Osteoporosis International, 7*(5), 407-413.

Haff, G.G., & Triplett, N.T. (Eds.). (2015). *Essentials of strength and conditioning* (4th ed.). Champaign, IL: Human Kinetics.

Hanson, A.M., Padua, D.A., Blackburn, J.T., Prentice, W.E., & Hirth, C.J. (2008). Muscle activation during side-step cutting maneuvers in male and female soccer athletes. *Journal of Athletic Training, 43*(2), 133-143.

Harrison, A.J., & Gaffney, S. (2001). Motor development and gender effects on stretch-shortening cycle performance. *Journal of Science and Medicine in Sport, 4*(4), 406-415.

Hay, J.G. (1993). *The biomechanics of sports techniques* (4th ed.). Englewood Cliffs, NJ: Prentice Hall.

Haywood, K.M., & Getchell, N. (2014). *Life span motor development* (6th ed.). Champaign, IL: Human Kinetics.

Hazari, A., Warsi, M., & Agouris, I. (2016). Electromyography analysis of shoulder and wrist muscles in semi-professional cricket fast bowlers during bouncer and Yorker delivery. A cross-sectional comparative study. *International Journal of Physical Education, Sports and Health, 3*(6), 77-87.

Hewett, T.E., Myer, G.D., Ford, K.R., Heidt, R.S., Jr., Colosimo, A.J., McLean, S.G., . . . Succop, P. (2005). Biomechanical measures of neuromuscular control and valgus loading of the knee predict anterior cruciate ligament injury risk in female athletes: A prospective study. *American Journal of Sports Medicine, 33*(4), 492-501.

Hollander, K., Argubi-Wollesen, A., Reer, R., & Zech, A. (2015). Comparison of minimalist footwear strategies for simulating barefoot running: A randomized crossover study. *PLoS One, 10*(5), E0125880.

Hotson, G., McMullen, D.P., Fifer, M.S., Johannes, M.S., Katyal, K.D., Para, M.P., . . . Crone, N.E. (2016). Individual finger control of the modular prosthetic limb using high-density electrocorticography in a human subject. *Journal of Neural Engineering, 13*(2), 026017.

Houglum, P.A., & Bertoti, D.B. (2012). *Brunnstrom's clinical kinesiology* (6th ed.) Philadelphia: Davis.

Hudson, J.L. (1986). Coordination of segments in the vertical jump. *Medicine & Science in Sports & Exercise, 18*(2), 242-251.

Huffman, K.D., Sanford, B.A., Zucker-Levin, A.R., Williams, J.L., & Mihalko, W.M. (2015). Increased hip abduction in high body mass index subjects during sit-to-stand. *Gait Posture, 41*(2), 640-645.

Inman, V.T., Ralston, H.J., & Todd, F. (1981). *Human walking*. Baltimore: Williams & Wilkins.

International Ergonomics Association. (2017). Definition and domains of ergonomics. Retrieved from www.iea.cc/whats

International Wheelchair Basketball Federation. (2014). *Official player classification manual*. Retrieved from https://iwbf.org/wp-content/uploads/2017/09/CLASSIFICATION-MANUAL-2014-2018-ENGLISH-FINAL.pdf

Jacobs, R., Bobbert, M.F., & van Ingen Schenau, G.J. (1996). Mechanical output from individual muscles during explosive leg extensions: The role of biarticular muscles. *Journal of Biomechanics, 29*(4), 513-523.

Jerome, J. (1980). *The sweet spot in time*. New York: Summit.

Jobe, F.W., Moynes, D.R., & Antonelli, D.J. (1986). Rotator cuff function during a golf swing. *American Journal of Sports Medicine, 14*(5), 388-392.

Jobe, F.W., Moynes, D.R., Tibone, J.E., & Perry, J. (1984). An EMG analysis of the shoulder in pitching: A second report. *American Journal of Sports Medicine, 12*(3), 218-220.

Jobe, F.W., Perry, J., & Pink, M. (1989). Electromyographic shoulder activity in men and women professional golfers. *American Journal of Sports Medicine, 17*(6), 782-787.

Jobe, F.W., Stark, H., & Lombardo, S.J. (1986). Reconstruction of the ulnar collateral ligament in athletes. *Journal of Bone and Joint Surgery, 68A*, 1158-1163

Johnell, O., & Kanis, J.A. (2006). An estimate of the worldwide prevalence and disability associated with osteoporotic fractures. *Osteoporosis International, 17*(12), 1726-1733.

Jorge, M., & Hull, M.L. (1986). Analysis of EMG measurements during bicycle pedalling. *Journal of Biomechanics, 19*(9), 683-694.

Judge, J.O., Ounpuu, S., & Davis, R.B. (1996). Effects of age on the biomechanics and physiology of gait. *Clinics in Geriatric Medicine, 12*(4), 659-678.

Kamen, G., & Gabriel, D.A. (2009). *Essentials of electromyography.* Champaign, IL: Human Kinetics.

Kannus, P., & Jozsa, L. (1991). Histopathologic changes preceding spontaneous rupture of a tendon. *Journal of Orthopaedic Trauma, 5,* 395-402.

Kao, J.T., Pink, M., Jobe, F.W., & Perry, J. (1995). Electromyographic analysis of the scapular muscles during a golf swing. *American Journal of Sports Medicine, 23*(1), 19-23.

Kawchuk, G.N., Fryer, J., Jaremko, J.L., Zeng, H., Rowe, L., & Thompson, R. (2015). Real-time visualization of joint cavitation. *PLOS One,* 1-11.

Kellis, E., Arabatzi, F., & Papadopoulos, C. (2003). Muscle co-activation around the knee in drop jumping using the co-contraction index. *Journal of Electromyography & Kinesiology, 13,* 229-238.

Kellis, E., & Katis, A. (2007). Biomechanical characteristics and determinants of instep soccer kick. *Journal of Sports Science and Medicine, 6,* 154-165.

Kelly, B.T., Backus, S.I., Warren, R.F., & Williams, R.J. (2002). Electromyographic analysis and phase definition of the overhead football throw. *American Journal of Sports Medicine, 30*(6), 837-844.

Kibler, W.B., Chandler, T.J., Shapiro, R., & Conuel, M. (2007). Muscle activation in coupled scapulohumeral motions in the high performance tennis serve. *British Journal of Sports Medicine, 41*(11), 745-749.

King, M.A., Kentel, B.B., & Mitchell, S.R. (2012). The effects of ball impact location and grip tightness on the arm, racquet and ball for one-handed tennis backhand groundstrokes. *Journal of Biomechanics, 45*(6), 1048-1052.

Knudson, D.V., & Blackwell, J. (1997). Upper extremity angular kinematics of the one-handed backhand drive in tennis players with and without tennis elbow. *International Journal of Sports Medicine, 18*(2), 79-82.

Koerte, I.K., Nichols, E., Tripodis, Y., Schultz, V., Lehner, S., Igbinoba, R., & Sereno, A.B. (2017). Impaired cognitive performance in youth athletes exposed to repetitive head impacts. *Journal of Neurotrauma, 34*(16), 2389-2395.

Komi, P.V., Fukashiro, S., & Järvinen, M. (1992). Biomechanical loading of Achilles tendon during normal locomotion. *Clinics in Sports Medicine, 11*(3), 521-531.

Kontos, A.P., Braithwaite, R., Chrisman, S.P.D., McAllister-Deitrick, J., Symington, L., Reeves, V.L., & Collins, M.W. (2017). Systemic review and meta-analysis of the effects of football heading. *British Journal of Sports Medicine, 51*(15), 1118-1124.

Krasnow, D., Wilmerding, V., Stecyk, S., Wyon, M., & Koutedakis, Y. (2011). Biomechanical research in dance: A literature review. *Medical Problems of Performing Artists, 26*(1), 3-23.

Kraus, K., Schütz, E., Taylor, W.R., & Doyscher, R. (2014). Efficacy of the functional movement screen: A review. *Journal of Strength and Conditioning Research, 28*(12), 3571-3584.

Lamontagne, A., Richards, C.L., & Malouin, F. (2000). Coactivation during gait as an adaptive behavior after stroke. *Journal of Electromyography & Kinesiology, 10*(6), 407-415.

Langendorfer, S.J., & Roberton, M.A. (2002). Individual pathways in the development of forceful throwing. *Research Quarterly in Exercise and Sport, 73*(3), 245-256.

Latash, M.L. (2016). Biomechanics as a window into the neural control of movement. *Journal of Human Kinetics, 52,* 7-20.

Latash, M.L., Levin, M.F., Scholz, J.P., & Schöner, G. (2010). Motor control theories and their applications. *Medicina (Kaunas), 46*(6), 382-392.

Lauer, J., Figueiredo, P., Vilas-Boas, J.P., Fernandes, R.J., & Rouard, A.H. (2013). Phase-dependence of elbow muscle coactivation in front crawl swimming. *Journal of Electromyography and Kinesiology, 23*(4), 820-825.

Leblanc, H., Seifert, L., Baudry, L., & Chollet, D. (2005). Arm-leg coordination in flat breaststroke: A comparative study between elite and non-elite swimmers. *International Journal of Sports Medicine, 26*(9), 787-797.

Levangie, P.K., & Norkin, C.C. (2011). *Joint structure and function: A comprehensive analysis* (5th ed.). Philadelphia: FA Davis.

Li, F., Harmer, P., Fisher, K.J., & McAuley, E. (2004). Tai Chi: Improving functional balance and predicting subsequent falls in older persons. *Medicine & Science in Sports & Exercise, 36*(12), 2046-2052.

Li, F., Harmer, P., Fisher, K.J., McAuley, E., Chaumeton, N., Eckstrom, E., & Wilson, N.L. (2005). Tai Chi and fall reductions in older adults: A randomized controlled trial. *Journals of Gerontology, Series A, Biological Sciences and Medical Sciences, 60*(2), 187-194.

Lieber, R.L. (2009). *Skeletal Muscle Structure, Function, and Plasticity: The Physiological Basis of Rehabilitation* (3rd ed.). Philadelphia: Lippincott Williams & Wilkins, 2009.

Lin, M.R., Hwang, H.F., Wang, Y.W., Chang, S.H., & Wolf, S.L. (2006). Community-based tai chi and its effect on injurious falls, balance, gait, and fear of falling in older people. *Physical Therapy, 86*(9), 1189-1201.

Ling, H., Hardy, J., & Zetterberg, H. (2015). Neurological consequences of traumatic brain injuries in sports. *Molecular and Cellular Neuroscience, 66*(Pt B), 114-122.

Lloyd, R.S., Cronin, J.B., Faigenbaum, A.D., Haff, G.G., Howard, R., Kraemer, W.J., . . . Oliver, J.L. (2016). National Strength and Conditioning Association position statement on long-term athletic development. *Journal of Strength and Conditioning Research, 30*(6), 1491-1509.

Lo, S.L., Raskin, K., Lester, H., & Lester, B. (2002). Carpal tunnel syndrome: A historical perspective. *Hand Clinics, 18*(2), 211-217.

Luhtanen, P., & Komi, R.V. (1978). Segmental contribution to forces in vertical jump. *European Journal of Applied Physiology and Occupational Physiology, 38*(3), 181-188.

Maffet, M.W., Jobe, F.W., Pink, M.M., Brault, J., & Mathiyakom, W. (1997). Shoulder muscle firing patterns during the windmill softball pitch. *American Journal of Sports Medicine, 25*(3), 369-374.

Maher, M.E., Hutchison, M., Cusimano, M., Comper, P., & Schweizer, T.A. (2014). Concussions and heading in soccer: A review of the evidence of incidence, mechanisms, biomarkers and neurocognitive outcomes. *Brain Injury, 28*(3), 271-285.

Mak, M.K., Levin, O., Mizrahi, J., & Hui-Chan, C.W. (2003). Joint torques during sit-to-stand in healthy subjects and people with Parkinson's disease. *Clinical Biomechanics, 18*(3), 197-206.

Maki, B.E., & McIlroy, W.E. (1996). Postural control in the older adult. *Clinics in Geriatric Medicine, 12*(4), 635-658.

Mandelbaum, B.R., Silvers, H.J., Watanabe, D.S., Knarr, J.F., Thomas, S.D., Griffin, L.Y., . . . Garrett, W., Jr. (2005). Effectiveness of a neuromuscular and proprioceptive training program in preventing anterior cruciate ligament injuries in female athletes: 2-year follow-up. *American Journal of Sports Medicine, 33*(7), 1003-1010.

Marcus, R., Cann, C., Madvig, P., Minkoff, U., Goddard, M., Bayer, M., . . . Genant, H. (1985). Menstrual function and bone mass in elite women distance runners: Endocrine and metabolic features. *Annals of Internal Medicine, 102*(2), 158-163.

Marta, S., Silva, L., Castro, M.A., Pezarat-Correia, P., & Cabri, J. (2012). Electromyography variables during the golf swing: A literature review. *Journal of Electromyography and Kinesiology, 22*(6), 803-813.

Marta, S., Silva, L., Vaz, J.R., Castro, M.A., Reinaldo, G., & Pezarat-Correia, P. (2016). Electromyographic analysis of the lower limb muscles in low- and high-handicap golfers. *Research Quarterly for Exercise and Sport, 87*(3), 318-324.

Martens, J., Figueiredo, P., & Daly, D. (2015). Electromyography in the four competitive swimming strokes: A systematic review. *Journal of Electromyography and Kinesiology, 25*(2), 273-291.

Martin, C., Bideau, B., Delamarche, P., & Kulpa, R. (2016). Influence of a prolonged tennis match play on serve biomechanics. *PLoS ONE, 11*(8), e0159979.

Martin, R.B., Burr, D.B., Sharkey, N.A., & Fyhrie, D.P. (2015). *Skeletal tissue mechanics* (2nd ed.). New York: Springer.

Martin, V., Scholz, J.P., & Schöner, G. (2009). Redundancy, self-motion and motor control. *Neural Computing, 21*(5), 1371-1414.

Martland, H. (1928). Punch drunk. *JAMA, 91*, 1103-1107.

Maurer, M.S., Burcham, J., & Cheng, H. (2005). Diabetes mellitus is associated with an increased risk of falls in elderly residents of a long-term care facility. *Journals of Gerontology, Series A, Biological Sciences and Medical Sciences, 60*(9), 1157-1162.

McGill, S. (2002). *Low back disorders: Evidence-based prevention and rehabilitation*. Champaign, IL: Human Kinetics.

McGill, S. (2016). *Low back disorders: Evidence-based prevention and rehabilitation* (3rd ed.). Champaign, IL: Human Kinetics.

McGinnis, P. (2013). *Biomechanics of sport and exercise* (3rd ed.). Champaign, IL: Human Kinetics.

McKee, A.C., Cantu, R.C., Nowinski, C.J., Hedley-Whyte, E.T., Gavett, B.E., Budson, A.E., . . . Stern, R.A. (2009). Chronic traumatic encephalopathy in athletes: Progressive tauopathy after repetitive head injury. *Journal of Neuropathology & Experimental Neurology, 68*(7), 709-735.

Mero, A., & Komi, P.V. (1986). Force-, EMG-, and elasticity-velocity relationships at submaximal, maximal and supramaximal running speeds in sprinters. *European Journal of Applied Physiology, 55*(5), 553-561.

Minick, K.I., Kiesel, K.B., Burton, L., Taylor, A., Plisky, P., & Butler, R.J. (2010). Interrater reliability of the functional movement screen. *Journal of Strength and Conditioning Research, 24*(2), 479-486.

Moran, R.W., Schneiders, A.G., Mason, J., & Sullivan S.J. (2017). Do functional movement screen (FMS) composite scores predict subsequent injury? A systematic review with meta-analysis. *British Journal of Sports Medicine*, doi: 10.1136/bjsports-2016-096938.

Moreno, M.A., Zamunér, A.R., Paris, J.V., Teodori, R.M., & Barros, R.M.L. (2012). Effects of wheelchair sports on respiratory muscle strength and thoracic mobility of individuals with spinal cord injury. *American Journal of Physical Medicine & Rehabilitation, 91*(6), 470-477.

Morris, M., Jobe, F.W., Perry, J., Pink, M., & Healy, B.S. (1989). Electromyographic analysis of elbow function in tennis players. *American Journal of Sports Medicine, 17*(2), 241-247.

Morrison, S., Colberg, S.R., Mariano, M., Parson, H.K., & Vinik, A.I. (2010). Balance training reduces falls risk in older individuals with type 2 diabetes. *Diabetes Care, 33*(4), 748-750.

Mullen, S., & Roby, E.B. (2013). Adolescent runners: The effect of training shoes on running kinematics. *Journal of Pediatric Orthopaedics, 33*(4), 453-457.

Murray, M.P., Guten, G.N., Mollinger, L.A., & Gardner, G.M. (1993). Kinematic and electromyographic patterns of Olympic race walkers. *American Journal of Sports Medicine, 11*(2), 68-74.

Naito, K., Fukui, Y., & Maruyama, T. (2010). Multijoint kinetic chain analysis of knee extension during the soccer instep kick. *Human Movement Science, 29*(2), 259-276.

Nakata, H., Miura, A., Yoshie, M., Kanosue, K., & Kudo, K. (2013). Electromyographic analysis of lower limbs during baseball batting. *Journal of Strength and Conditioning Research, 27*(5), 1179-1187.

Narazaki, K., Berg, K., Stergiou, N., & Chen, B. (2009). Physiological demands of competitive basketball. *Scandinavian Journal of Medicine & Science in Sports, 19*(3), 425-432.

National Institute for Occupational Safety and Health (NIOSH). (2007). *Ergonomic guidelines for manual material handling*. San Francisco: California Department of Industrial Relations.

Neumann, D.A. (2016). *Kinesiology of the musculoskeletal system: Foundations for rehabilitation* (3rd ed.). St. Louis: Mosby.

Newell, K.M. (1986). Constraints on the development of coordination. In M.G. Wade & H.T.A. Whiting (Eds.), *Motor development in children: Aspects of coordination and control* (pp. 341-361). Amsterdam: Martin Nijhoff.

O'Bryan, S.J., Brown, N.A.T., Billaut, F., & Rouffet, D.M. (2014). Changes in muscle coordination and power output during sprint cycling. *Neuroscience Letters, 576*, 11-16.

O'Donoghue, D.H. (1984). *Treatment of injuries to athletes* (4th ed.). Philadelphia: Saunders.

O'Kane, J.W. (2016). Is heading in youth soccer dangerous play? *Physician and Sportsmedicine, 44*(2), 190-194.

Okazaki, V.H.A., Rodacki, A.L.F., & Satern, M.N. (2015). A review of the basketball jump shot. *Sports Biomechanics, 14*(1), 1-16.

Oliver, G.D., Plummer, H.A., & Keeley, D.W. (2011). Muscle activation patterns of the upper and lower extremity during the windmill softball pitch. *Journal of Strength and Conditioning Research, 25*(6), 1653-1658.

Olstad, B.H., Vaz, J.R., Zinner, C., Cabri, J.M.H., Kjendlie, P-L. (2017). Muscle coordination, activation and kinematics of world-class and elite breaststroke swimmers during submaximal and maximal efforts. *Journal of Sports Sciences, 35*(11), 1107-1117.

Olstad, B.H., Zinner, C., Vaz, J.R., Cabri, J.M.M., Kjendlie, P-L. (2017). Muscle activation in world-champion, world-class, and national breaststroke swimmers. *International Journal of Sports Physiology and Performance, 12*(4), 538-547.

Osternig, L.R., Hamill, J., Lander, J.E., & Robertson, R. (1986). Co-activation of sprinter and distance runner muscles in isokinetic exercise. *Medicine & Science in Sports & Exercise, 18*(4), 431-435.

Ostrosky, K.M., VanSwearingen, J.M., Burdett, R.G., & Gee, Z. (1994). A comparison of gait characteristics in young and old subjects. *Physical Therapy, 74*(7), 637-646.

Parkkari, J., Kannus, P., Palvanen, M., Natri, A., Vainio, J., Aho, H., . . . Järvinen, M. (1999). Majority of hip fractures occur as a result of a fall and impact on the greater trochanter of the femur: A prospective controlled hip fracture study with 206 consecutive patients. *Calcified Tissue International, 65*(3), 183-187.

Parchmann, C.J., & McBride, J.M. (2011). Relationship between functional movement screen and athletic performance. *Journal of Strength and Conditioning Research, 25*(12), 3378-3384.

Pascal, B., & Krailsheimer, A.J. (1995). *Pensees*. New York: Penguin.

Perrine, J.J., & Edgerton, V.R. (1978). Muscle force-velocity and power-velocity relationships under isokinetic loading. *Medicine & Science in Sports & Exercise, 10*(3), 159-166.

Perry, J. (1992). *Gait analysis: Normal and pathological function*. Thorofare, NJ: Slack.

Perry, J., & Burnfield, J. (2010). *Gait analysis: Normal and pathological function* (2nd ed.). Thorofare, NJ: Slack.

Peterson, L., & Renström, P. (2001). *Sports injuries: Their prevention and treatment*. Champaign, IL: Human Kinetics.

Pink, M., Jobe, F.W., & Perry, J. (1990). Electromyographic analysis of the shoulder during the golf swing. *American Journal of Sports Medicine, 18*(2), 137-140.

Pink, M., Perry, J., & Jobe, F.W. (1993). Electromyographic analysis of the trunk in golfers. *American Journal of Sports Medicine, 21*(3), 385-388.

Pink, M., Jobe, F.W., Perry, J., Browne, A., Scovazzo, M.L., & Kerrigan, J. (1993). The painful shoulder during the butterfly stroke. An electromyographic and cinematographic analysis of twelve muscles. *Clinical Orthopaedics and Related Research, 288*, 60-72.

Pink, M., Jobe, F.W., Perry, J., Kerrigan, J., Browne, A., & Scovazzo, M.L. (1992). The normal shoulder during the backstroke: An EMG and cinematographic analysis of 12 muscles. *Clinical Journal of Sport Medicine, 2*, 6-12.

Pink, M., Jobe, F.W., Perry, J., Kerrigan, J., Browne, A., & Scovazzo, M.L. (1993). The normal shoulder during the butterfly swim stroke. An electromyographic and cinematographic analysis of twelve muscles. *Clinics in Orthopaedic and Related Research, 288*, 48-59.

Pink, M., Perry, J., Browne, A., Scovazzo, M.L., & Kerrigan, J. (1991). The normal shoulder during freestyle swimming. An electromyographic and cinematographic analysis of twelve muscles. *American Journal of Sports Medicine, 19*(6), 569-576.

Plamondon, A., Delisle, A., Bellefeuille, S., Denis, D., Gagnon, D., & Larivière, C. (2014). Lifting strategies of expert and novice workers during a repetitive palletizing task. *Applied Ergonomics, 45*(3), 471-481.

Plamondon, A., Larivière, C., Denis, D., Mecheri, H., & Nastasia, I. (2017). Difference between male and female workers lifting the same relative load when palletizing boxes. *Applied Ergonomics, 60*, 93-102.

Plamondon, A., Larivière, C., Denis, D., St-Vincent, M., & Delisle, A. (2014). Sex differences in lifting strategies during a repetitive palletizing task. *Applied Ergonomics, 45*, 1558-1569.

Pojskić, H., Šeparović, V., Užičanin, E., Muratović, M., & Mačković, S. (2015). Positional role differences in the aerobic and anaerobic power of elite basketball players. *Journal of Human Kinetics, 49*(1), 219-227.

Preston, D.C., & Shapiro, B.E. (2012). *Electromyography and neuromuscular disorders: Clinical-electrophysiologic correlations* (3rd ed.). Philadelphia: Saunders.

Prilutsky, B.I., & Zatsiorsky, V.M. (1994). Tendon action of two-joint muscles: Transfer of mechanical energy between joints during jumping, landing, and running. *Journal of Biomechanics, 27*(1), 25-34.

Puniello, M.S., McGibbon, C.A., & Krebs, D.E. (2001). Lifting strategy and stability in strength-impaired elders. *Spine, 26*(7), 731-737.

Putnam, C.A. (1991). A segment interaction analysis of proximal-to-distal sequential segment motion patterns. *Medicine & Science in Sports & Exercise, 23*(1), 130-144.

Ramsey, V.K., Miszko, T.A., & Horvat, M. (2004). Muscle activation and force production in Parkinson's patients during sit to stand transfers. *Clinical Biomechanics, 19*(4), 377-384.

Reeser, J.C., Fleisig, G.S., Bolt, B., & Ruan, M. (2010). Upper limb biomechanics during the volleyball serve and spike. *Sports Health, 2*(5), 368-374.

Roberton, M.A., & Halverson, L.E. (1984). *Developing children: Their changing movement*. Philadelphia: Lea & Febiger.

Robertson, D. & Mosher, R. (1985). Work and power of the leg muscles in soccer kicking. In D. Winter (Ed.), *Biomechanics IX-B* (pp. 533-538). Champaign, IL: Human Kinetics.

Rodrigues, A.C., Lasmar, R.P., & Caramelli, P. (2016). Effects of soccer heading on brain structure and function. *Frontiers in Neurology, 7*(38), 1-11.

Rojas, I.L., Provencher, M.T., Bhatia, S., Foucher, K.C., Bach, B.R., Jr., Romeo, A.A., . . . Verma, N.N. (2009). Biceps activity during windmill softball pitching. *American Journal of Sports Medicine, 37*(3), 558-565.

Rokito, A.S., Jobe, F.W., Pink, M.M., Perry, J., & Brault, J. (1998). Electromyographic analysis of shoulder function during the volleyball serve and spike. *Journal of Shoulder and Elbow Surgery, 7*(3), 256-263.

Rose, J., & Gamble, J.G. (2005). *Human walking* (3rd ed.). Philadelphia: Lippincott Williams & Wilkins.

Roston, J.B., & Wheeler Haines, R. (1947). Cracking in the metacarpo-phalangeal joint. *Journal of Anatomy, 81*(2), 165-173.

Rota, S., Morel, B., Saboul, D., Rogowski, I., & Hautier, C. (2014). Influence of fatigue on upper limb muscle activity and performance in tennis. *Journal of Electromyography and Kinesiology, 24*(1), 90-97.

Rouard, A.H., & Clarys, J.P. (1995). Cocontraction in the elbow and shoulder muscles during rapid cyclic movements in an aquatic environment. *Journal of Electromyography and Kinesiology, 5*(3), 177-183.

Ruwe, P.A., Pink, M., Jobe, F.W., Perry, J., & Scovazzo, M.L. (1994). The normal and the painful shoulders during the breaststroke. Electromyographic and cinematographic analysis of twelve muscles. *American Journal of Sports Medicine, 22*(6), 789-796.

Ryschon, T.W., Fowler, M.D., Wysong, R.E., Anthony, A., & Balaban, R.S. (1997). Efficiency of human skeletal muscle in vivo: Comparison of isometric, concentric, and eccentric muscle action. *Journal of Applied Physiology, 83*(3), 867-874.

Ryu, R.K.N., McCormick, J., Jobe, F.W., Moynes, D.R., & Antonelli, D.J. (1988). An electromyographic analysis of shoulder function in tennis players. *American Journal of Sports Medicine, 16*(5), 481-485.

Santos, S., Krishnan, C., Alonso, A.C., & Greve, J.M.D. (2017). Trunk function correlates positively with wheelchair basketball player classification. *American Journal of Physical Medicine & Rehabilitation, 96*(2), 101-108.

Schmidt, R.A., & Lee, T.D. (2011). *Motor control and learning: A behavioral emphasis* (5th ed.). Champaign, IL: Human Kinetics.

Schmitz, C., Martin, N., & Assaiante, C. (1999). Development of anticipatory postural adjustments in a bimanual load-lifting task in children. *Experimental Brain Research, 126*(2), 200-204.

Schoenfeld, B.J. (2010). Squatting kinematics and kinetics and their application to exercise performance. *Journal of Strength and Conditioning Research, 24*(12), 3497-3506.

Schwartz, A.V., Hillier, T.A., Sellmeyer, D.E., Resnick, H.E., Gregg, E., Ensrud, K.E., . . . Cummings, W.R. (2002). Older women with diabetes have a higher risk of falls: A prospective study. *Diabetes Care, 25*(10), 1749-1754.

Scott, S.H., & Winter, D.A. (1990). Internal forces at chronic running injury sites. *Medicine & Science in Sports & Exercise, 22*(3), 357-369.

Scovazzo, M.L., Browne, A., Pink, M., Jobe, F.W., & Kerrigan, J. (1991). The painful shoulder during freestyle swimming. An electromyographic cinematographic analysis of twelve muscles. *American Journal of Sports Medicine, 19*(6), 577-582.

Scurr, J.C., Abbott, V., & Ball, N. (2011). Quadriceps EMG muscle activation during accurate soccer instep kicking. *Journal of Sports Sciences, 29*(3), 247-251.

Serrien, B., Ooijen, J., Goossens, M., & Baeyens, J.-P. (2016). A motion analysis in the volleyball spike—Part 1: Three-dimensional kinematics and performance. *International Journal of Human Movement and Sports Sciences, 4*(4), 70-82.

Shaffer, B., Jobe, F.W., Pink, M., & Perry, J. (1993). Baseball batting. An electromyographic study. *Clinical Orthopaedics and Related Research, 292*, 285-293.

Shan, G., & Westerhoff, P. (2005). Full-body kinematic characteristics of the maximal instep soccer kick by male soccer players and parameters related to kick quality. *Sports Biomechanics, 4*(1), 59-72.

Sheppard, J.M., & Triplett, N.T. (2015). Program design for resistance training. In G.G. Haff & N.T. Triplett (Eds.), *Essentials of strength and conditioning* (4th ed., pp. 439-470). Champaign, IL: Human Kinetics.

Shih, Y., Lin, K-L., & Shiang, T-Y. (2013). Is the foot striking pattern more important than barefoot or shod conditions in running? *Gait & Posture, 38*(3), 490-494.

Sibella, F., Galli, M., Romei, M., Montesano, A., & Crivellini, M. (2003). Biomechanical analysis of sit-to-stand movement in normal and obese subjects. *Clinical Biomechanics, 18*(8), 745-750.

Sinclair, J., Fewtrell, D., Taylor, P.J., Atkins, S., Bottoms, L., & Hobbs, S.J. (2014). Three-dimensional kinematic differences between the preferred and non-preferred limbs during maximal instep soccer kicking. *Journal of Sports Sciences, 32*(20), 1914-1923.

Sinclair, J., Fewtrell, D., Taylor, P.J., Bottoms, L., Atkins, S., & Hobbs, S.J. (2014). Three-dimensional kinematic correlates of ball velocity during maximal instep soccer kicking in males. *European Journal of Sport Science, 14*(8), 799-805.

Sinclair, J., & Hobbs, S.J. (2016). Bilateral differences in knee and ankle loading of the support limb during maximal instep soccer kicking. *Science & Sports, 31*(4), e73-e78.

Sisto, D.J., Jobe, F.W., Moynes, D.R., & Antonelli, D.J. (1987). An electromyographic analysis of the elbow in pitching. *American Journal of Sports Medicine, 15*(3), 260-263.

Smith, S.M., Heer, M.A., Shackelford, L.C., Sibonga, J.D., Ploutz-Snyder, L., & Zwart, S.R. (2012). Benefits for bone from resistance exercise and nutrition in long-duration space-flight: Evidence from biochemistry and densitometry. *Journal of Bone and Mineral Research, 27*(9), 1896-1906.

Smith, L.B., & Thelen, E. (2003). Development as a dynamic system. *TRENDS in Cognitive Sciences, 7*(8), 343-348.

Smith, L.K., Weiss, E.L., & Lehmkuhl, L.D. (1996). *Brunnstrom's clinical kinesiology*. Philadelphia: Davis.

Squadrone, R., Rodano, R., Hamill, J., & Preatoni, E. (2015). Acute effect of different minimalist shoes on foot strike pattern and kinematics in rearfoot strikers during running. *Journal of Sports Sciences, 33*(11), 1196-1204.

Stanislavski, C. (1984). *An actor prepares*. New York: Theatre Arts Books.

Sterne, L. (1980). *Tristram shandy*. New York: Norton.

Stewart, W.F., Kim, N., Ifrah, C.S., Lipton, R.B., Bachrach, T.A., Zimmerman, M.E., . . . Lipton, M.L. (2017). Symptoms from repeated intentional and unintentional head impact in soccer players. *Neurology, 88*(9), 901-908.

Sungkarat, S., Boripuntakul, S., Chattipakorn, N., Watcharasaksilp, K., & Lord, S.R. (2017). Effects of Tai Chi on cognition and fall risk in older adults with mild cognitive impairment: A randomized controlled trial. *Journal of the American Geriatrics Society, 65*(4), 721-727.

Sutherland, D.H. (2001). The evolution of clinical gait analysis. Part I: Kinesiological EMG. *Gait and Posture, 14*(1), 61-70.

Sutherland, D.H. (2002). The evolution of clinical gait analysis. Part II: Kinematics. *Gait and Posture, 16*(2), 159-179.

Sutherland, D.H. (2005). The evolution of clinical gait analysis. Part III: Kinetics and energy assessment. *Gait and Posture, 21*(4), 447-461.

Toffler, A. (1990). *Powershift*. New York: Bantam.

Tosi, V. (1992). Marey and Muybridge: How modern biolocomotion analysis started. In A. Cappozzo, M. Marchetti, & V. Tosi (Eds.), *Biolocomotion: A century of research using moving pictures* (pp. 51-69). Rome: Promograph.

Trepman, E., Gellman, R.E., Micheli, L.J., & De Luca, C.J. (1998). Electromyographic analysis of grand-plié in ballet and modern dancers. *Medicine & Science in Sports & Exercise, 30*(12), 1708-1720.

U.S. Bureau of Labor Statistics. (2016a). *2015 nonfatal occupational injuries and illnesses: Cases with days away from work.* Retrieved from www.bls.gov/iif/oshwc/osh/case/osch0058.pdf

U.S. Bureau of Labor Statistics. (2016b). Illnesses, injuries, and fatalities: Occupational safety and health definitions. Retrieved from https://data.bls.gov/cgi-bin/print.pl/iif/oshdef.htm

U.S. Department of Labor, Occupational Safety and Health Administration. (n.d.) Solutions to control hazards. Retrieved from www.osha.gov/SLTC/ergonomics/controlhazards.html

Unnithan, V.B., Dowling, J.J., Frost, G., & Bar-Or, O. (1999). Role of mechanical power estimates in the O_2 cost of walking in children with cerebral palsy. *Medicine & Science in Sports & Exercise, 31*(12), 1703-1708.

Unnithan, V.B., Dowling, J.J., Frost, G., Volpe Ayub, B., & Bar-Or, O. (1996). Cocontraction and phasic activity during GAIT in children with cerebral palsy. *Electromyography & Clinical Neurophysiology, 36*(8), 487-494.

Unsworth, A., Dowson, D., & Wright, V. (1971). 'Cracking joints': A bioengineering study of cavitation in the metacarpophalangeal joint. *Annals of the Rheumatic Diseases, 30*(4), 348-358.

Van der Ploeg, H.P., Chey, T., Korda, R.J., Banks, E., & Bauman, A. (2012). Sitting time and all-cause mortality risk in 222,497 Australian adults. *Archives of Internal Medicine, 172*(6), 494-500.

Vaz, J.R., Olstad, B.H., Cabri, J., Kjendlie, P.L., Pezarat-Correia, P., & Hug, F. (2016). Muscle coordination during breaststroke swimming: Comparison between elite swimmers and beginners. *Journal of Sports Sciences, 34*(20), 1941-1948.

Verma, R., Hansen, E.A., de Zee, M., & Madeleine, P. (2016). Effect of seat positions on discomfort, muscle activation, pressure distribution and pedal force during cycling. *Journal of Electromyography and Kinesiology, 27*(2), 78-86.

Voight, M.L., Hoogenboom, B.J., Cook, G., & Rose, G. (2014). Functional training and advanced rehabilitation. In B.J. Hoogenboom, M.L. Voight, & W.E. Prentice (Eds.), *Musculoskeletal interventions: Techniques for therapeutic exercises* (3rd ed., pp. 513-546). New York: McGraw-Hill.

Wakeling, J.M., Blake, O.M., & Chan, H.K. (2010). Muscle coordination is key to the power output and mechanical efficiency of limb movements. *Journal of Experimental Biology, 213*(3), 487-492.

Wang, Y.T., Chen, S., Limroongreungrat, W., & Change, L-S. (2005). Contributions of selected fundamental factors to wheelchair basketball performance. *Medicine & Science in Sports & Exercise, 37*(1), 130-137.

Waters, T.R., Putz-Anderson, V., Garg, A., & Fine, L.J. (1993). Revised NIOSH equation for the design and evaluation of manual lifting tasks. *Ergonomics, 36*(7), 749-776.

Watkins, R.G., Uppal, G.S., Perry, J., Pink, M., & Dinsay, J.M. (1996). Dynamic electromyographic analysis of trunk musculature in professional golfers. *American Journal of Sports Medicine, 24*(4), 535-538.

Wei, S-H., Chiang, J-Y., Shiang, T-Y., & Chang, H-Y. (2006). Comparison of shock transmission and forearm electromyography between experienced and recreational tennis players during backhand strokes. *Clinical Journal of Sport Medicine, 16*(2), 129-135.

Werner, S.L., Guido, J.A., McNeice, R.P., Richardson, J.L., Delude, N.A., & Stewart, G.W. (2005). Biomechanics of youth windmill softball pitching. *American Journal of Sports Medicine, 33*(4), 552-560.

Werner, S.L., Jones, D.G., Guido, J.A., Jr., & Brunet, M.E. (2006). Kinematics and kinetics of elite windmill softball pitching. *American Journal of Sports Medicine, 34*(4), 597-603.

Whiteside, D., Deneweth, J.M., Pohorence, M.A., Sandoval, B., Russell, J.R., McLean, S.G., Zernicke, R.F., & Goulet, G.C. (2016). Grading the functional movement screen: A comparison of manual (real-time) and objective methods. *Journal of Strength and Conditioning Research, 34*(4), 924-933.

Whiting, W.C., & Rugg, S. (2006). *Dynatomy: Dynamic human anatomy.* Champaign, IL: Human Kinetics.

Whiting, W.C., & Zernicke, R.F. (2008). *Biomechanics of musculoskeletal injury* (2nd ed.) Champaign, IL: Human Kinetics.

Williams, J.H., Akogyrem, E., & Williams, J.R. (2013). A meta-analysis of soccer injuries on artificial turf and natural grass. *Journal of Sports Medicine.* doi:10.1155/2013/380523

Williams, K., Haywood, K., & VanSant, A. (1991). Throwing patterns of older adults: A follow-up investigation. *International Journal of Aging and Human Development, 33*(4), 279-294.

Williams, K., Haywood, K., & VanSant, A. (1998). Changes in throwing by older adults: A longitudinal investigation. *Research Quarterly for Exercise and Sport, 69*(1), 1-10.

Yildirim, N.U., Comert, E., & Ozengin, N. (2010). Shoulder pain: A comparison of wheelchair basketball players with trunk control and without trunk control. *Journal of Back and Musculoskeletal Rehabilitation, 23*(2), 55-61.

Zhu, W., & Owen, N. (Eds.). (2017). *Sedentary behavior and health.* Champaign, IL: Human Kinetics.

Index

Note: The italicized *f* and *t* following page numbers refer to figures and tables, respectively.

D

E

F

G

N

O

P

T

U

V

W

X

Y

Z

About the Author

William C. Whiting, PhD, is a professor and codirector of the biomechanics laboratory in the department of kinesiology at California State University at Northridge, where he has won both the Distinguished Teaching Award and Scholarly Publication Award. Whiting earned his PhD in kinesiology at UCLA. He has taught courses in biomechanics and human anatomy for more than 35 years and has published more than 40 articles and 30 research abstracts. He is coauthor of *Biomechanics of Musculoskeletal Injury*.

Whiting currently serves on the editorial board of NSCA's *Journal of Strength and Conditioning Research* and serves as a reviewer for a number of scholarly journals. Whiting is a fellow of the American College of Sports Medicine (ACSM) and has served as president of the Southwest Regional Chapter of ACSM. He is also a member of the American Society of Biomechanics; the International Society of Biomechanics; and the National Strength and Conditioning Association.

In his leisure time, Whiting enjoys playing basketball and volleyball, reading, camping, and hiking. He lives in Glendale, California, with his wife, Marji; sons, Trevor and Tad; and daughter, Emmi.

HUMAN KINETICS

Books
Ebooks